PIER 1

Progress In Electromagnetics Research

PIER 1

Progress In Electromagnetics Research

Chief Editor

Jin Au Kong

Massachusetts Institute of Technology
Cambridge, MA, U.S.A.

Elsevier
New York • Amsterdam • London

Elsevier Science Publishing Co., Inc.
655 Avenue of the Americas, New York, New York 10010

Sole distributors outside the United States and Canada:
Elsevier Science Publishers B.V.
P.O. Box 211, 1000 AE Amsterdam, The Netherlands

This book is printed on acid-free paper.

Library of Congress Cataloging-in-Publication Data

ISBN 0-444-01490-X

Current printing (last digit):
10 9 8 7 6 5 4 3 2 1

Manufactured in the United States of America

PREFACE

PROGRESS IN ELECTROMAGNETICS RESEARCH (PIER) is a book series devoted to reporting advancements and progresses in the modern development of electromagnetics and its new and innovative applications. It serves as an international forum for the publication of the state of the art review articles on new theories, methodologies and computational techniques, and interpretations of both theoretical and experimental results. The scope is very wide and covers the spectrum from statics to optical frequencies and beyond.

In this first volume PIER I, five chapters are included. The first chapter is authored by Dr. Wait on the topic of complex resistivity of the Earth which covers the complex resistivity concept and induced polarization response of spheroidal models. The second chapter is written by Dr. Rappaport who addresses the issue of synthesis of optimum microwave antenna applicators for use in treating deep localized tumors. In Chapter 3, Ding and Tsang study effective propagation constants in media with densely distributed dielectric particles of multiple sizes and permittivities by using pair distribution functions. Chapter 4 is authored by Yueh, Shin, and Kong who consider scattering from randomly perturbed periodic and quasiperiodic surfaces. In Chapter 5, Shin and Kong apply radiative transfer theory for active remote sensing of two-layer random media with planar and rough interfaces.

All contributions to the series are reviewed and prospective contributors are welcome to contact the editor and members of the Advisory Board. The Advisory Board members serve on three-year rotation terms and advise on new developments and trends in electromagnetics research. It is our hope that this series will serve as useful references for scientists and engineers in the electromagnetics profession, as sources of new topics for researchers, and as advanced text books for students in electromagnetics.

J. A. Kong

Cambridge, Massachusetts
December 1988

CONTRIBUTORS

J. R. Wait, *University of Arizona*
Tucson, Arizona 85721, USA

C. M. Rappaport, *Northeastern University*
Boston, Massachusetts 02115, USA

K. H. Ding, *University of Washington*
Seattle, Washington 98195, USA

L. Tsang, *University of Washington*
Seattle, Washington 98195, USA

H. A. Yueh, *Massachusetts Institute of Technology*
Cambridge, Massachusetts 02139, USA

R. T. Shin, *Massachusetts Institute of Technology*
Cambridge, Massachusetts 02139, USA

J. A. Kong, *Massachusetts Institute of Technology*
Cambridge, Massachusetts 02139, USA

CONTENTS

PIER 1

Progress In Electromagnetics Research

1

COMPLEX RESISTIVITY OF THE EARTH

J. R. Wait

1.1 Preliminary Theory

a. D.C. Current Flow

As a starting point we consider a point source of direct current I in a homogeneous medium of conductivity σ. The radial current density J_r, in amps/m, is given by the expression

$$J_r = I/(4\pi r^2) \tag{1}$$

where r is the radial distance to the point of observation from the source point. The corresponding radial electric field E_r is obtained simply from Ohm's law:

$$E_r = J_r/\sigma = I/(4\pi\sigma r^2) \tag{2}$$

By symmetry there is no other components of the electric field. Thus, in vector notation, the electric field is written

$$\overline{E} = \hat{r}E_r = \hat{r}I/(4\pi\sigma r^2) \tag{3}$$

where $\hat{r}$ is a unit vector in the r direction.

We now introduce the concept of potential by postulating that $\overline{E}$ can be derived from the gradient of a scalar function V. Thus we write

$$\overline{E} = -\text{grad}\, V \tag{4}$$

or, in the present example,

$$E_r = -\partial V/\partial r \tag{5}$$

If we stipulate the V vanishes as $r \to \infty$, it is clear that

$$V = I/(4\pi\sigma r) \tag{6}$$

We are now in the position to derive the field expressions for a pair of current point sources of equal and opposite signs. The situation is indicated in Fig. 1.1.1 where, for convenience, we have chosen a cylindrical coordinate system (ρ, ϕ, z). The positive current source $+I$ is located at $z = +L/2$ while the negative current source $-I$ (i.e. a

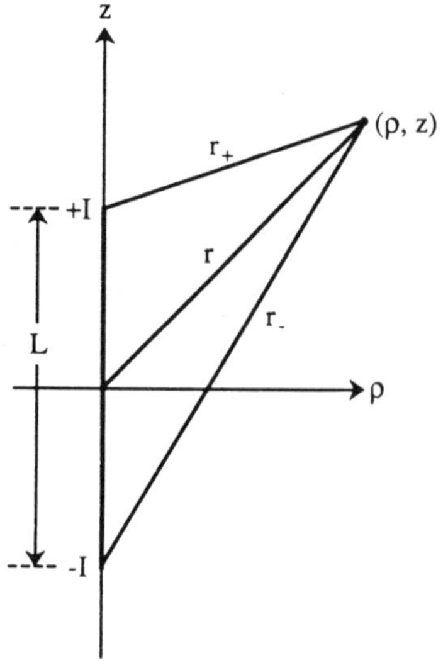

Figure 1.1.1 A current point source and a current point sink.

point sink) is located at $z = -L/z$. Both are located on the z axis. The observer at (ρ, z) is independent of the azimuthal angle ϕ.

Making use of (6), it is a simple matter to show that the resultant potential V at (ρ, z) is given by

$$V = (I/4\pi\sigma)(r_+^{-1} - r_-^{-1}) \tag{7}$$

where

$$r_+ = \left[\rho^2 + (z - L/2)^2\right]^{\frac{1}{2}} \tag{8}$$

and

$$r_- = \left[\rho^2 + (z + L/2)^2\right]^{\frac{1}{2}} \tag{9}$$

The corresponding electric field components are obtained from (4). Thus we easily deduce that

$$E_\rho = -\frac{\partial V}{\partial \rho} = \frac{I\rho}{4\pi\sigma}\left[\frac{1}{r_+^3} - \frac{1}{r_-^3}\right] \tag{10}$$

and

$$E_z = -\frac{\partial V}{\partial z} = \frac{I}{4\pi\sigma}\left[\frac{z - L/2}{r_+^3} - \frac{z + L/2}{r_-^3}\right] \tag{11}$$

Exercise. Show that in rectangular coordinates

$$E_x = \frac{Ix}{4\pi\sigma}\left[\frac{1}{r_+{}^3} - \frac{1}{r_-{}^3}\right] \tag{12}$$

and

$$E_y = \frac{Iy}{4\pi\sigma}\left[\frac{1}{r_+{}^3} - \frac{1}{r_-{}^3}\right] \tag{13}$$

where now

$$r_+ = \left[x^2 + y^2 + (z - L/2)^2\right]^{\frac{1}{2}} \tag{14}$$

and

$$r_- = \left[x^2 + y^2 + (z + L/2)^2\right]^{\frac{1}{2}} \tag{15}$$

Then verify that V, E_x, E_y and E_z satisfy Laplace's equation, i.e.

$$\nabla^2 V \quad \text{or} \quad \frac{\partial^2 V}{\partial x^2} + \frac{\partial^2 V}{\partial y^2} + \frac{\partial^2 v}{\partial z^2} = 0 \tag{16}$$

b. D.C. Magnetic Fields Simply Obtained

To obtain expressions for the magnetic field $\overline{H}$ of the two current point sources, we need to be a bit more specific about the configuration of the circuit connections from the generator (e.g. battery) to the electrodes. For simplicity we will consider a linear thin wire carrying a constant current I as depicted in Fig. 1.1.2. For this situation we can assert that the magnetic field has only an azimuthal component H_ϕ which itself is independent of ϕ.

A simple statement of Ampere's law for the present configuration is

$$\int_0^{2\pi} H_\phi \rho d\phi = \int_{\phi'=0}^{2\pi} \int_{\rho'=0}^{\rho} J_z(\rho', \phi', z)\rho' d\phi' d\rho' \tag{17}$$

which is a statement that the integral of the tangential magnetic field around a closed circuit of radius ρ is equal to the *total* enclosed current in the vertical (i.e. z) direction. Because of symmetry

$$2\pi\rho H_\phi = 2\pi \int_0^{\rho} J_z(\rho', z)\rho' d\rho' \tag{18}$$

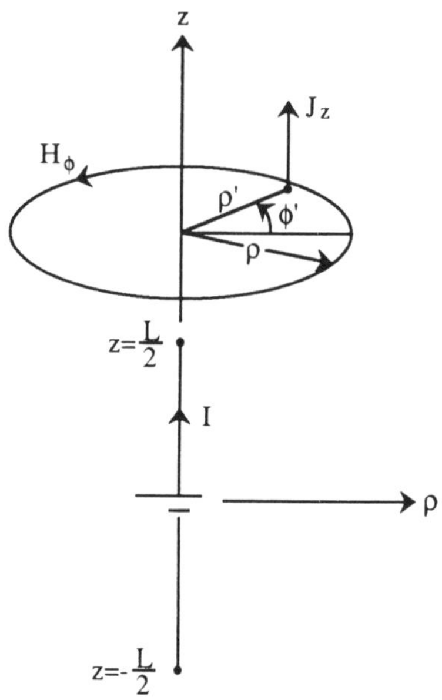

Figure 1.1.2 Concentric circular path to deduce magnetic field.

where, for the case $|z| > L/2$,

$$J_z(\rho', z) = \frac{I}{4\pi}\left[\frac{z - L/2}{(r_+')^3} + \frac{z + L/2}{(r_-')^3}\right] \tag{19}$$

and

$$r_+' = \left[(\rho')^2 + (z - (L/2))^2\right]^{\frac{1}{2}} \tag{20}$$

$$r_-' = \left[(\rho')^2 + (z + (L/2))^2\right]^{\frac{1}{2}} \tag{21}$$

The integration over ρ' can be carried out to give

$$H_\phi = \frac{1}{4\pi\rho}\left(\frac{z - (L/2)}{\left\{\rho^2 + [z + (L/2)]^2\right\}^{\frac{1}{2}}} - \frac{z - (L/2)}{\left\{\rho^2 + [z - (L/2)]\right\}^{\frac{1}{2}}}\right) \tag{22}$$

which is the resultant magnetic field at (ρ, z) for the linear current I grounded at $z = \pm L/2$.

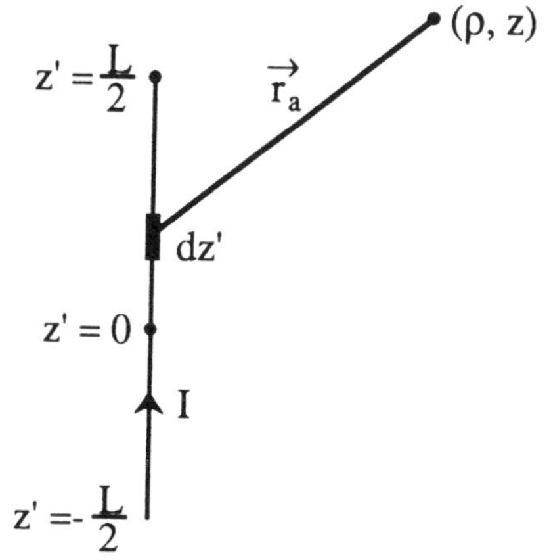

Figure 1.1.3 Geometry for applying the Biot-Savart law.

We can deal with the case $|z| < L/2$ by noting that (18) is modified to read

$$2\pi\rho H_\phi = I + 2\pi \int_0^\rho J_z(\rho', z)\rho' d\rho' \tag{23}$$

because the current on the wire now intersects the plane of the loop. Then, for example, if $z > L/2$ we can see that in the limit $\rho' \to 0$

$$\frac{I}{2}\left(z - \frac{L}{2}\right)\left(-\frac{1}{r_+{}'}\right) = \frac{I}{2}\left(z - \frac{L}{2}\right)\left(-\frac{1}{\left[(\rho')^2 + \left(z - \frac{L}{2}\right)^2\right]^{\frac{1}{2}}}\right) \to \frac{I}{2} \tag{24}$$

bearing in mind that $r'_+ > 0$ always. The up shot is that the same expression for the magnetic field H_ϕ is obtained as before. Thus it is valid for all values of z.

Exercise Show that (22) may also be derived by an application of the Biot-Savart law

Solution The relevant form is

$$\overline{H} = \frac{I}{4\pi} \int_{-L/2}^{L/2} \frac{d\overline{z}' \times \overline{r}_a}{r_a{}^3} \tag{25}$$

where

$$r_a = \left[\rho^2 + (z - z')^2\right]^{\frac{1}{2}} \tag{26}$$

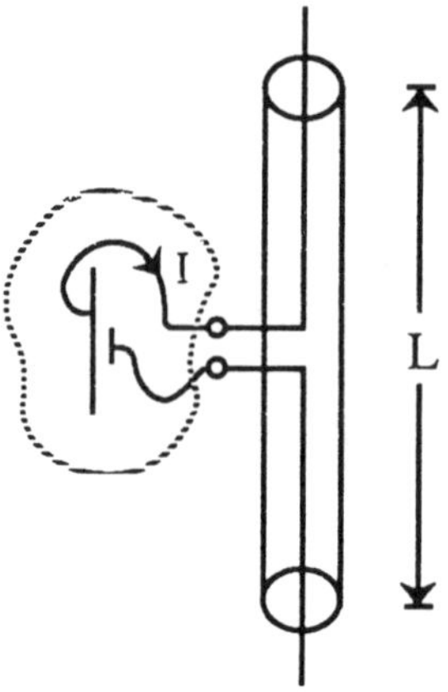

Figure 1.1.4 Idealization of an insulated current carrying wire with bare ends.

and where $\overline{r}_a$ is the vector of magnitude r_a from the element $d\overline{z}'$ to the observer at (ρ, z). See Fig. 1.1.3 . The details of the solution are given elsewhere. [Wait 1985, p. 36]

Exercise Show that yet still another method to obtain (22) is to use the vector potential $\overline{A}$ for the problem.

Solution Noting that $\overline{H}$ = curl $\overline{A}$ it follows quite easily from the symmetry of the problem that

$$H_\phi = -\frac{\partial A_z}{\partial \rho} \tag{27}$$

where

$$A_z = \frac{I}{4\pi} \int_{-L/2}^{L/2} \frac{1}{r_a} dz' \tag{28}$$

is the only component of A. This solution is also detailed in the referenced text book [Wait 1985, p. 37].

c. *Current Injection and Voltage Pick Up*

Perhaps it is desirable if we indicate that a linear current-carrying wire in a conducting medium can be realized by a physical structure such as illustrated in Fig. 1.1.4. The filamental wire of length L is fed by the battery or zero-frequency generator at two terminals. The structure is covered by a concentric insulated cover throughout its length but the ends are exposed. Thus the current I is conveyed to the medium at the bare end points. As we have indicated, the structure is represented analytically as a point source and point sink of current separated by a distance L. The electric fields so produced are deduced on this basis. However, the magnetic field is dependent on the assumption of symmetry about the axis of the structure. Here we make no attempt to say that the magnetic field is produced by the current along the wire or whether it is a manifestation of the current flow within the external medium. But as we have indicated in the preceding exercises, consistent results for H_ϕ are obtained by either supposition.

Another point of interest is that we ignore the presence of the insulating cover in deducing the magnetic field using the Biot-Savart law or the vector potential method. A possible justification for this idealization is to note that the derived field expression for H_ϕ given by (22) reduces to $I/(2\pi\rho)$ as ρ becomes vanishingly small provided $|z| < L/2$. Thus $2\pi\rho H_\phi$ can be associated with the total current I whether we choose $\rho =$ wire radius or $\rho =$ insulation radius.

Other physical approximations involve the precise method that we feed the structure and how it is terminated in the medium. In some applications, both of these questions are important but in geophysical prospecting schemes the voltage drop at the electrodes is not crucial because the injected current I into the medium is monitored. Also, the connecting leads from the battery or generator to the structure are configured such that the resulting fields are minimized (e.g. by using twisted pair lead wire of minimum length).

A few words might also be said about how the fields are to be measured within the conductor. The ideal probe for the electric field is again an insulated linear wire of say length ℓ with open ends. The induced voltage v is then observed at the terminal pairs as suggested by the sketch in Fig. 1.1.5. Now it is clear, under the assumption that the linear wire is perfectly conducting, the full induced voltage will appear at the terminals. Thus

$$v = -\int_{-\ell/2}^{\ell/2} E_z(z')dz' \tag{29}$$

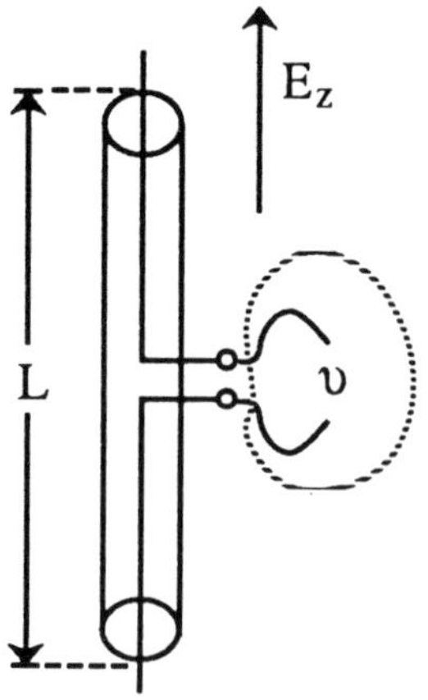

Figure 1.1.5 Idealization of an insulated linear wire with bare ends for use as an electric field probe.

which is the open circuit voltage for the case when the linear wire probe is aligned in the z direction. If E_z was substantially constant over the length ℓ we could write $v \simeq -E_z \cdot \ell$ indicating that the electric field is directly proportional to the voltage v at the detector. It is clear that the probe system requires the ends of the linear wire have electrical contact with the medium but the precise nature of the interface is not crucial provided negligible current is drawn by the voltage detector.

When E_z is not a constant along the linear wire (i.e. probe is "near" the source) it is desirable to express v in terms of the potentials at the ends of the structure. Thus in the context of Fig. 1.1.5, we note that

$$E_z = -\frac{\partial V}{\partial z}$$

so that

$$v = V(z' = \frac{\ell}{2}) - V(z' = -\frac{\ell}{2}) \tag{30}$$

is the expression for the induced potential.

A specific four electrode measuring scheme is illustrated in Fig. 1.1.6 where again the surrounding medium is assumed to be homogeneous of conductivity σ. The "current electrodes" A and B are fed by the battery or DC source so that essentially A is a point source of current I and B is a point sink. The voltage v or difference of potential between the electrodes M and N is then measured by the

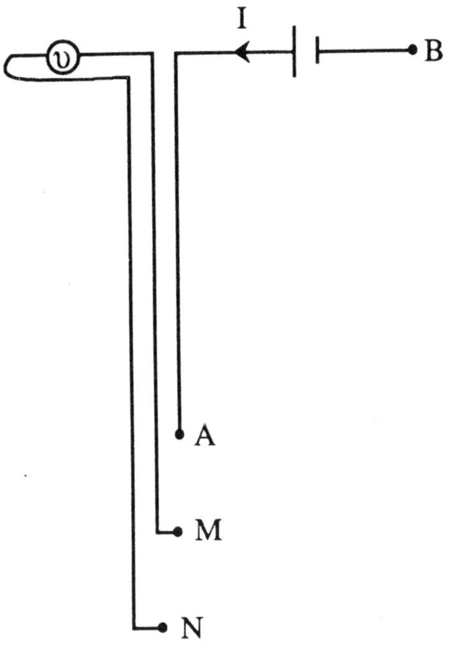

Figure 1.1.6 Four electrode array showing voltage and current circuits.

detector.

From what we have said above, it is clear that the so called transfer resistance R of the four electrode array is given by

$$R = \frac{v}{I} = \frac{1}{4\pi\sigma}\left[\frac{1}{AM} + \frac{1}{BN} - \frac{1}{AN} - \frac{1}{BM}\right] \tag{31}$$

where AM, BN, AN and BM are linear distances. This particular configuration is relevant to boreholes or well logging. Here, for example, A, M and N are located on the axis of the hole and they make contact with the adjacent medium. Obviously, we have ignored the influence of the hole, itself including the effect of any conductive fluids therein. More about this type of measurement scheme will be brought up later when more realistic conditions are considered.

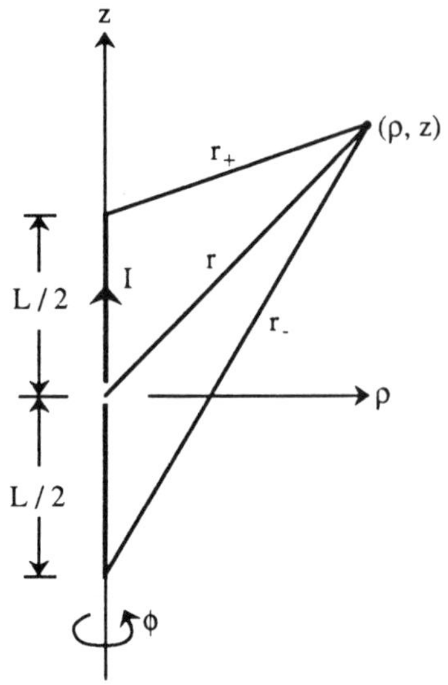

Figure 1.1.7 Linear wire of length L.

d. The Dipole Concept

In many instances the fields of the linear wire of length L, carrying a current I, are observed at distances r large compared with L. The situation is illustrated in Fig. 1.1.7 where $r = (\rho^2 + z^2)^{\frac{1}{2}}$ is measured from the center point to the observer at (ρ, z) again using cylindrical coordinates. Now if $r \gg L$ it is clear that

$$\begin{aligned} r_+^{-1} &= \left[\rho^2 + (z - L/2)^2\right]^{-\frac{1}{2}} \\ &= \left[\rho^2 + z^2 - zL + L^2/4\right]^{-\frac{1}{2}} \\ &\simeq (r^2 - zL)^{-\frac{1}{2}} = r^{-1}(1 - zL/r^2)^{-\frac{1}{2}} \\ &\simeq r^{-1}(1 + zL/2r^2) = r^{-1} + zL/2r^3 \end{aligned} \tag{32}$$

Similarly

$$r_-^{-1} \simeq r^{-1} - zL/2r^3 \tag{33}$$

The potential as given by (7) is now approximated as follows:

$$V = \frac{I}{4\pi\sigma}\left[\frac{1}{r_+} - \frac{1}{r_-}\right] \simeq \frac{ILz}{4\pi\sigma r^3} \tag{34}$$

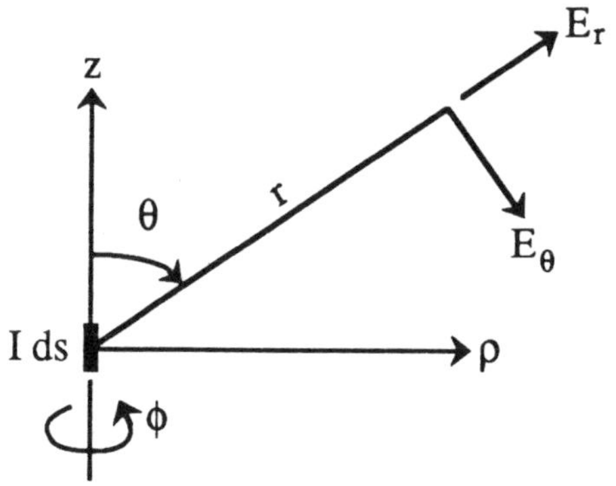

Figure 1.1.8 Electric dipole or short current element.

which is proportional to the product IL called the current moment of the resulting electric dipole. Clearly an alternative form of (34) is

$$V \simeq \frac{IL}{4\pi\sigma r^2}\cos\theta \qquad (35)$$

in terms of spherical coordinates (r, θ, ϕ) as indicated in Fig. 1.1.8. In both (34) and (35) above we now replace IL by Ids to signify that the current element has an infinitesimal length ds. Thus, we write (35) as an equality

$$V = \frac{Ids}{4\pi\sigma r^2}\cos\theta \qquad (36)$$

The corresponding electric fields for the dipole source are obtained from (4) so that we arrive at the explicit expressions:

$$E_r = -\frac{\partial V}{\partial r} = \frac{Ids}{2\pi\sigma r^3}\cos\theta \qquad (37)$$

and

$$E_\theta = -\frac{1}{r}\frac{\partial V}{\partial \theta} = \frac{Ids}{4\pi\sigma r^3}\sin\theta \qquad (38)$$

The inverse distance dependence, to the inverse third power, is characteristic of the static fields of an elementary dipole.

Exercise Obtain the electric field components of the electric dipole Ids expressed in cylindrical coordinates.

Solution We proceed directly by working with (4) and (34). Thus

$$E_\rho = -\frac{\partial V}{\partial \rho} = \frac{3Ids\rho z}{4\pi\sigma r^5} = \frac{Ids}{4\pi\sigma}\cdot\frac{3\rho z}{(\rho^2+z^2)^{5/2}} \tag{39}$$

and

$$E_z = -\frac{\partial V}{\partial z} = \frac{Ids}{4\pi\sigma}\left(\frac{-1}{r^3}+\frac{3z^2}{r^5}\right) = \frac{Ids}{4\pi\sigma}\frac{2z^2-\rho^2}{(\rho^2+z^2)^{5/2}} \tag{40}$$

where we have noted that $\partial r/\partial z = z/r$ and $\partial r/\partial\rho = \rho/r$.

The static magnetic field of the electric dipole Ids can be obtained directly from (22) in the limiting case where $\rho^2+z^2 \gg L^2$ and IL is replaced by Ids. Then

$$H_\phi = \frac{Ids\rho}{4\pi r^3} = \frac{Ids}{4\pi r^2}\sin\theta \tag{41}$$

which, of course, is valid for either cylindrical or spherical coordinates.

Our next observation is to note that the fields of the dipole can be conveniently derived from a vector potential with only a z component A_z. Thus, we readily confirm that if

$$A_z = Ids/4\pi r \tag{42}$$

then

$$\sigma E_\rho = \partial^2 A_z/\partial\rho\partial z \tag{43}$$

$$\sigma E_z = \partial^2 A_z/\partial z^2 \tag{44}$$

and

$$H_\phi = -\partial A_z/\partial\rho \tag{45}$$

lead back to (39), (40) and (41), respectively.

These relations are consistent with the general forms

$$\overline{E} = -\text{grad } V \tag{46}$$

and

$$\overline{H} = \text{curl } \overline{A} \tag{47}$$

where $\overline{A}$ and V are related by $\sigma V = -\text{div } \overline{A}$.

e. Dynamic Fields

Up to this point we have been dealing with purely static fields in a homogeneous conductor of infinite extent. We now wish to generalize the results to time varying fields. For convenience we specify that the fields vary harmonically with an *angular frequency* ω. Thus, for example, the component E_x is a phasor or complex quantity with an *amplitude* $|E_x|$ and phase ϕ_x. The actual physical quantity is then

$$e_x(t) = \text{Re}\,\{E_x e^{j\omega t}\} = |E_x| \cos(\omega t + \phi_x) \tag{48}$$

Correspondingly, the complex vector $\overline{E}$ is written explicitly in the form

$$E = \hat{x}E_x + \hat{y}E_y + \hat{z}E_z \tag{49}$$

when E_x, E_y and E_z are phasors. The physical time-varying vector is then

$$\overline{e}(t) = \hat{x}e_x(t) + \hat{y}e_y(t) + \hat{z}e_z(t) \tag{50}$$

where

$$e_y(t) = \text{Re}\,\{E_y e^{j\omega t}\} \quad \text{and} \quad e_z(t) = \text{Re}\,\{E_z e^{j\omega t}\} \tag{51}$$

It is important that the reader not confuse the meanings of vectors and phasors. In what follows we will deal almost exclusively with complex phasors corresponding to a single *angular frequency* ω. There is no loss of generality here because general time varying quantities can be handled by Fourier synthesis involving spectra over all values of ω.

At this juncture we write down Maxwell's equations for *a source-free* region characterized by a conductivity σ, (electric) permittivity ϵ, and (magnetic) permeability μ. For the implied time factor $\exp(j\omega t)$:

$$-j\mu\omega\overline{H} = \text{curl}\,\overline{E} \tag{52}$$

$$(\sigma + j\epsilon\omega) = \text{curl}\,\overline{H} \tag{53}$$

where $\overline{E}$ and $\overline{H}$ are the vector complex electric and magnetic fields, respectively. If the region is homogeneous, we readily deduce that

$$\text{curl curl}\,\overline{E} + \gamma^2\overline{E} = 0 \tag{54}$$

and

$$\text{curl curl}\,\overline{H} + \gamma^2\overline{H} = 0 \tag{55}$$

where

$$\gamma^2 = j\mu\omega(\sigma + j\epsilon\omega)$$

In the case of rectangular coordinates, these two equations reduce to the well known scalar wave equation such that each field component (e.g. E_x) satisfies

$$\left(\frac{\partial^2}{\partial x^2} + \frac{\partial^2}{\partial y^2} + \frac{\partial^2}{\partial z^2} - \gamma^2\right) E_x = 0 \tag{56}$$

However, the situation can be more complicated in cylindrical and spherical coordinates. Thus, it is more convenient to utilize the vector potential $\overline{A}$ defined such that

$$\overline{H} = \text{curl}\,\overline{A} \tag{57}$$

Then from (53) we see that

$$\overline{E} = \frac{1}{\sigma + j\epsilon\omega}\text{curl curl}\,\overline{A} \tag{58}$$

In particular if we choose $\overline{A} = \hat{z}A_z$ (i.e. z directed vector potential) it follows that

$$H_\phi = -\partial A_z/\partial\rho \tag{59}$$

and

$$E_\rho = \frac{1}{\sigma + j\epsilon\omega}\frac{\partial^2}{\partial\rho\partial z}A_z \tag{60}$$

$$E_z = \frac{1}{\sigma + j\epsilon\omega}\left(-\gamma^2 + \frac{\partial^2}{\partial z^2}\right) A_z \tag{61}$$

where we have made use of elementary results from vector calculus (Wait 1986). We have also assumed azimuthal symmetry (i.e. $\partial/\partial\phi = 0$). Then we can show that

$$(\nabla^2 - \gamma^2)A_z = 0 \tag{62}$$

where in cylindrical coordinates

$$\nabla^2 = \frac{1}{\rho}\frac{\partial}{\partial\rho}(\rho\frac{\partial}{\partial\rho}) + \frac{\partial^2}{\partial z^2} \tag{63}$$

is the Laplacian operator.

Now in the case of zero frequency, it is clear that $\gamma^2 = 0$ and $\nabla^2 A_z = 0$ which is Laplace's equation. For the elementary source dipole of current moment Ids we showed that the static (i.e. $\omega = 0$) solution for the problem was given by

$$A_z = \frac{Ids}{4\pi r} \tag{64}$$

where

$$r = (\rho^2 + z^2)^{\frac{1}{2}}$$

The appropriate solution of (62) that reduces to (64) is clearly given by

$$A_z = \frac{Ids}{4\pi r} e^{-\gamma r} \tag{65}$$

where $\gamma = [j\mu\omega(\sigma + j\epsilon\omega)]^{\frac{1}{2}}$ is defined such that Re $\gamma > 0$. Using (59) we readily deduce that

$$\begin{aligned} H_\phi &= \frac{Ids}{4\pi r^3}(1 + \gamma r)e^{-\gamma r}\rho \\ &= \frac{Ids}{4\pi r^2}(1 + \gamma r)e^{-\gamma r}\sin\theta \end{aligned} \tag{66}$$

which reduces to (41) as it should when $\gamma r \to 0$. The corresponding electric field components are most easily obtained from (53). In spherical coordinates

$$E_r = \frac{Ids}{2\pi(\sigma + j\epsilon\omega)r^3}(1 + \gamma r)e^{-\gamma r}\cos\theta \tag{67}$$

and

$$E_\theta = \frac{Ids}{4\pi(\sigma + j\epsilon\omega)r^3}(1 + \gamma r + \gamma^2 r^2)e^{-\gamma r}\sin\theta \tag{68}$$

We can observe that (67) and (68) reduce to the corresponding static forms (37) and (38), respectively, in the limit $\gamma r \to 0$ and $\omega \to 0$.

It is useful to note here that, if $|\gamma r| \ll 1$, (without requiring that $\omega \to 0$), (67) and (68) are approximated by the "so-called" quasi-static forms

$$E_r = \frac{Ids}{2\pi(\sigma + j\epsilon\omega)r^3}\cos\theta \tag{69}$$

and

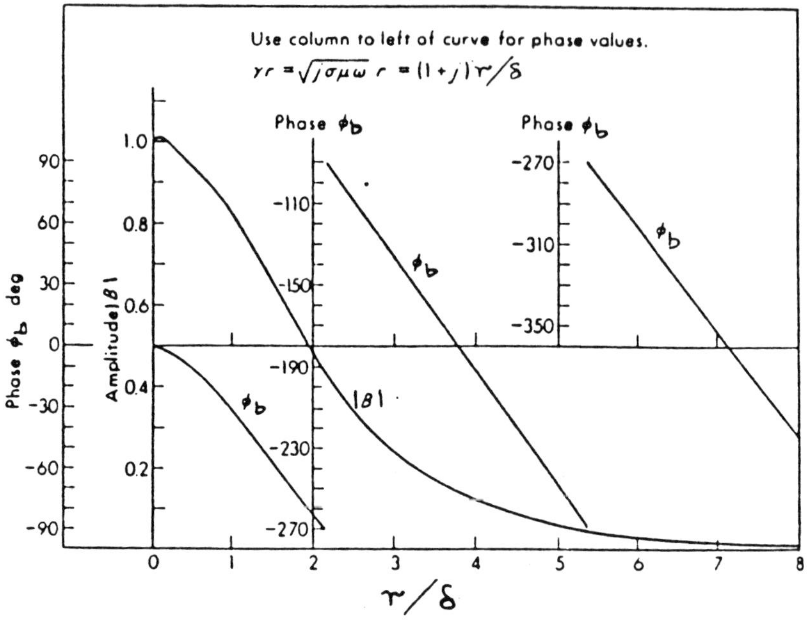

Figure 1.1.9 Amplitude and phase of the generic field function B as a function of normalized distance to source.

$$E_\theta = \frac{Ids}{4\pi(\sigma + j\epsilon\omega)r^3}\sin\theta \tag{70}$$

Now, if we let $\omega \to 0$, the static forms given by (37) and (38) are recovered.

f. Generic Field Structure

To give some insight to the nature of the field structure of the dipole we will assume that $\sigma \gg \epsilon\omega$. In this case the conduction currents dominate the displacement currents. Then we may approximate the propagation constant as follows

$$\gamma = [j\mu\omega(\sigma + j\epsilon\omega)]^{\frac{1}{2}} \simeq (j\mu\omega\sigma)^{\frac{1}{2}} = (1+j)/\delta \tag{71}$$

where

$$\delta = [2/(\sigma\mu\omega)]^{\frac{1}{2}}$$

is the classical skin depth. Now we may write (66), (67) and (68) in the forms

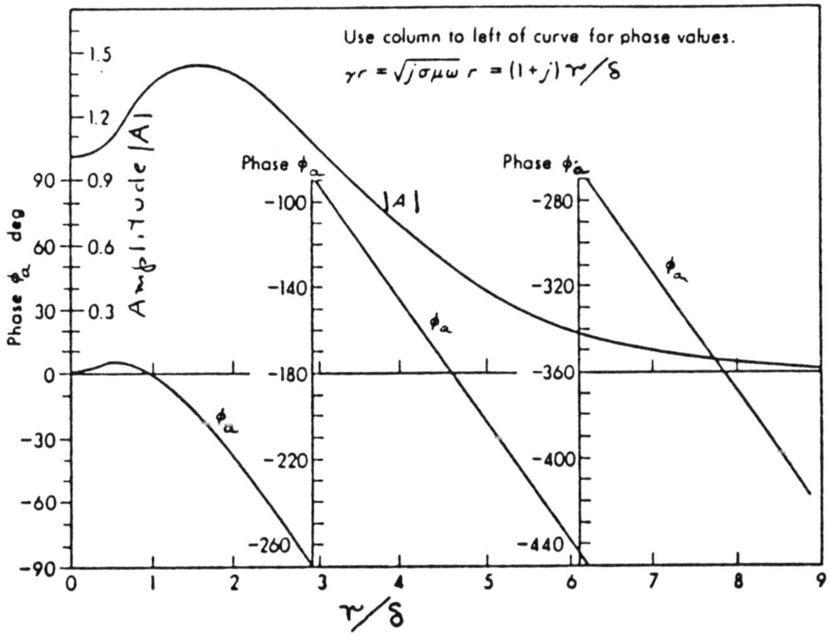

Figure 1.1.10 Amplitude and phase of the generic field function A as a function of normalized distance to source.

$$H_\phi = \frac{Ids}{4\pi r^2} B(r/\delta) \sin\theta \tag{72}$$

$$E_r = \frac{Ids}{2\pi\sigma r^3} B(r/\delta) \cos\theta \tag{73}$$

$$E_\theta = \frac{Ids}{4\pi\sigma r^3} A(r/\delta) \sin\theta \tag{74}$$

where

$$B(r/\delta) = [1 + (1+j)r/\delta] \exp[-(1+j)r/\delta] \tag{75}$$

and

$$A(r/\delta) = \left[1 + (1+j)r/\delta + 2jr^2/\delta^2\right] \exp[-(1+j)r/\delta] \tag{76}$$

The dimensionless complex functions $B = |B|e^{j\phi_b}$ and $A = |A|e^{j\phi_a}$ represent the normalized actual fields at a normalized distance r/δ from the source dipole. The amplitude and phase values of B and A

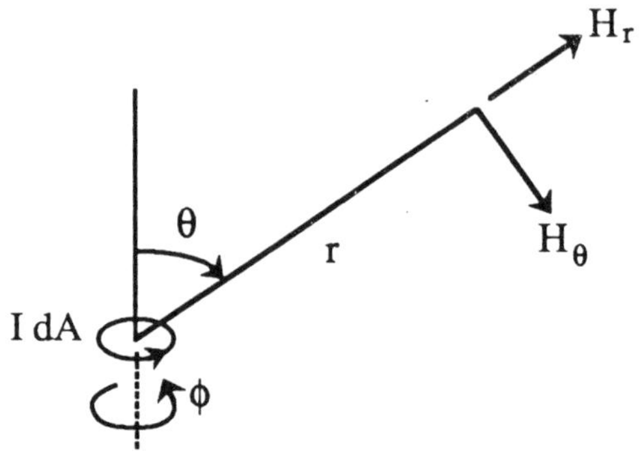

Figure 1.1.11 Magnetic dipole or small current carrying loop.

are plotted in Figs. 1.1.9 and 1.1.10, respectively. As indicated both A and B approach 1 as $r/\delta \to 0$ which, of course, is the static limit. On the other hand, if $r/\delta \gg 1$, the functions are exceptionally damped and the phase lag behaves as r/δ radians.

An interesting property of the A function is that its magnitude is actually 1.4 for $r/\delta \simeq 1.8$. Thus, contrary to simple intuition, the field $|E_\theta|$ is much greater than that predicted on the basis of simple plane wave theory.

g. The Magnetic Dipole

While we have been dealing exclusively with an electric dipole source, the generic functions A and B are actually relevant to a magnetic dipole source. We call attention to this fact for future reference.

A magnetic dipole is here regarded as a small loop of area dA carrying a current I as indicated in Fig. 1.1.11. Again, the surrounding medium is considered to be homogeneous with conductivity σ, permittivity ϵ, and magnetic permeability, μ. We refer the reader to standard texts for the derivations of the field expressions. For example, the principle of duality may be exploited which shows that the oscillating magnetic dipole source is analogous to the electric dipole source if the role of $\vec{E}$ and $\vec{H}$ are suitably interchanged [Wait 1985, p. 39].

Using spherical coordinates (r, θ, ϕ) with the magnetic dipole (i.e. small loop with vertical axis) located at the origin we can express the non-zero field components as follows

$$E_\phi = -\frac{j\mu\omega I dA}{4\pi r^2}(1 + \gamma r)e^{-\gamma r}\sin\theta \tag{77}$$

$$H_r = \frac{IdA}{2\pi r^3}(1+\gamma r)e^{-\gamma r}\cos\theta \tag{78}$$

and

$$H_\theta = \frac{IdA}{4\pi r^3}(1+\gamma r+\gamma^2 r^2)e^{-\gamma r}\sin\theta \tag{79}$$

These expressions are analogous to the electric dipole forms given by (66), (67) and (68). In particular, we note that in the static limit $\omega \to 0$

$$H_r = \frac{IdA}{2\pi r^3}\cos\theta \tag{80}$$

$$H_\theta = \frac{IdA}{4\pi r^3}\sin\theta \tag{81}$$

which are the classical forms for the DC magnetic field of the static magnetic dipole source. Of course, in this limit $E_\phi \to 0$. But, in the quasi-static sense where $\omega \neq 0$ and $|\gamma r| \ll 1$, we may approximate (77) by

$$E_\phi \simeq -\frac{j\mu\omega IdA}{4\pi r^2}\sin\theta \tag{82}$$

When the surrounding medium is a good conductor (i.e. $\sigma \gg \epsilon\omega$) we can express (77), (78) and (79) in the useful forms

$$E_\phi = -\frac{j\mu\omega IdA}{4\pi r^2}B(r/\delta)\sin\theta \tag{83}$$

$$H_r = \frac{IdA}{2\pi r^3}B(r/\delta)\cos\theta \tag{84}$$

$$H_\theta = \frac{IdA}{4\pi r^3}A(r/\delta)\sin\theta \tag{85}$$

where the complex functions B and A are given by (75) and (76), respectively. Thus, the plots in Figs. (1.1.9) and (1.1.10) are still applicable.

h. Dispersion of σ and ϵ

At this stage, one could extend the calculation of the fields to include displacement currents. Thus, one could utilize the forms given by (66), (67) and (68) for the field components with γ defined by

$$\gamma = [j\mu\omega(\sigma + j\epsilon\omega)]^{\frac{1}{2}}$$

Typically such calculations assign a fixed value for σ and ϵ. Unfortunately, such results are very artificial because in actual geological media, both σ and (particularly) ϵ vary significantly with frequency. We prefer to deal with the complex resistivity function $\rho(j\omega)$ which is defined as

$$\rho(j\omega) = [\sigma(\omega) + j\omega\epsilon(\omega)]^{-1}$$

in terms of the frequency dependent real conductivity $\sigma(\omega)$ and real permittivity $\epsilon(\omega)$.

We defer further discussion of the influence of dispersion (i.e. frequency dependence) of the conductivity and permittivity. Obviously, it is an important topic.

References

[1] Wait, J. R., *Electromagnetic Wave Theory*, Harper and Row, 1985.

[2] Wait, J. R., *Introduction to Antennas and Propagation*, Peter Peregrinus Ltd., Stevenage, UK, 1986.

1.2 Complex Resistivity Concept

a. Ohm's Law

The essential property we exploit in the induced polarization method of geophysical is the frequency dispersion of the medium. From a macroscopic point of view, we are saying that the real conductivity $\sigma(\omega)$ and the real permittivity $\epsilon(\omega)$ are functions of the angular frequency ω. Thus, for a particular point in the medium, we may assert that electric field $\overline{E}(j\omega)$ and the associated current density $\overline{J}(j\omega)$ are connected by Ohm's law

$$\overline{E}(j\omega) = \rho(j\omega)\overline{J}(j\omega) \tag{1}$$

where

$$\rho(j\omega) = [\sigma(\omega) + j\epsilon(\omega) \cdot \omega]^{-1} \tag{2}$$

is the complex resistivity*. It is assumed here and in most of what follows that $\rho(j\omega)$ does not depend on the magnitude of the current density $\overline{J}$. That is, we are dealing with linear phenomena. The implied time factor is $\exp(j\omega t)$.

b. Basic Dipole Model

While (1) is valid for an inhomogeneous region, we find it convenient to deal with piecewise homogeneous regions. In fact, in order to deal with basic concepts at the most elementary level, it is desirable to imagine that the source of the excitation is an electric dipole with current moment $I(j\omega)ds$. The surrounding medium is homogeneous with complex resistivity $\rho(j\omega)$. We may now write down expressions for the field components with reference to spherical coordinates [Wait 1986a]:

$$E_r(j\omega) = \frac{I(j\omega)ds\,\rho(j\omega)}{2\pi r^3}B(j\omega)\cos\theta \tag{3}$$

$$E_\theta(j\omega) = \frac{I(j\omega)ds\,\rho(j\omega)}{4\pi r^3}A(j\omega)\sin\theta \tag{4}$$

and

---

* The use of the lower case rho to represent complex resistivity should not cause confusion with the radial cylindrical coordinate.

$$H_\phi(j\omega) = \frac{1}{4\pi r^2} I(j\omega) ds\, B(j\omega) \sin\theta \tag{5}$$

where

$$B(j\omega) = [1 + \gamma(j\omega)r] \exp[-\gamma(j\omega)r] \tag{6}$$

and

$$A(j\omega) = \left[1 + \gamma(j\omega)r + \gamma^2(j\omega)r^2\right] \exp[-\gamma(j\omega)r] \tag{7}$$

The complex propagation constant is here written in terms of the complex resistivity as follows

$$\gamma(j\omega) = \{j\omega\mu[\sigma(\omega) + j\omega\epsilon(\omega)]\}^{\frac{1}{2}} = [j\omega\mu/\rho(j\omega)]^{\frac{1}{2}}$$

Here we can see the complexity of the problem resulting from the dispersion in the medium. For example, if we were to measure the ratio $E_\theta(j\omega)/I(j\omega)$ for some fixed point (r, θ), the resulting frequency dependence would be related to the frequency dependence of $\rho(j\omega)$ in a rather complicated manner. This fact means that any inverse procedure to deduce the complex resistivity of the medium from the measured field is not simple. We shall consider various aspects of this problem below and in the following chapters.

c. *Quasi-Static Limit*

There is a great simplification when the angular frequency ω is low and/or the radial distance r is sufficiently small such that

$$|\gamma(j\omega)r| \ll 1$$

Then $B(j\omega)$ and $A(j\omega)$ can both be replaced by 1. This restriction is equivalent to the quasi-static assumption discussed earlier. We can now say that an electric field component $\Psi(j\omega)$ is linearly related to the complex resistivity in the manner

$$\Psi(j\omega) = I(j\omega) F \rho(j\omega) \tag{8}$$

when F depends only on the geometry. More explicitly, we would write

$$E_r(j\omega) \simeq I(j\omega)\left[\frac{ds}{2\pi r^3}\cos\theta\right]\rho(j\omega) \tag{9}$$

and

$$E_\theta(j\omega) \simeq I(j\omega)\left[\frac{ds}{4\pi r^3}\sin\theta\right]\rho(j\omega) \qquad (10)$$

On the other hand, $H_\phi(j\omega)$ is independent of $\rho(j\omega)$ at least within the quasi-static idealization. Thus

$$H_\phi(j\omega) \simeq I(j\omega)\left[\frac{ds}{4\pi r^2}\sin\theta\right] \qquad (11)$$

Linear system theory [Wait 1984] can be employed to discuss the time domain or transient response of the electric field response $\Psi(t)$ (i.e. $e_r(t)$ or $e_\theta(t)$) for a suddenly applied current $i(t)$ at $t = 0$. Using Laplace transform notation, we can pose the problem as follows: Given the source current, deduce the response; thus, begin by taking the Laplace transform

$$I(s) = \int_0^\infty i(t)e^{-st}dt = \mathcal{L}i(t) \qquad (12)$$

where formally we identify s with $j\omega$ — but we can allow $\mathrm{Re}\{s\} > 0$ to guarantee convergence of (12). Then clearly

$$\Psi(t) = \mathcal{L}^{-1}\Psi(s) = \frac{1}{2\pi j}\int_{\delta-j\infty}^{\delta+j\infty} e^{st}\psi(s)ds \qquad (\text{where } \delta > 0) \qquad (13)$$

or, to be explicit,

$$\Psi(t) = F\mathcal{L}^{-1}I(s)\rho(s) \qquad (14)$$

In the interests of conciseness, we have used the symbols $\mathcal{L}$ and $\mathcal{L}^{-1}$ to denote the direct and inverse Laplace transform operations, respectively.

While the Laplace transformation really involves all frequencies here, we argue that the procedure is valid in an approximate sense even though we have invoked the limitation that $|\gamma(s)r| \ll 1$. The early part of the transient response would be modified if propagation effects were allowed for. This question is related to "electromagnetic coupling" which we discuss later.

d. Apparent Resistivity

When the medium is inhomogeneous, we need to revise our approach. To illustrate the problem, consider the situation shown in Fig.

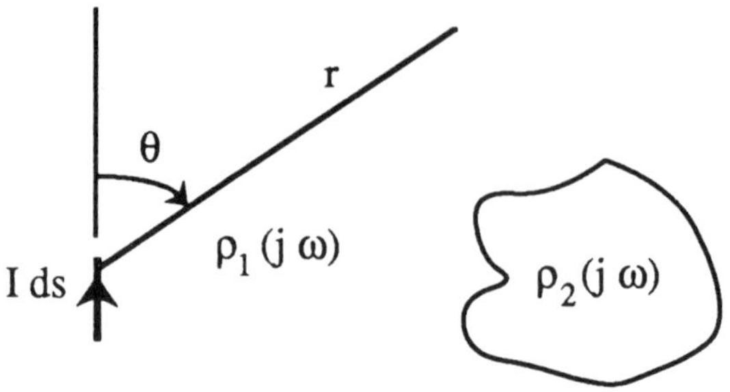

Figure 1.2.1 Current element source in homogeneous region (1) in presence of target region (2).

1.2.1. An electric dipole of current moment $I(j\omega)ds$ is located in a homogeneous region of resistivity $\rho_1(j\omega)$ which is unbounded except for an adjacent region of resistivity $\rho_2(j\omega)$. The symmetry of the problem has been destroyed but it is useful to introduce the concept of apparent (complex) resistivity $\rho_a(j\omega)$ with reference to the homogeneous model [i.e. where $\rho_2(j\omega) = \rho_1(j\omega)$]. For example, in the homogeneous case, we might deal with the radial field $E_r(j\omega)$ at $\theta = 0$. Thus

$$E_r(j\omega) = I(j\omega)ds\rho_1(j\omega)/2\pi r^3 \qquad \text{for } \rho_2 = \rho_1 \text{ and } \theta = 0 \quad (15)$$

Now, in the presence of the inhomogeneity, we can write

$$E_r(j\omega) = I(j\omega)ds\rho_a(j\omega)/2\pi r^3 \qquad \text{for } \rho_2 \neq \rho_1 \text{ and } \theta = 0 \quad (16)$$

In other words, the apparent complex resistivity for the configuration indicated is an effective parameter that is normalized by the homogeneous medium parameter.

In a functional sense, we can write

$$\rho_a(j\omega) = \rho_a\left[\rho_1(j\omega), \rho_2(j\omega)\right] \quad (17)$$

for a specific configuration of the source dipole and receiving circuit. However, such a statement is only valid in the quasi-static sense where all propagation effects are ignored. Some interesting properties can be discussed using this model. In fact, as pointed out by Seigel (1985), it is generally true that for any dimensionless factor λ

$$\rho_a(\lambda\rho_1, \lambda\rho_2) = \lambda\rho_a(\rho_1, \rho_2) \quad (18)$$

In other words, if the resistivities are increased by a factor λ, then the electric field components will also be increased by λ. Clearly now, if we differentiate both sides of (18) with respect to λ we get

$$\frac{\partial \rho_a(\lambda\rho_1, \lambda\rho_2)}{\partial(\lambda\rho_1)}\rho_1 + \frac{\partial \rho_a(\lambda\rho_1, \lambda\rho_2)}{\partial(\lambda\rho_2)}\rho_2 = \rho_a(\rho_1, \rho_2) \tag{19}$$

Now, we set $\lambda = 1$ to give

$$\rho_1 \frac{\partial \rho_a}{\partial \rho_1} + \rho_2 \frac{\partial \rho_a}{\partial \rho_2} = \rho_a \tag{20}$$

Clearly, this equation is equivalent to

$$\frac{\partial \ln \rho_a}{\partial \ln \rho_1} + \frac{\partial \ln \rho_a}{\partial \ln \rho_2} = 1 \tag{21}$$

which is true for any frequency — again within the limits of the quasi-static approximation. An alternative statement of (20) or (21) is

$$\rho_a = K \rho_1{}^{B_1} \rho_2{}^{B_2} \tag{22}$$

where

$$B_1 = \frac{\rho_1}{\rho_a}\frac{\partial \rho_a}{\partial \rho_1} \quad \text{and} \quad B_2 = \frac{\rho_2}{\rho_a}\frac{\partial \rho_a}{\partial \rho_2} \tag{23}$$

and K is a constant. The functional frequency dependence of ρ_a, K, ρ_1 and ρ_2 is understood. In fact, we can write

$$\frac{\rho_a(j\omega)}{\rho_a(0)} = \frac{K(j\omega)}{K(0)} \left(\frac{\rho_1(j\omega)}{\rho_1(0)}\right)^{B_1(j\omega)} \left(\frac{\rho_2(j\omega)}{\rho_2(0)}\right)^{B_2(j\omega)} X(j\omega) \tag{24}$$

where we have normalized all factors by the zero-frequency values and where

$$X(j\omega) = \rho_1(0)^{B_1(j\omega)-B_1(0)} \rho_2(0)^{B_2(j\omega)-B_2(0)} \tag{25}$$

By making use of the identity

$$B_1(j\omega) + B_2(j\omega) = 1 \tag{26}$$

it is possible to write the above expression in the form

$$X(j\omega) = [\rho_2(0)/\rho_1(0)]^{B_2(j\omega)-B_2(0)} \tag{27}$$

which agrees with Song and Vozoff [1985]. They argue that a useful approximation to (24) is to set $[K(j\omega)/K(0)]X(j\omega) = 1$ whence

$$\frac{\rho_a(j\omega)}{\rho_a(0)} \simeq \left[\frac{\rho_1(j\omega)}{\rho_1(0)}\right]^{B_1(j\omega)} \left[\frac{\rho_2(j\omega)}{\rho_2(0)}\right]^{B_2(j\omega)} \tag{28}$$

The applicability of this type of approximation has been confirmed by Song and Vozoff [1985] who carried out extensive numerical experiments using relevant geophysical parameters for both a layered earth model and a buried spherical target. They show extensive plots of the frequency dependent dilution factors B_1 and B_2 using a Cole-Cole model for the complex resistivity functions $\rho_1(j\omega)$ and $\rho_2(j\omega)$. Presumably, they use the real parts of B_1 and B_2 although this point is not made clear.

A closely related approach is to return to (22) and take logarithms of both slices. Thus,

$$\ln \rho_a = \ln K + B_1 \ln \rho_1 + B_2 \ln \rho_2 \tag{29}$$

If we write

$$\rho_a = |\rho_a| e^{j\phi_a}, \qquad \rho_1 = |\rho_1| e^{j\phi_1}, \quad \text{and} \quad \rho_2 = |\rho_2| e^{j\phi_2}$$

it follows that

$$\begin{aligned} \ln|\rho_a| = \ln|K| &+ \mathrm{Re}\{\mathrm{B}_1\}\ln|\rho_1| - \mathrm{Im}\{\mathrm{B}_1\}\phi_1 \\ &+ \mathrm{Re}\{\mathrm{B}_2\}\ln|\rho_2| - \mathrm{Im}\{\mathrm{B}_2\}\phi_2 \end{aligned} \tag{30}$$

and

$$\begin{aligned} \phi_a = \phi_k &+ \mathrm{Re}\{\mathrm{B}_1\}\phi_1 + \mathrm{Im}\{\mathrm{B}_1\}\ln|\rho_1| \\ &+ \mathrm{Re}\{\mathrm{B}_2\}\phi_2 + \mathrm{Im}\{\mathrm{B}_2\}\ln|\rho_2| \end{aligned} \tag{31}$$

If we now focus our attention on the frequency dependence of $\ln|\rho_a|$ and ϕ_d and neglect higher order terms, we obtain Guptasarma's (1984) approximations. They read

$$\frac{d\ln|\rho_a|}{d\ln\omega} \simeq \mathrm{Re}\{\mathrm{B}_1\}\frac{d\ln|\rho_1|}{d\ln\omega} + \mathrm{Re}\{\mathrm{B}_2\}\frac{d\ln|\rho_2|}{d\ln\omega} \tag{32}$$

and

$$\phi_a \simeq \mathrm{Re}\{\mathrm{B}_1\}\phi_1 + \mathrm{Re}\{\mathrm{B}_2\}\phi_2 \tag{33}$$

where we note that $\mathrm{Re}\{\mathrm{B}_1\} = 1 - \mathrm{Re}\{\mathrm{B}_2\}$.

In order for (32) and (33) be valid approximations to (30) and (31), it is evident that $B_2(j\omega)$ should not vary too rapidly with ω and also the imaginary part of $B_2(j\omega)$ should be small compared with the real part. The neglect of ϕ_k would be justified because $\phi_a \to 0$ when both ϕ_1 and $\phi_2 \to 0$. It is also evident that (32) and (33) also follow from (28) if the imaginary parts of B_1 and B_2 are ignored. Thus, in spite of widely differing derivations, Song and Vozoff's [1985] final working formulas are very similar to those presented by Guptasarma [1984]. In both cases, frequency dependent dilution factors are needed. While the authors present convincing evidence that their methods work for specific situations, it is not clear if such approximate procedures have wider applications. Also, the extension to the time domain would raise some difficulties. We discuss this aspect of the problem below.

e. Dilution and Distortion

We now recall that

$$\rho_a(j\omega) = \rho_a\left[\rho_1(j\omega), \rho_2(j\omega)\right] \tag{34}$$

Then we adopt the view that the *D.C.* or zero frequency solutions are a convenient reference. Thus, set

$$\rho_1(j\omega) = \rho_1(0)\left[1 + \delta_1(j\omega)\right] \tag{35}$$

$$\rho_2(j\omega) = \rho_2(0)\left[1 + \delta_2(j\omega)\right] \tag{36}$$

and

$$\rho_3(j\omega) = \rho_a(0)\left[1 + \delta_a(j\omega)\right] \tag{37}$$

where δ_1, δ_2 and δ_a can be regarded as dispersion functions which vanish as $\omega \to 0$. We now employ a MacLaurin series expansion so that in effect

$$\rho_a(1 + \delta_a) = \sum_{n=0}^{N-1} \frac{1}{n!}\left[\rho_1\delta_1\frac{\partial}{\partial\rho_1} + \rho_2\delta_2\frac{\partial}{\partial\rho_2}\right]^n \rho_a(\rho_1, \rho_2) + R_N \tag{38}$$

where R_N is a suitably defined remainder term [Sokolnikoff 1939]. It is to be noted that the derivatives in (38) involve only *DC* resistivities. To be explicit, we should note that

$$\frac{\partial\rho_a}{\partial\rho_1} = \lim_{\substack{\delta_1\to 0\\ \delta_2\to 0}}\left[\frac{\partial\rho_a(\rho_1(1+\delta_1), \rho_2(1+\delta_2))}{\partial(\rho_1(1+\delta_1))}\right] \tag{39}$$

with a similar convention for the other derivatives.

In accordance with the form of (38) we may write

$$\delta_a(j\omega) = B_1\delta_1(j\omega) + B_2\delta_2(j\omega) + B_{1,1}\delta_1{}^2(j\omega) + B_{2,2}\delta_2{}^2(j\omega) + B_{1,2}\delta_1(j\omega)\delta_2(j\omega) + \cdots \tag{40}$$

where, to be explicit,

$$B_1 = \frac{\rho_1(0)}{\rho_a(0)}\frac{\partial\rho_a(0)}{\partial\rho_1(0)} = \frac{\rho_1}{\rho_a}\frac{\partial\rho_a}{\partial\rho_1} = \frac{\partial\ln\rho_a}{\partial\ln\rho_1} \tag{41}$$

is, by definition, a dilution factor. Similarly

$$B_2 = \frac{\rho_2}{\rho_a}\frac{\partial\rho_a}{\partial\rho_2} = \frac{\partial\ln\rho_a}{\partial\ln\rho_2} \tag{42}$$

is also a dilution factor. Both B_1 and B_2 as defined involve only DC resistivity functions and such a convention conforms with the original works of Siegel [1959]. As indicated, the dilution concept refers to the first order diminuition of the responses of the constitutive regions * On the other hand, the higher order terms in (42) produce a more complicated change of the apparent dispersion function [Wait 1981]. Thus, we call these coefficients distortion factors and the second order forms are defined by

$$B_{1,1} = \frac{1}{2}\frac{\rho_1{}^2}{\rho_a}\frac{\partial^2\rho_a}{\partial\rho_1{}^2} \tag{43}$$

$$B_{2,2} = \frac{1}{2}\frac{\rho_2{}^2}{\rho_a}\frac{\partial^2\rho_a}{\partial\rho_2{}^2} \tag{44}$$

and

$$B_{1,2} = \frac{\rho_1\rho_2}{\rho_a}\frac{\partial^2\rho_a}{\partial\rho_1\partial\rho_2} \tag{45}$$

where again only DC (real) resistivity functions are used.

---

* From Webster: Dilute — to make thinner, to diminish,.....
Distort — to twist out of regular shape,.....

Exercise Using the "homogeneity" property stated by (18), show that the following identities exist [Wait 1986b]

$$B_1 = 1 - B_2 \tag{46}$$

$$B_{1,1} + B_{1,2} + B_{2,2} = 0 \tag{47}$$

and

$$B_{1,1} = B_{2,2} = -\frac{1}{2}B_{1,2} \tag{48}$$

Also show that

$$B_{1,2} = \frac{\partial B_2}{\partial \ln \rho_1} + B_1 B_2 \tag{49}$$

f. Extension to Time Domain

The series expansion of the dispersion function, involving only DC resistivities, is particularly suitable for dealing with the time-domain or transient response [Wait 1982, 1986b]. We discuss this problem here in the context of the quasi-static theory.

When a step current $I_0 u(t)$ is injected into the current electrodes, the corresponding voltage response $v(t)$ is expressible in the form

$$v(t) = I_0 \rho_a(0) A(t) \tag{50}$$

where $\rho_a(0)$ is the DC apparent resistivity and $A(t)$ is the normalized step response (i.e. $A(t) \to 1$ as $t \to \infty$). Then from linear system theory

$$A(t) = \frac{1}{\rho_a(0)} \mathcal{L}^{-1} \frac{\rho_a(s)}{s} \tag{51}$$

where $\mathcal{L}^{-1}$ is the inverse of the Laplace transform operator. Alternatively, we can write

$$\rho_a(s)/s = \rho_a(0) \mathcal{L} A(t) \tag{52}$$

Clearly, if $\rho_a(s)$ could be replaced by $\rho_a(0)$, corresponding to non-dispersion, $A(t)$ would simply be the unit step function $u(t)$.

If we now use (51) in conjunction with (40), it is evident that the step function response is given by

$$\begin{aligned} A(t) =& u(t) + B_1 \mathcal{L}^{-1} \frac{\delta_1(s)}{s} + B_2 \mathcal{L}^{-1} \frac{\delta_2(s)}{s} + B_{1,1} \mathcal{L}^{-1} \frac{{\delta_1}^2(s)}{s} \\ &+ B_{2,2} \mathcal{L}^{-1} \frac{{\delta_2}^2(s)}{s} + B_{1,2} \mathcal{L}^{-1} \frac{\delta_1(s)\delta_2(s)}{2} + \cdots \end{aligned} \tag{53}$$

The functions $B_1, B_2, B_{1,1} \cdots$ etc are real and time independent.

To discuss the time domain responses in a more explicit manner, it is desirable to introduce the "decay function" $M(t)$ which for $t > 0$ is defined by

$$M(t) = 1 - A(t) \tag{54}$$

Clearly, this corresponds to the response at the potential electrodes *following* the cessation of a steady current I_0 applied to the current electrodes. Then we may write

$$\begin{aligned} M(t) =& B_1 m_1(t) + B_2 m_2(t) + B_{1,1} m_{1,1}(t) + B_{2,2} m_{2,2}(t) \\ &+ B_{1,2} m_{1,2}(t) + \cdots \end{aligned} \tag{55}$$

where

$$m_1(t) = -\mathcal{L}^{-1}\delta_1(s)/s \tag{56}$$
$$m_2(t) = -\mathcal{L}^{-1}\delta_2(s)/s \tag{57}$$
$$m_{1,1}(t) = -\mathcal{L}^{-1}\delta_1{}^2(s)/s \tag{58}$$
$$m_{2,2}(t) = -\mathcal{L}^{-1}\delta_2{}^2(s)/s \tag{59}$$
$$m_{1,2}(t) = -\mathcal{L}^{-1}\delta_1(s)\delta_2(s)/s \tag{60}$$

It is certainly evident from the form of (55) that the first two terms on the R.H.S. represent pure "dilution" of the response. The succeeding second order terms lead to "distortion" or mutilation of the decay curve shapes.

By a simple application of the convolution theorem for Laplace transforms, we may express (55) in the explicit form [Wait, 1986b]:

$$\begin{aligned} M(t) =& B_1 m_1(t) + B_2 m_2(t) \\ &- B_{1,1}\left[\int_0^t m_1{}'(t-\tau)m_1(\tau)d\tau + m_1(0)m_1(t)\right] \\ &- B_{2,2}\left[\int_0^t m_2{}'(t-\tau)m_2(\tau)d\tau + m_2(0)m_2(t)\right] \\ &- B_{1,2}\left[\int_0^t m_1{}'(t-\tau)m_2(\tau)d\tau + m_1(0)m_2(t)\right] + \cdots \end{aligned} \tag{61}$$

Clearly higher order terms involve multiple convolution integrals that represent further distortion of the decay curves.

Exercise Choose $m_1(t) = m_1 e^{-\alpha t}u(t)$ and $m_2(t) = m_2 e^{-\beta t}u(t)$ where

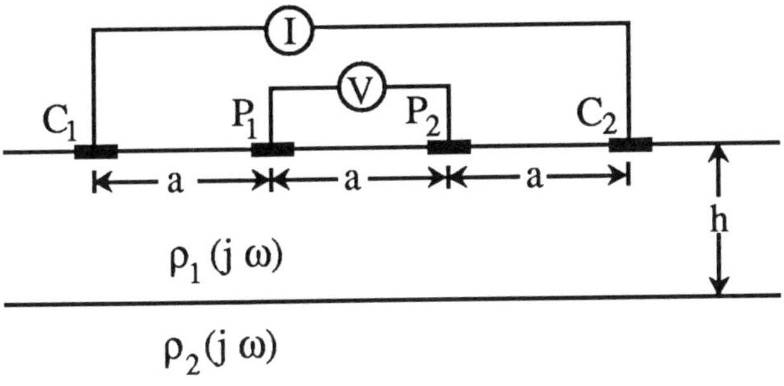

Figure 1.2.2 Four electrode Wenner array over a two layer polarizable earth.

$m_1 = m_1(0)$ and $m_2 = m_2(0)$ and where α and β are decay constants for the assumed exponential responses. Then show that [Wait 1986b]:

$$\begin{aligned} M(t) =& B_1 m_1 e^{-\alpha t} + B_2 m_2 e^{-\beta t} - B_{1,1} m_1^2 (1 - \alpha t) e^{-\alpha t} \\ & - B_{2,2} m_2^2 (1 - \beta t) e^{-\beta t} - B_{1,2} m_1 m_2 \frac{1}{\beta - \alpha} \left[\beta e^{-\beta t} - \alpha e^{-\alpha t} \right] \\ & + \cdots \end{aligned} \tag{62}$$

g. Two Layer Polarizable Earth Model

To illustrate the concept of dilution and distortion, it is desirable to adopt a concrete model which involves two well-defined regions. Such an example is a two layer flat earth model as illustrated in Fig. 1.2.2. For convenience, a four electrode array is shown with the current electrodes, C_1 and C_2, outside the potential electrodes, P_1 and P_2. The common spacing between the electrodes of this so-called Wenner array is a. The current injected into the electrodes, C_1 and C_2, is $I \exp(j\omega t)$. The resulting open circuit voltage at the potential electrodes is $V \exp(j\omega t)$. The apparent complex resistivity $\rho_a(j\omega)$ is defined according to

$$\frac{V}{I} = \frac{\rho_a(j\omega)}{2\pi a} \tag{63}$$

Assuming the validity of potential theory (i.e. electromagnetic coupling is negligible), it follows that

$$\frac{\rho_a(j\omega)}{\rho_1(j\omega)} = 1 + 2y \sum_{m=1}^{\infty} [N(j\omega)]^m \Omega(y,m) \tag{64}$$

where

$$y = a/h = \text{electrode spacing/upper layer thickness}$$
$$N(j\omega) = \frac{\rho_2(j\omega) - \rho_1(j\omega)}{\rho_2(j\omega) + \rho_1(j\omega)} \tag{65}$$

and

$$\Omega(y,m) = \frac{1}{m}\left[\frac{1}{[1+(y^2/4m^2)]^{\frac{1}{2}}} - \frac{1}{[1+(y^2/m^2)]^{\frac{1}{2}}}\right] \tag{66}$$

Exercise Derive the preceding expression for ρ_a/ρ_1 assuming *DC* theory is relevant using an image method. Note, C_1 can be treated as a current point source and C_2 a current point sink [e.g. Wait, 1982, Chap. 1].

In order to show explicit results, we need to specify a dispersion model for two polarizable regions. We follow current practice and adopt the Cole-Cole representation which is a versatile four-parameter model. In the present context,

$$\rho_1(j\omega) = \rho_1(\infty) + \frac{\rho_1(0) - \rho_1(\infty)}{1 + (j\omega\tau_1)^{k_1}} \tag{67}$$

and

$$\rho_2(j\omega) = \rho_2(\infty) + \frac{\rho_2(0) - \rho_2(\infty)}{1 + (j\omega\tau_2)^{k_2}} \tag{68}$$

where $\rho_i(0)$ and $\rho_i(\infty)$ are the zero and infinite frequency limits, respectively, of the complex resistivity of the ith layer ($i = 1, 2$), τ_i is a relaxation time constant, and k_i is a dispersion index. The index k_j would be 1 for a pure polar liquid but for geological materials it would range typically from 0.6 to 0.1.

Using the basic definitions for the dilution factors as given by (41) and (42), we may deduce the "first order" approximation

$$\rho_a(j\omega) \cong \rho_a(0)[1 + \delta_a(j\omega)] \tag{69}$$

where

$$\delta_a(j\omega) = B_1\delta_1(j\omega) + B_2\delta_2(j\omega) \tag{70}$$

The "second order" approximation is given by

$$\begin{aligned}\delta_a(j\omega) =&B_1\delta_1(j\omega) + B_2\delta_2(j\omega) + B_{1,1}{\delta_1}^2(j\omega) \\ &+ B_{2,2}{\delta_2}^2(j\omega) + B_{1,2}\delta_1(j\omega)\delta_2(j\omega)\end{aligned} \tag{71}$$

where the indicated distortion factors are deduced from (43), (44) and (45). Explicit expressions for the dilution and distortion factors are given by Gruszka and Wait [1985] for this particular configuration.

For the numerical example, we choose the "chargeability" $m_2 = [\rho_2(0) - \rho_2(\infty)]/\rho_2(0) = 0.8$ and $\rho_1(j\omega) = \rho_1(0)$. In other words, the lower region is polarizable but the upper layer (i.e. the overburden) is taken to be non-polarizable. To be representative of an actual field situation, the dispersion index k_2 for the lower region is taken to be $1/2$. Then, we specify three values of the $D.C.$ resistivity ratios of the two regions; namely,

$$C = \rho_2(0)/\rho_1(0) = 0.1,\ 1.0 \text{ and } 10$$

For the conditions indicated above, we graph the complex values of $\rho_a(j\omega)/\rho_1(0)$ in the plane in Fig. 1.2.3. The "exact" version corresponds to using (64) with the corresponding Cole-Cole form for $\rho_2(j\omega)$ given by (68). The parameter $\omega\tau$ ranges through all values from 0 to ∞ for each value of C indicated. For purpose of comparison, the curves for the first approximation given by (69) and (70) are also shown. In addition, we show the second order approximation given by (69) and (71). All curves are arcs or approximately arcs of circles in accordance with the expected behavior for a Cole-Cole model [Daniel 1949, Wait 1984]. But more significantly, it is seen that the first order and even the second order approximated curves may diverge significantly from the exact form. This discrepancy is most notable for the case where $C = 0.1$ corresponding to a relatively well-conducting overburden. When the lower region is better conducting than the upper layer, the divergence of the curves is almost insignificant.

Actually, the numerical example we have chosen is rather extreme in the sense that the polarization (i.e. dispersion) in the lower layer is very high. When m_2 is reduced from 0.8 to say 0.2, the first order approximation, involving just the real dilution factors, B_1 and B_2, is very good for all values of the resistivity contrast. Some further examples are given by Gruszka and Wait [1985].

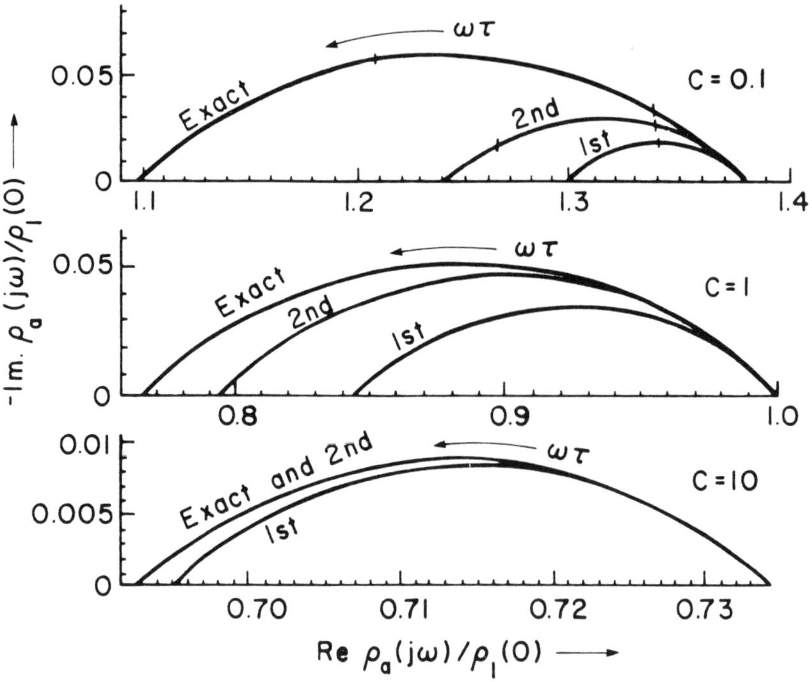

Figure 1.2.3 Complex resistivity trajectories in Argand plane for two layer polarizable earth model.

Exercise Consider a multi-region model [Wait, 1982] where

$$\rho_a(j\omega) = \rho_a[\rho_1(j\omega),\ \rho_2(j\omega),\ \rho_3(j\omega),\ \cdots \rho_J(j\omega)] \qquad (72)$$

where $\rho_i(j\omega)$ is the complex resistivity of the ith homogeneous region (and $i = 1, 2, 3 \cdots J$). In analogy, to the two region cases, show that

$$\rho_a(j\omega) = \rho_a(0)\,[1 + \delta_a(j\omega)] \qquad (73)$$

where

$$\delta_a(j\omega) \simeq \sum_{i=1}^{J} B_i \delta_i(j\omega) \qquad (74)$$

when the dilution factors are defined in terms of the $D.C.$ resistivities such that

$$B_i = \frac{\rho_i}{\rho_a}\frac{\partial \rho_a}{\partial \rho_i} = \frac{\partial \ln \rho_a}{\partial \ln \rho_i} \tag{75}$$

and

$$\sum_{i=1}^{J} B_i = 1 \tag{76}$$

h. An Alternative Approach

A further approximate representation follows from the fact that for a two-region problem, we can write

$$\rho_a(j\omega) = \rho_1(j\omega) f\left(\frac{\rho_2(j\omega)}{\rho_1(j\omega)}\right) \tag{77}$$

where f is a function of the complex ratio of the two complex resistivities. An equivalent statement of (77) is

$$\frac{\rho_a(0)[1+\delta_a(j\omega)]}{\rho_1[1+\delta_1(j\omega)]} = f\left(\frac{\rho_2}{\rho_1}[1+\delta_r(j\omega)]\right) \tag{78}$$

where

$$1+\delta_r(j\omega) = \frac{1+\delta_2(j\omega)}{1+\delta_1(j\omega)} \tag{79}$$

and where it is understood that $\rho_1 = \rho_1(0)$ and $\rho_2 = \rho_2(0)$ are the $D.C.$ resistivity values.

Following Gruszka [1987], we now expand the right hand side of (78) about $\delta_r = 0$. Thus

$$\begin{aligned}\frac{\rho_a(0)[1+\delta_a(j\omega)]}{\rho_1[1+\delta_1(j\omega)]} =& f\left(\frac{\rho_2}{\rho_1}\right) + \frac{\rho_2}{\rho_1} f^{(1)}\delta_r + \frac{1}{2!}\left(\frac{\rho_2}{\rho_1}\right)^2 f^{(2)}\delta_r^2 \\ &+ \frac{1}{3!}\left(\frac{\rho_2}{\rho_1}\right)^3 f^{(3)}\delta_r^3 + \cdots + \frac{1}{n!}\left(\frac{\rho_2}{\rho_1}\right)^n f^{(n)}\delta_r^n + \cdots\end{aligned} \tag{80}$$

where

$$f^{(n)} = f^{(n)}\left(\frac{\rho_2}{\rho_1}\right) = \frac{d^n f(x)}{dx^n}\bigg|_{x=\rho_2/\rho_1} \tag{81}$$

A similar expansion was also employed by Guptasarma [1984]. As Gruszka [1987] points out, we can solve for $\delta_a(j\omega)$ by inserting the expression for $\delta_r(j\omega)$, given by (79), into (80) to yield

$$\delta_a(j\omega) = \delta_1(j\omega) + \beta_1\left(\frac{\rho_2}{\rho_1}\right)(\delta_2 - \delta_1) + \beta_2\left(\frac{\rho_2}{\rho_1}\right)\frac{(\delta_2-\delta_1)^2}{1+\delta_1} + \cdots + \beta_n\left(\frac{\rho_2}{\rho_1}\right)\frac{(\delta_2-\delta_1)^n}{(1+\delta_1)^{n-1}} \tag{82}$$

where

$$\beta_n(x) = \frac{1}{n!}\, x^n\, \frac{f^{(n)}(x)}{f(x)} \tag{83}$$

$$\delta_1 = \delta_1(j\omega) \qquad \text{and} \qquad \delta_2 = \delta_2(j\omega)$$

Actually, (82) is a compact version of (71) as can be verified by expanding the terms of the form $(1+\delta_1)^{-m}$ about δ_1 and then collecting terms in the same order of δ_1 and δ_2. This process yields the identities

$$B_1 = 1 - \beta_1,\ B_2 = \beta_1,\ B_{11} = B_{22} = \beta_2, \text{ and } B_{12} = -2\beta_2$$

While (82) is more compact than (71), it is not as convenient for application to time domain responses as displayed by (61).

Another extension leads to expansion coefficients β_1, β_2, $\cdots$ which are complex. In this case, again following Gruszka [1987], we begin by noting that

$$\rho_a(j\omega) = |\rho_a(j\omega)|\ \exp(j\phi_a) \tag{84}$$

$$\rho_1(j\omega) = |\rho_1(j\omega)|\ \exp(j\phi_1) \tag{85}$$

and

$$\rho_2(j\omega) = |\rho_2(j\omega)|\ \exp(j\phi_2) \tag{86}$$

Thus, (78) with obvious contraction of notation can be written

$$\left|\frac{\rho_a(j\omega)}{\rho_1(j\omega)}\right| e^{j(\phi_a-\phi_1)} = f\left(\left|\frac{\rho_2(j\omega)}{\rho_1(j\omega)}\right| e^{j\phi_r}\right) \tag{87}$$

where

$$\phi_r = \phi_2 - \phi_1$$

We now take the natural logarithm of both sides of (83) to give

$$\ln\left|\frac{\rho_a(j\omega)}{\rho_1(j\omega)}\right| + j(\phi_a - \phi_1) = \ln f(\cdots) \tag{88}$$

Then, on expanding the right hand side of (88) about $\phi_r = 0$, we obtain

$$\begin{aligned}\ln\left|\frac{\rho_a(j\omega)}{\rho_1(j\omega)}\right| + j(\phi_a - \phi_1) &= \ln f(|z|) + j\beta_1(|z|)\phi_r \\ &-[\beta_2(|z|) + \frac{1}{2}\beta_1(|z|) - \frac{1}{2}\beta_1^2(|z|)]\phi_r^2 + \cdots\end{aligned} \tag{89}$$

where

$$z = \rho_2(j\omega)/\rho_1(j\omega)$$

$$\beta_1(x) = \frac{x}{f(x)}\frac{\partial f(x)}{\partial x} \quad \text{and} \quad \beta_2(x) = \frac{x^2}{2}\frac{\partial^2 f(x)}{\partial x^2}\frac{1}{f(x)}$$

where $x = |z|$ is real !

The definitions of $\beta_1(|z|)$ and $\beta_2(|z|)$, as a function of amplitude $|z|$, are consistent with (83), but note here $\beta_1(|z|)$ and $\beta_2(|z|)$ are frequency dependent real functions. If we now equate the real and imaginary parts of (89), it follows that the leading terms are

$$\begin{aligned}|\rho_a(j\omega)| \simeq &|\rho_1(j\omega)| f(|z|) \\ &\exp\left\{-\left[\beta_2(|z|) + \frac{1}{2}\beta_1(|z|) - \frac{1}{2}\beta_1{}^2(|z|)\right](\phi_2 - \phi_1)^2\right\}\end{aligned} \tag{90}$$

and

$$\phi_a(j\omega) \simeq \phi_1 + \beta_1(|z|)(\phi_2 - \phi_1) \tag{91}$$

where, as implied above, $\phi_1 = \phi_1(j\omega)$ and $\phi_2 = \phi_2(j\omega)$ and $|z| = |\rho_2(j\omega)/\rho_1(j\omega)|$ are real functions of frequency.

As Gruszka [1987] points out, (90) can be further simplified if β_1 is near 0 or 1 and β_2 is small, the exponential term can be replaced by 1. The resulting expressions for the amplitude and phase of $\rho_a(j\omega)$ can be called the "G approximation": it bears some similarity with the form given by Guptasarma [1984] but there are some subtle differences. For example, here β_1 is conveniently a *real* function of the frequency.

It is an interesting and instructive exercise to compare three approximations for the complex resistivity $\rho_a(j\omega)$ over a specific two

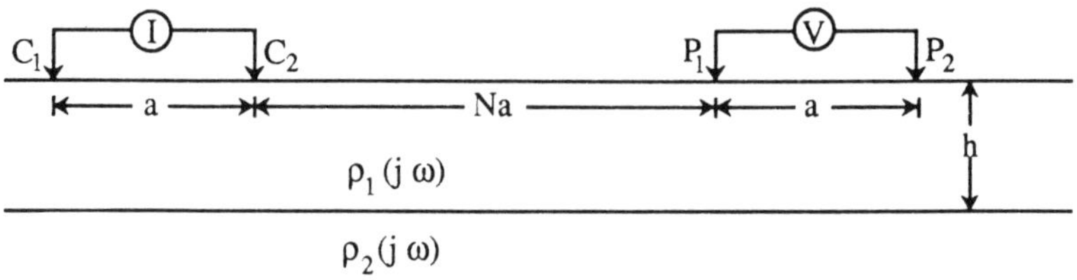

Figure 1.2.4 The dipole-dipole array over a two layer polarizable earth model.

layer model as shown in Fig. 1.2.4. Again, we are abstracting from Gruszka [1987] who used the dipole-dipole array. He gives explicit formulas for the functions β_1 and β_2 which we will not reproduce here* but they can be obtained from the prescription given by (83). The three approximations are summarized as follows

$$\rho_a(j\omega)/\rho_1 = \text{RHS (Right Hand Side)}$$

where, for the S (for Seigel) approximation,

$$\text{RHS} \simeq 1 + \delta_1(j\omega) + \beta_1\left(\frac{\rho_2}{\rho_1}\right)[\delta_2(j\omega) - \delta_1(j\omega)] \tag{92}$$

where, for the W (for Wait) approximation,

$$\begin{aligned}\text{RHS} \simeq & 1 + \delta_1(j\omega) + \beta_1\left(\frac{\rho_2}{\rho_1}\right)[\delta_2(j\omega) - \delta_1(j\omega)] \\ & + \beta_2\left(\frac{\rho_2}{\rho_1}\right)[\delta_2(j\omega) - \delta_1(j\omega)]^2\end{aligned} \tag{93}$$

and where, for the G (for Gruszka) approximation,

$$\begin{aligned}\text{RHS} \simeq & \left|\frac{\rho_1(j\omega)}{\rho_1}\right| \frac{f\left(|\rho_2(j\omega)/\rho_1(j\omega)|\right)}{f(\rho_2/\rho_1)} \\ & \exp\left[j\left(\phi_1(j\omega) + \beta_1\left|\frac{\rho_2(j\omega)}{\rho_1(j\omega)}\right|(\phi_2(j\omega) - \phi_1(j\omega))\right)\right]\end{aligned} \tag{94}$$

* Actually, $\beta_1 = B_2$ and $\beta_2 = B_{22}$, and explicit expressions are given for $B_2(x)$ and $B_{22}(x)$ by (101) and (103), respectively.

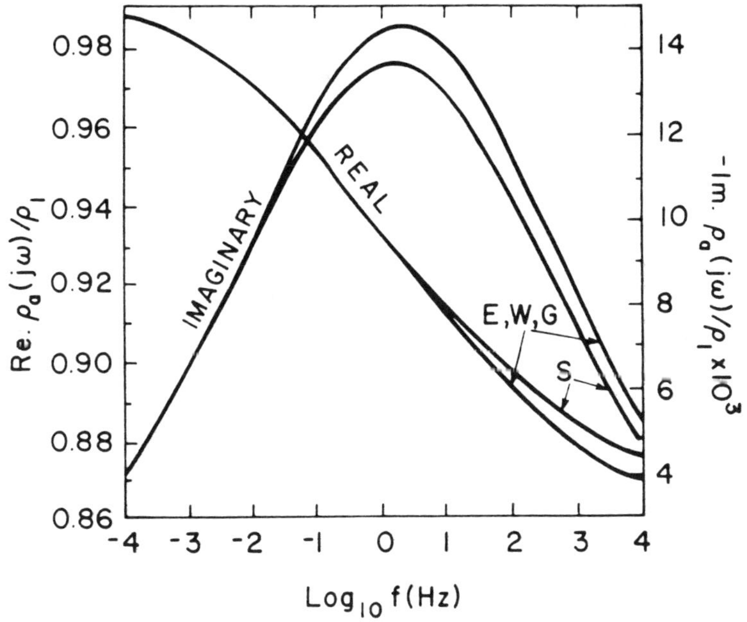

Figure 1.2.5 Real and imaginary parts of the normalized apparent resistivity for a dipole-dipole array over a two layer earth model where the lower layer is polarizable. Approximations S (Seigel), W (Wait) and G (Gruszka) are shown along with E (exact) curves. Parameters: $\rho_1 = 10^3$ ohm m, $m_2 = 0.2, \tau_2 = 0.1$ sec., $k_2 = 0.25, N = 2, a = 20$ m, and $h = 10$m .

We characterize the two layers by Cole-Cole forms; thus

$$\rho_i(j\omega) = \rho_i \left[1 - m_i \left(1 - \frac{1}{1 + (j\omega\tau_i)^k}\right)\right] \tag{95}$$

where $i = 1, 2$. For the example illustrated here, we set $m_1 = 0$ so the upper layer is non polarizable, i.e. $\rho_1(j\omega) = \rho_1$.

Using (92), (93) and (94), we plot, in . 1.2.5, the real and imaginary parts of the normalized apparent resistivity as a function of frequency. For comparison, the corresponding "exact" results denoted by E, from Gruszka and Wait [1985] are also shown. The specific parameter values are indicated in the caption to Fig. 1.2.5. As can be seen, only the Seigel or S approximation departs significantly from the exact calculations. Other more detailed comparisons have been given by Gruszka [1987].

Graphs of Dilution and Distortion Factors

We summarize here some results for the dilution factor B_2 and $B_{2,2}$ for dipole-dipole array located on the surface of a two-layer earth model. Electromagnetic coupling effects are not included so we may view one complication at a time. We also show analogous results for a borehole model.

The geometry of the two layer model is shown in Fig. 1.2.4 so the notation need not be discussed again. Also, the potential theory relevant to this model was already considered [Wait 1982]. Thus, we need only state the working formulas.

Following (77), we may write the $D.C.$ limiting form as

$$\rho_a = \rho_1 f(\rho_2/\rho_1) \tag{96}$$

where

$$f(x) = \frac{1}{2\pi h}\sum_{i=1}^{2}\sum_{j=1}^{2}(-1)^{i+j+1}\cdot\frac{h}{x_{ij}}F(x) \tag{97}$$

and, following Gruszka (1987),

$$F(x) = 1 + \frac{2\sum_{i=1}^{2}\sum_{j=1}^{2}(-1)^{i+j+1}\int_0^\infty \frac{Ke^{-2g}}{1-Ke^{-2g}}J_0\left(g\frac{x_{ij}}{h}\right)dg}{\sum_{i=1}^{2}\sum_{j=1}^{2}(-1)^{i+j+1}\cdot\frac{h}{x_{ij}}} \tag{98}$$

where

$$K = \frac{\rho_2 - \rho_1}{\rho_2 + \rho_1} \tag{99}$$

Then to obtain the dilution factor β_2, we perform the operation

$$B_2 = \frac{\rho_2}{\rho_a}\frac{\partial\rho_a}{\partial\rho_2} \tag{100}$$

to yield

$$B_2 = \frac{\dfrac{4x}{(1+x)^2}\displaystyle\sum_{i=1}^{2}\sum_{j=1}^{2}(-1)^{i+j+1}\int_0^\infty \frac{e^{-2g}}{(1-Ke^{-2g})^2}J_0\left(g\,\frac{x_{ij}}{h}\right)dg}{F(x)\displaystyle\sum_{i=1}^{2}\sum_{j=1}^{2}(-1)^{i+j+1}\,\frac{h}{x_{ij}}} \tag{101}$$

A further differentiation is needed to obtain an explicit expression for the distortion factor $B_{2,2}$. Thus we use

$$B_{2,2} = \frac{1}{2}\frac{\rho^2}{\rho_a}\frac{\partial^2\rho_a}{\partial{\rho_2}^2} \tag{102}$$

to yield

$$B_{2,2} = \frac{\dfrac{8x^2}{(1+x)^4}\displaystyle\sum_{i=1}^{2}\sum_{j=1}^{2}(-1)^{i+j+1}\int_0^\infty \frac{e^{-4g}}{(1-Ke^{-2g})^3}J_0\left(g\,\frac{x_{ij}}{h}\right)dg}{F(x)\displaystyle\sum_{i=1}^{2}\sum_{j=1}^{2}(-1)^{i+j+1}\frac{h}{x_{ij}}} \tag{103}$$

As indicated, B_2 and B_{22} are real functions. Then, the second order approximation for the complex apparent resistivity is given by

$$\rho_a(j\omega) = \rho_a(1+\delta_a(j\omega)) \tag{104}$$

where

$$\begin{aligned}\delta_a(j\omega) =& B_1\delta_1(j\omega) + B_2\delta_2(j\omega) + B_{11}{\delta_1}^2(j\omega)\\ &+ B_{12}\delta_1(j\omega)\delta_2(j\omega) + B_{22}{\delta_2}^2(j\omega)\end{aligned} \tag{105}$$

As we have indicated before $\delta_1(j\omega)$ and $\delta_2(j\omega)$ are defined in terms of the layer complex resistivities in accordance with

$$\rho_1(j\omega) = \rho_1[1+\delta_1(j\omega)] \tag{106}$$

and

$$\rho_2(j\omega) = \rho_2[1+\delta_2(j\omega)] \tag{107}$$

Also, as we indicated before,

$$B_1 = 1 - B_2 \tag{108}$$

and

$$B_{12} = -2B_{11} = -2\ B_{22} \tag{109}$$

Thus, computed data for B_2 and B_{22} provided by Gruszka [1987] are adequate to generate the complex (apparent) resistivity up to second order in δ_1 and δ_2.

The linear distances x_{ij} in (98), (101) and (103) are related to the electrode separations according to $x_{ij} = |P_j - C_i|$. Thus, for the dipole-dipole array shown in Fig. 1.2.4, $x_{11} = C_1P_1$, $x_{12} = C_1P_2$, $X_{21} = C_2P_1$ and $x_{22} = C_2P_2$ where $C_1P_1 = (N+1)a = C_2P_2$, $C_1P_2 = (N+2)a$ and $C_2P_1 = Na$ when the electrodes are all in line.

The graphical plots of B_2 and B_{22} as a function of the resistivity ration ρ_2/ρ_1 are shown in Figs. 1.2.6 and 1.2.7 for various a/h and N as indicated in the captions. Corresponding results for B_2 and B_{22}, as a function of a/h for various ρ_2/ρ_1 and N, are shown in Figs. 1.2.8 and 1.2.9. Then, the results for B_2 and B_{22}, as a function of N for various ρ_2/ρ_1 and a/h are shown in Figs. 1.2.10 and 1.2.11. These results allow one to estimate how the frequency dependence of the measured complex resistivity for a subsurface polarizable region are diluted and/or distorted by the presence of an upper layer which may or may not be polarizable.

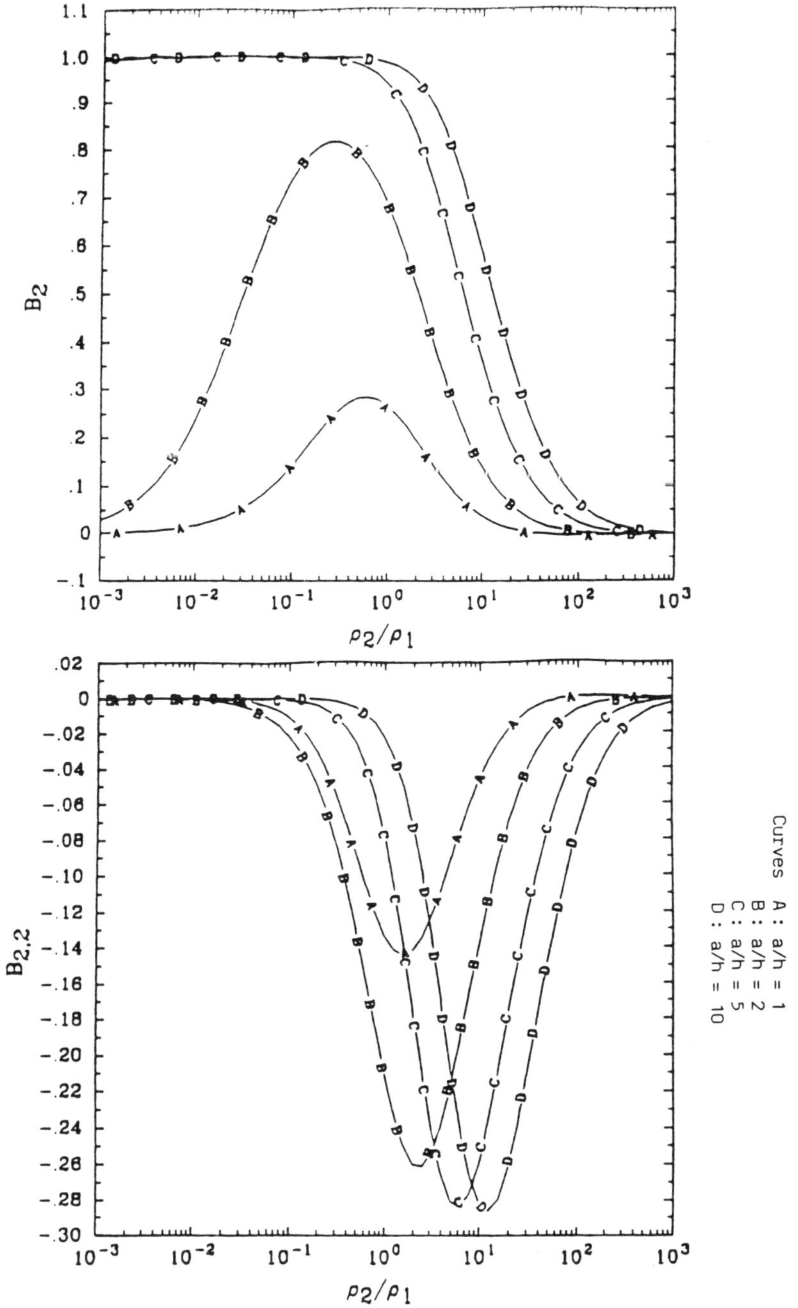

Figure 1.2.6 Dilution and distortion factors as a function of ρ_2/ρ_1 for various a/h for $N = 2$.

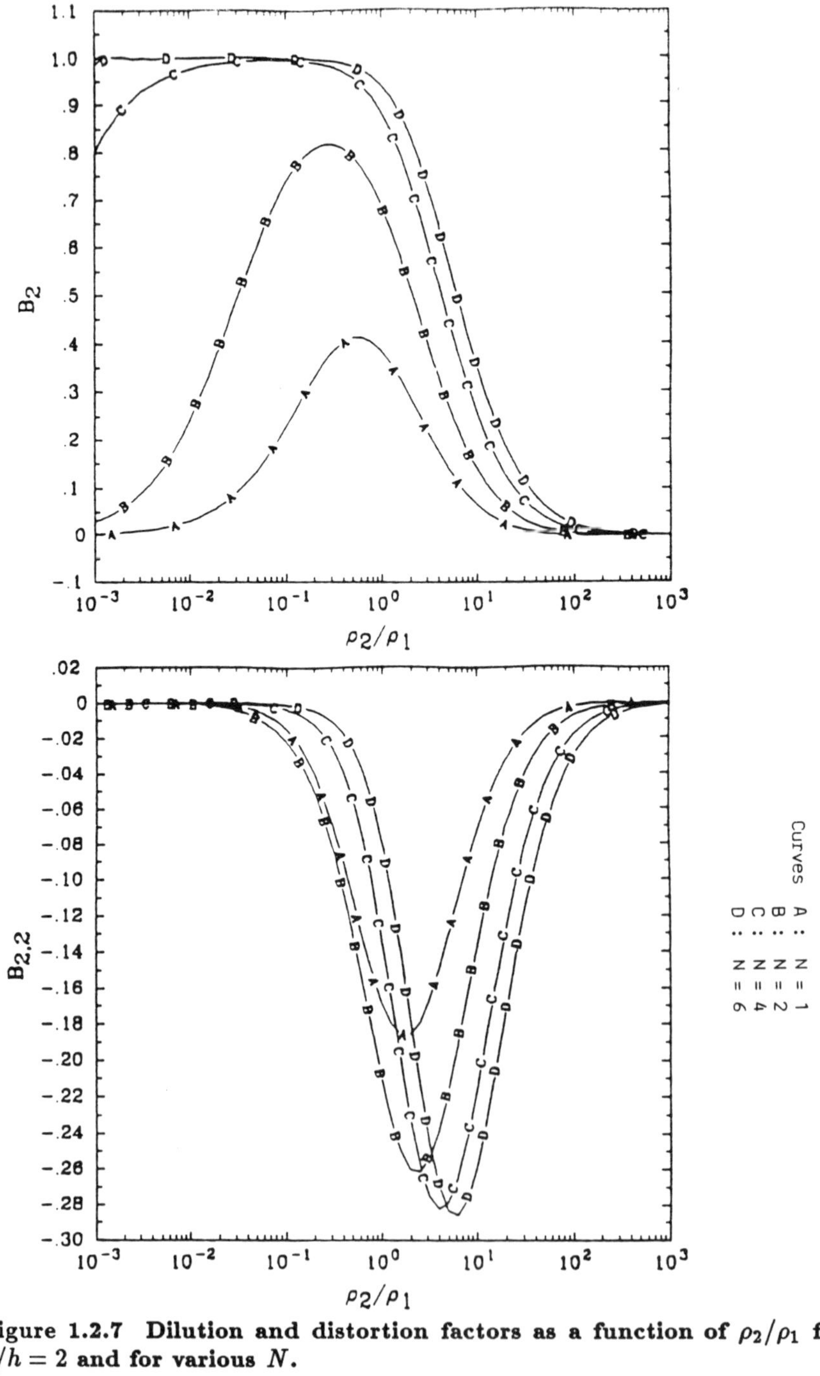

Figure 1.2.7 Dilution and distortion factors as a function of ρ_2/ρ_1 for $a/h = 2$ and for various N.

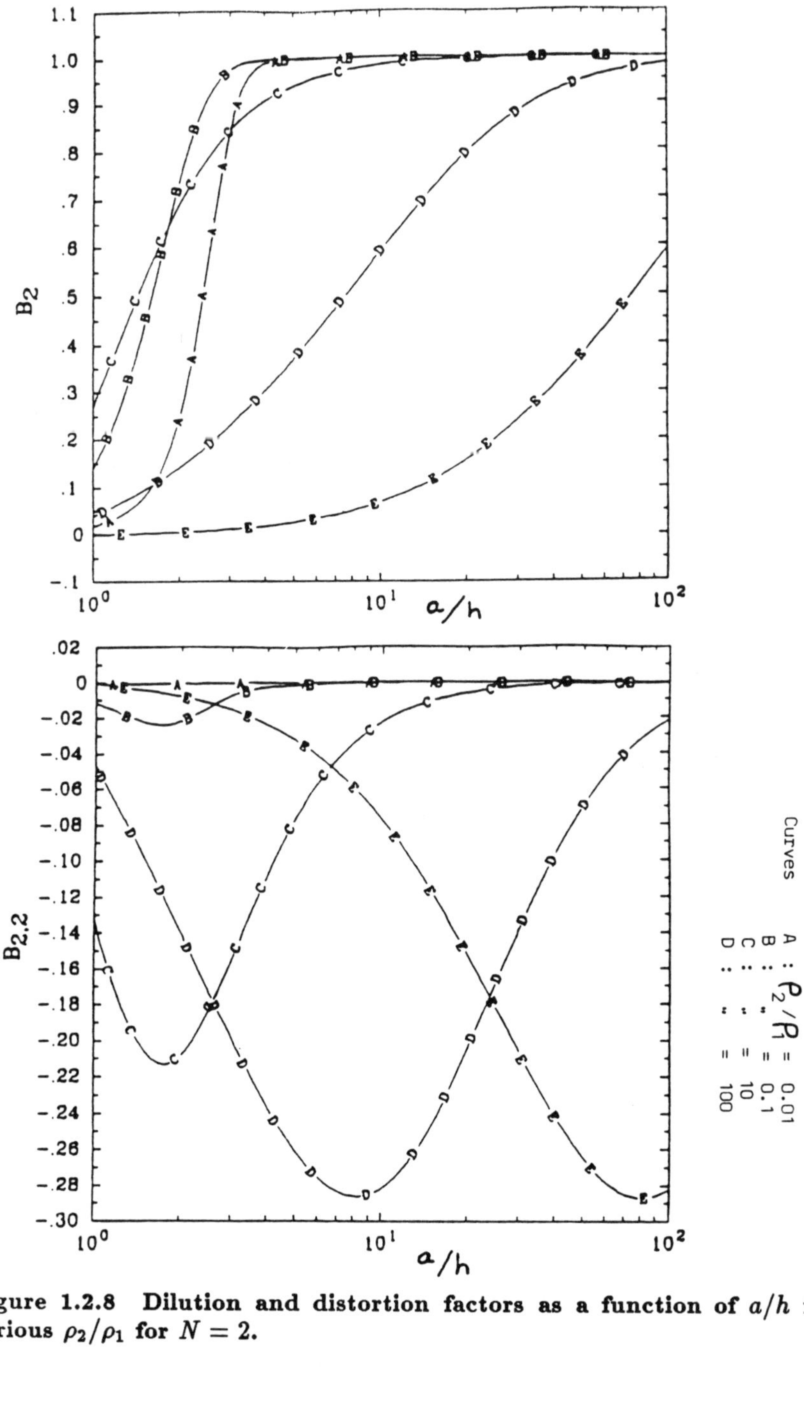

Figure 1.2.8 Dilution and distortion factors as a function of a/h for various ρ_2/ρ_1 for $N = 2$.

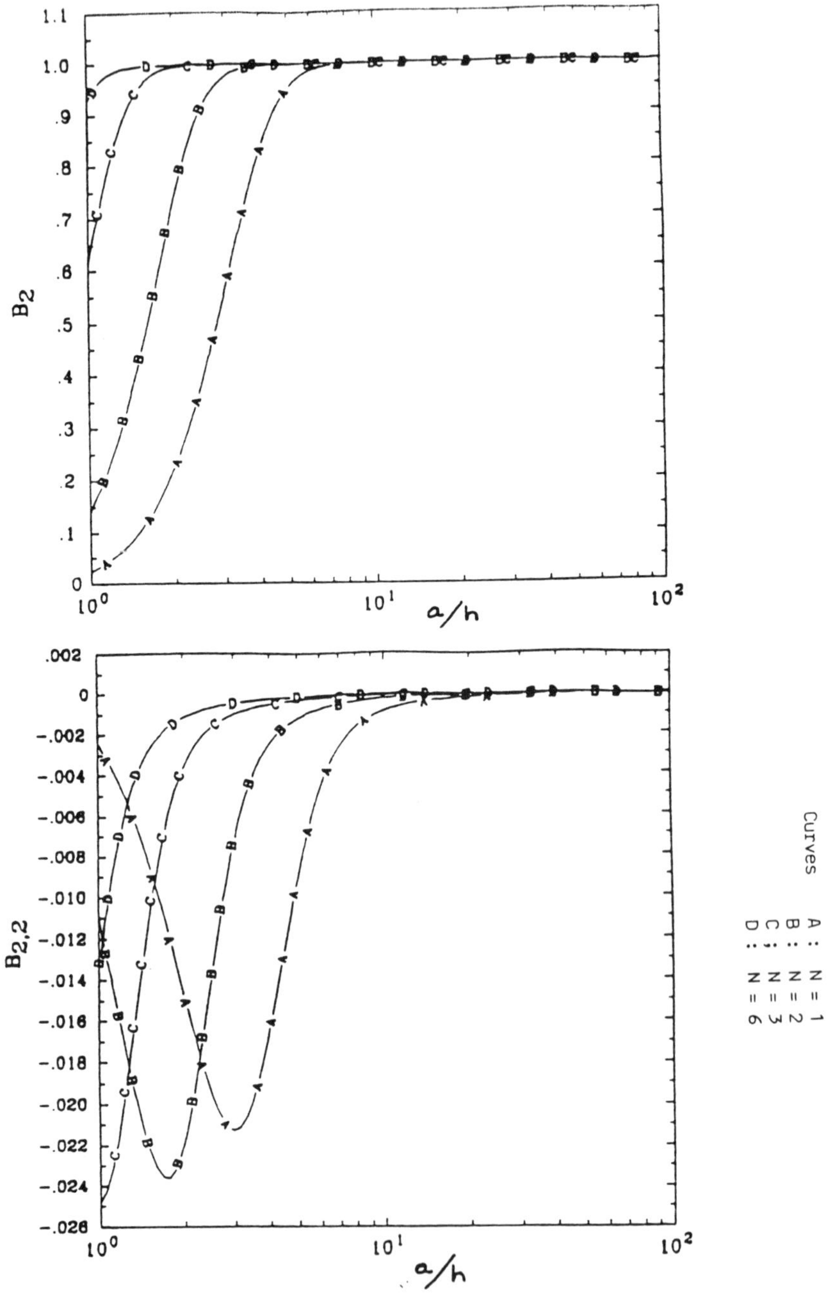

Figure 1.2.9 Dilution and distortion factors as a function of a/h for various N for ρ_2/ρ_1=0.1.

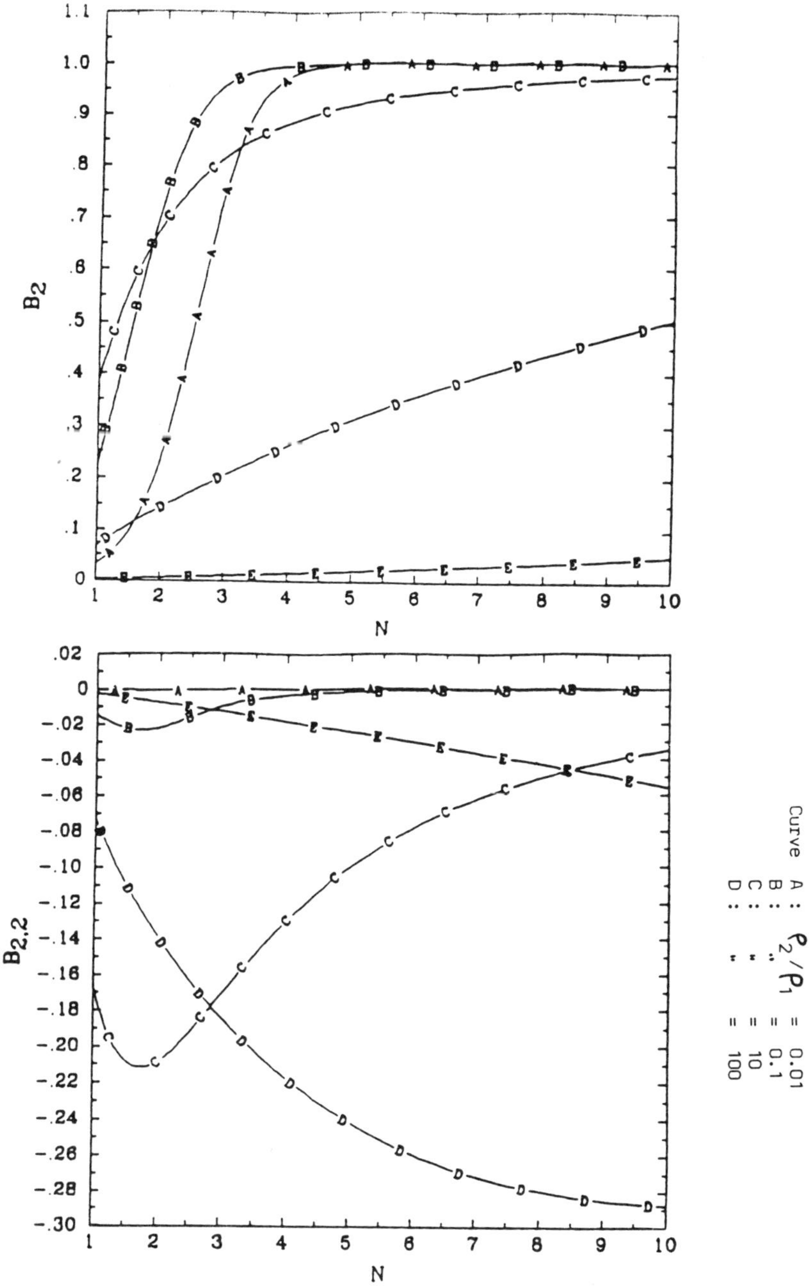

Figure 1.2.10 Dilution and distortion factors as a function of N for various ρ_2/ρ_1 for $a/h = 2$.

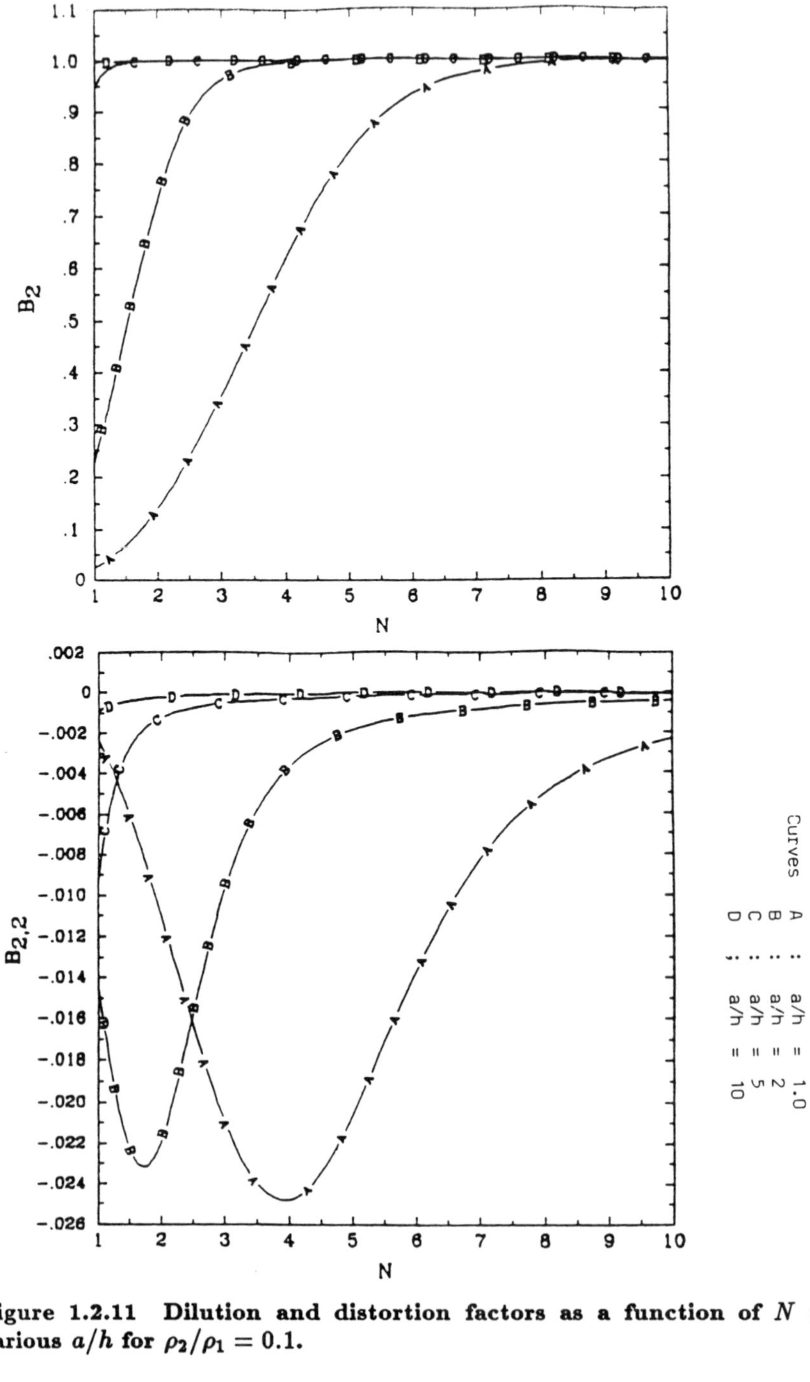

Figure 1.2.11 Dilution and distortion factors as a function of N for various a/h for $\rho_2/\rho_1 = 0.1$.

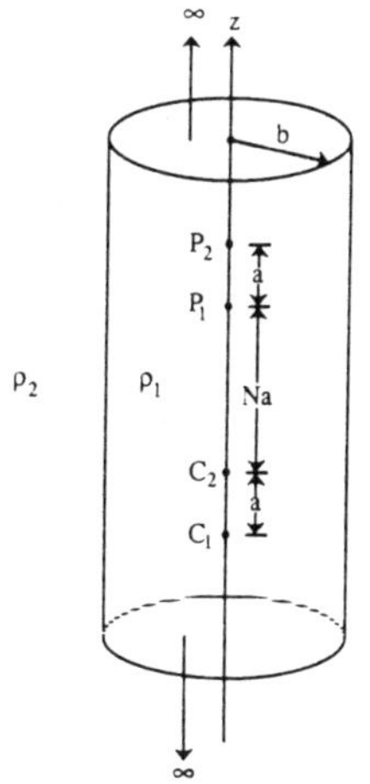

Figure 1.2.12 Dipole-dipole electrode array on the axis of a bore-hole.

We now locate the electrodes in a fluid filled bore-hole which is analogous to the two layer planar structure discussed above. The situation is shown in Fig. 1.2.12 where the electrodes are located on the axis of the cylindrical bore-hole of radius b. The resistivity of the fluid is ρ_1 and the resistivity of the external region or formation is ρ_2.

The relevant potential theory solution for this type of problem is outlined elsewhere [Chap. 1, Wait 1982, and Wait and Gruszka 1987] so we will state the working expressions without further ado. For the model shown in Fig. 1.2.12, the normalized apparent resistivity is given by

$$\frac{\rho_a}{\rho_1} = 1 + \left(\frac{2}{\pi} \sum_{j,i=1}^{2} (-1)^{i+j+1} \int_0^\infty A(\lambda b) \cos(\lambda x_{ij}) d\lambda \right) \Big/ \text{Den} \quad (110)$$

where

$$A(\lambda b) = -\frac{\lambda b(k-1) K_0(\lambda b) K_1(\lambda b)}{1 + \lambda b(k-1) I_0(\lambda b) K_1(\lambda b)} \quad (111)$$

where

$$k = \rho_1 / \rho_2 \quad (112)$$

and

$$\text{Den} = \sum_{i=1}^{2} \sum_{j=1}^{2} (-1)^{i+j+1} \frac{1}{z_{ij}} \quad (113)$$

For the dipole-dipole array:

$$z_{11} = C_1P_1 = z_{22} = C_2P_2 = (N+1)a$$
$$z_{12} = C_1P_2 = (N+2)a \text{ and } z_{21} = C_2P_1 = Na$$

The dilution factor is obtained from

$$B_2 = \frac{\rho_2}{\rho_a}\frac{\partial\rho_a}{\partial\rho_2}$$

giving

$$B_2 = \frac{\left[k\frac{2}{\pi}\sum_{i=1}^{2}\sum_{j=1}^{2}(-1)^{i+j+1}\int_0^\infty A'(\lambda b)\cos(\lambda z_{ij})d\lambda\right] \Big/ \text{Den}}{1+\left[\frac{2}{\pi}\sum_{i=1}^{2}\sum_{j=1}^{2}(-1)^{i+j+1}\int_0^\infty A(\lambda b)\cos(\lambda z_{ij})d\lambda\right] \Big/ \text{Den}} \tag{114}$$

where

$$A'(\lambda b) = \frac{\lambda b\, K_0(\lambda b)\, K_1(\lambda b)}{[1+\lambda b(k-1)I_0(\lambda b)\, K_1(\lambda b)]^2} \tag{115}$$

The distortion factor is obtained from

$$B_{22} = \frac{{\rho_2}^2}{\rho_a}\frac{\partial^2\rho_a}{\partial{\rho_2}^2} \tag{116}$$

which is conveniently written in the form

$$B_{22} = B_2 - 1 + \frac{\left[k^2\frac{2}{\pi}\sum_{i=1}^{2}\sum_{j=1}^{2}(-1)^{i+j+1}\int_0^\infty A''(\lambda b)\cos(\lambda z_{ij})d\lambda\right] \Big/ \text{Den}}{1+\left[\frac{2}{\pi}\sum_{i=1}^{2}\sum_{j=1}^{2}(-1)^{i+j+1}\int_0^\infty A(\lambda b)\cos(\lambda z_{ij})d\lambda\right] \Big/ \text{Den}} \tag{117}$$

where

$$A''(\lambda b) = \frac{(\lambda b)^2\, I_0(\lambda b)\, K_0(\lambda b)\, {K_1}^2(\lambda b)}{[1+\lambda b(k-1)\, I_0(\lambda b)\, K_1(\lambda b)]^3} \tag{118}$$

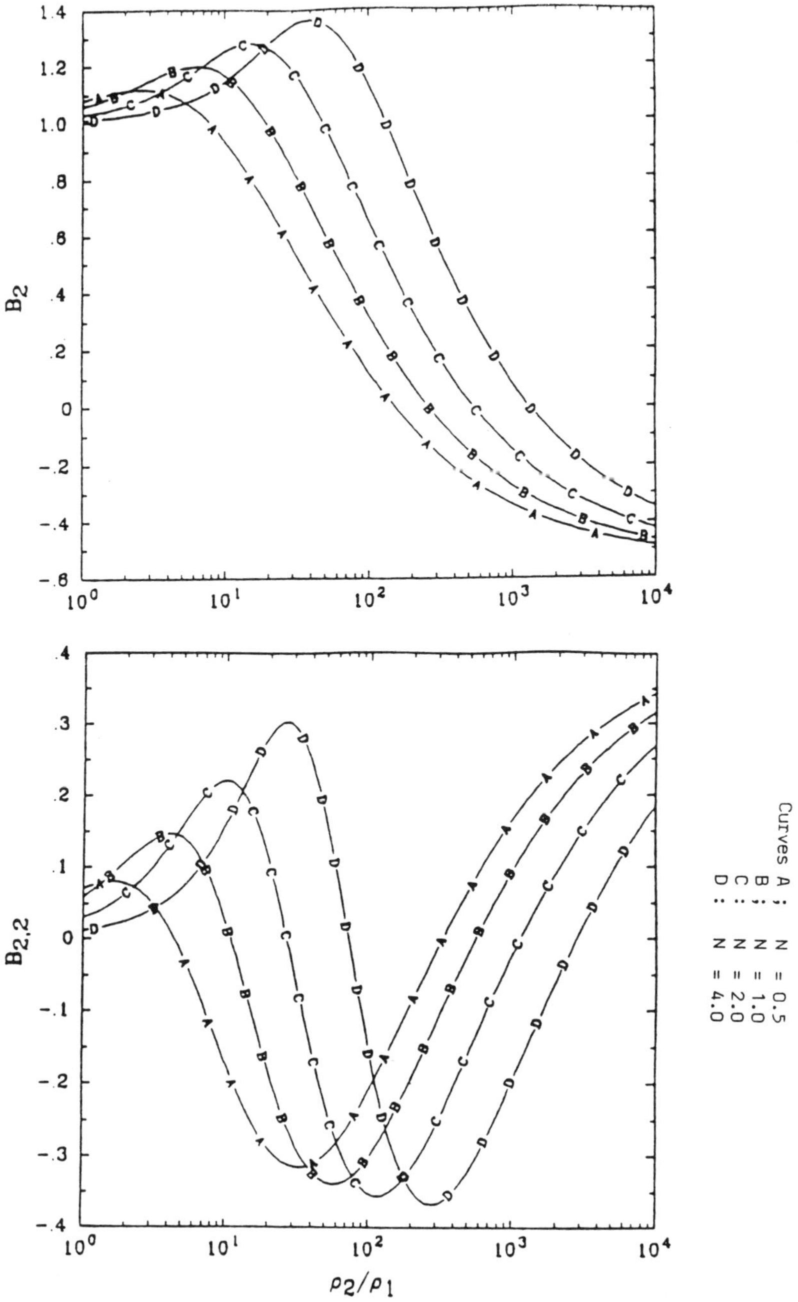

Figure 1.2.13 **Dilution and distortion factors as a function of ρ_2/ρ_1 for various N for $a/b = 10$.**

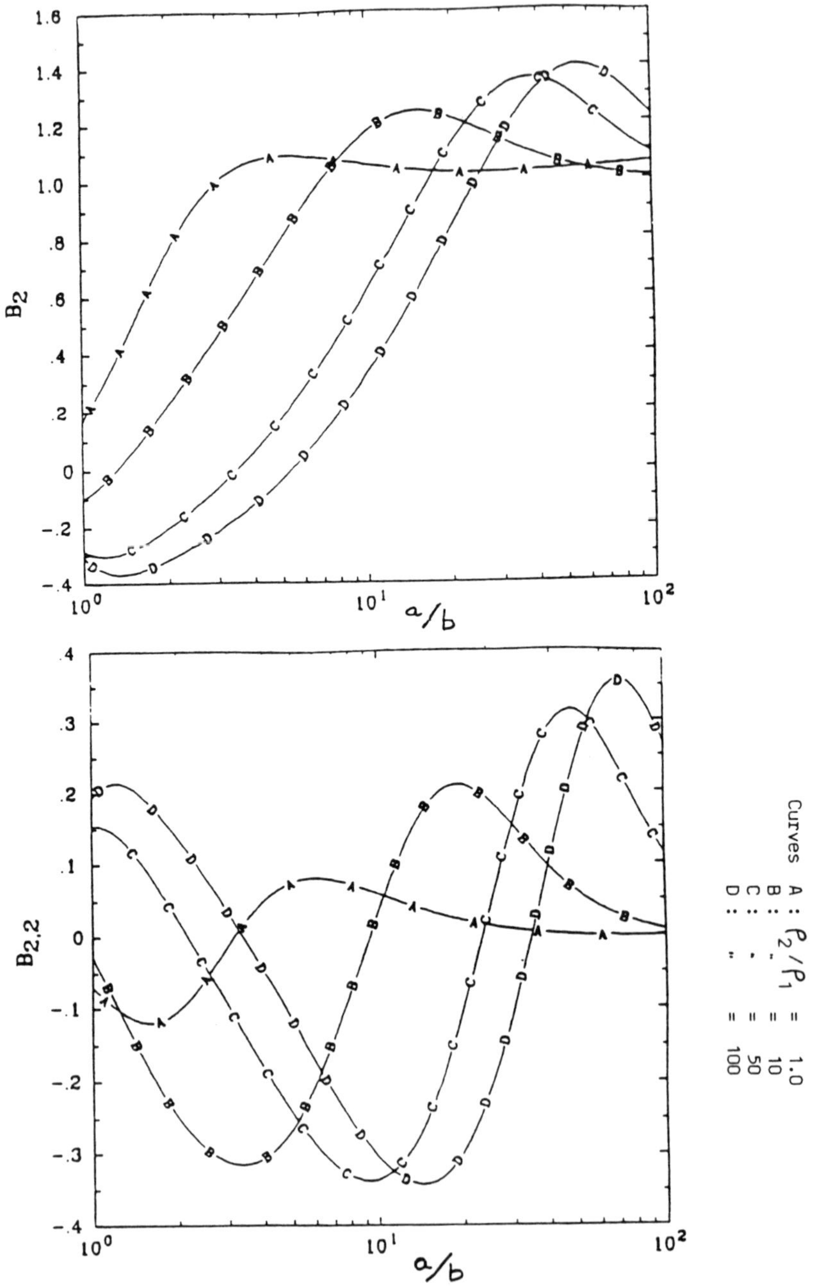

Figure 1.2.14 Dilution and distortion factors as a function of a/b for various ρ_2/ρ_1 for $N = 1.0$.

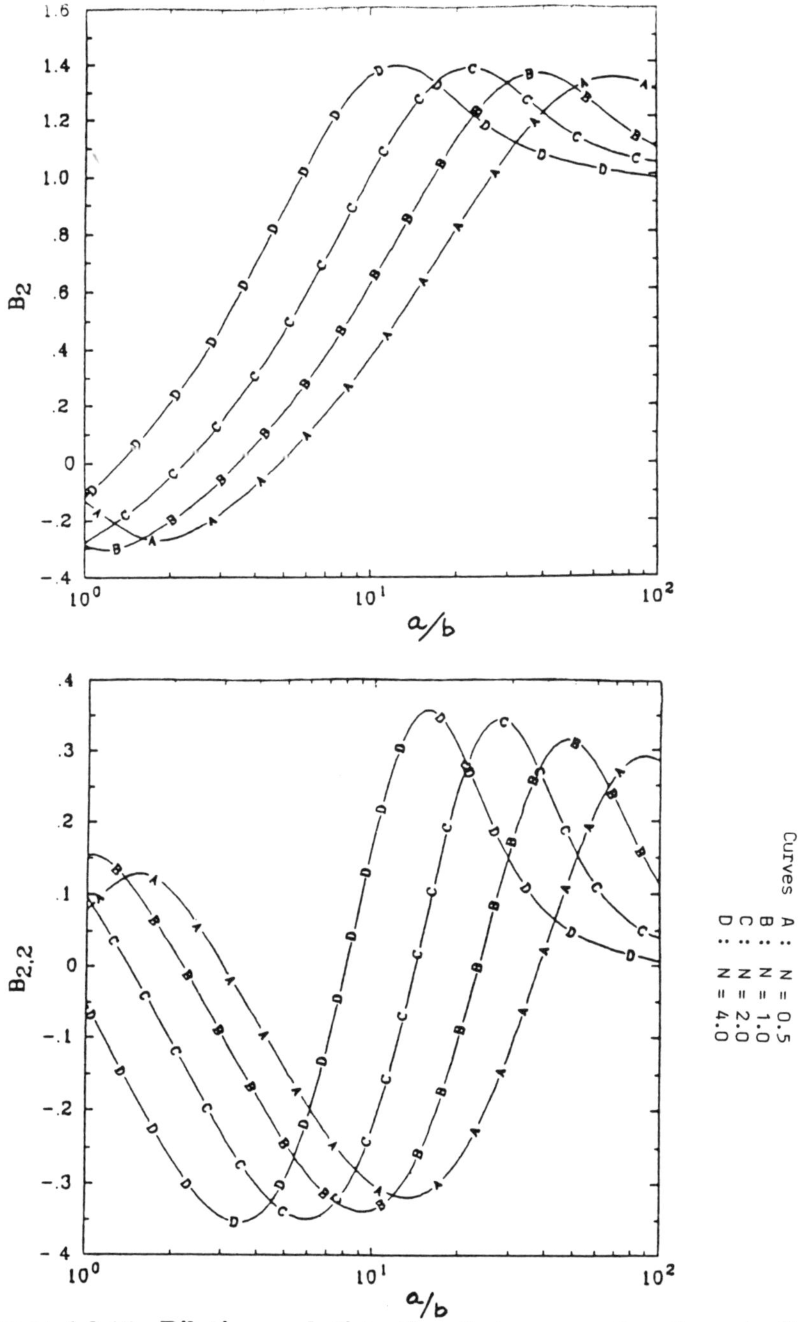

Figure 1.2.15 Dilution and distortion factors as a function of a/b for various N for $\rho_2/\rho_1 = 50$.

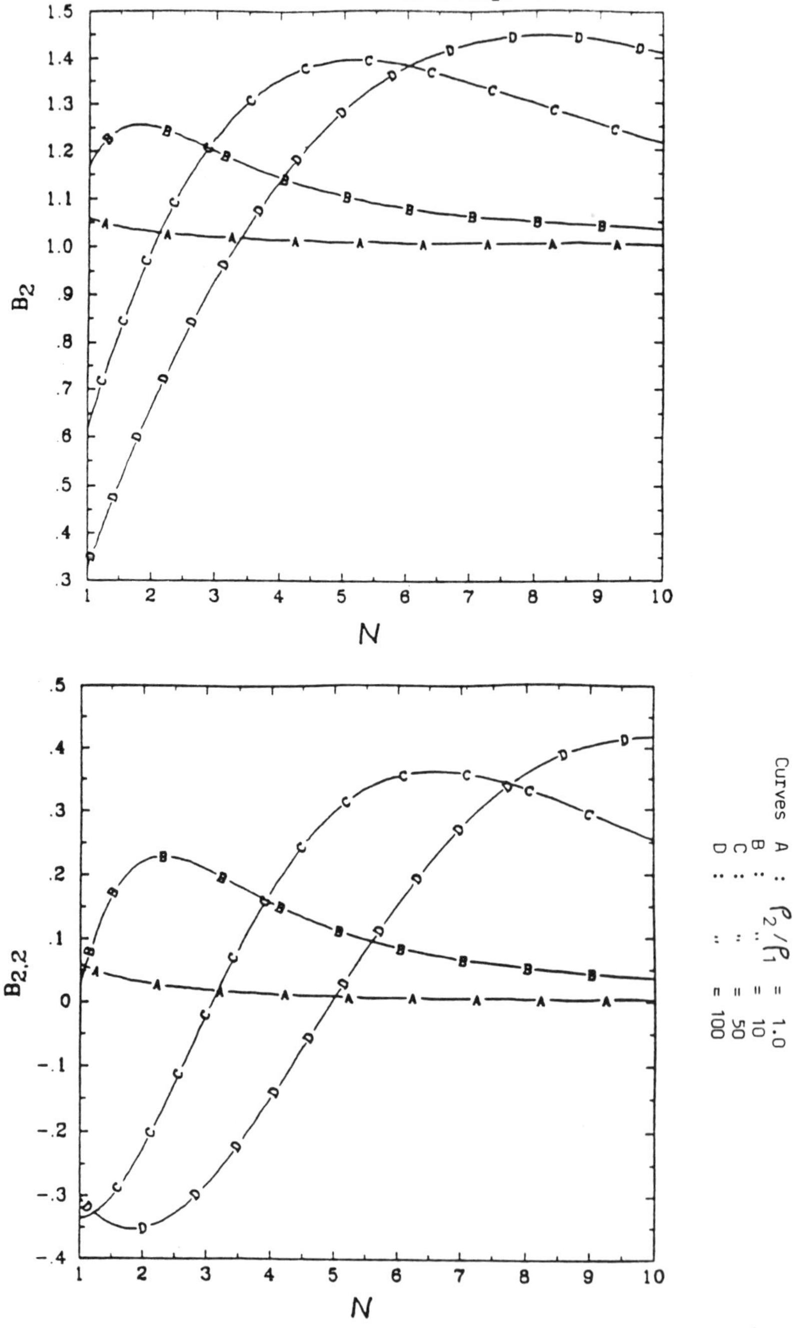

Figure 1.2.16 Dilution and distortion factors as a function of N for a/b =10 for various ρ_2/ρ_1.

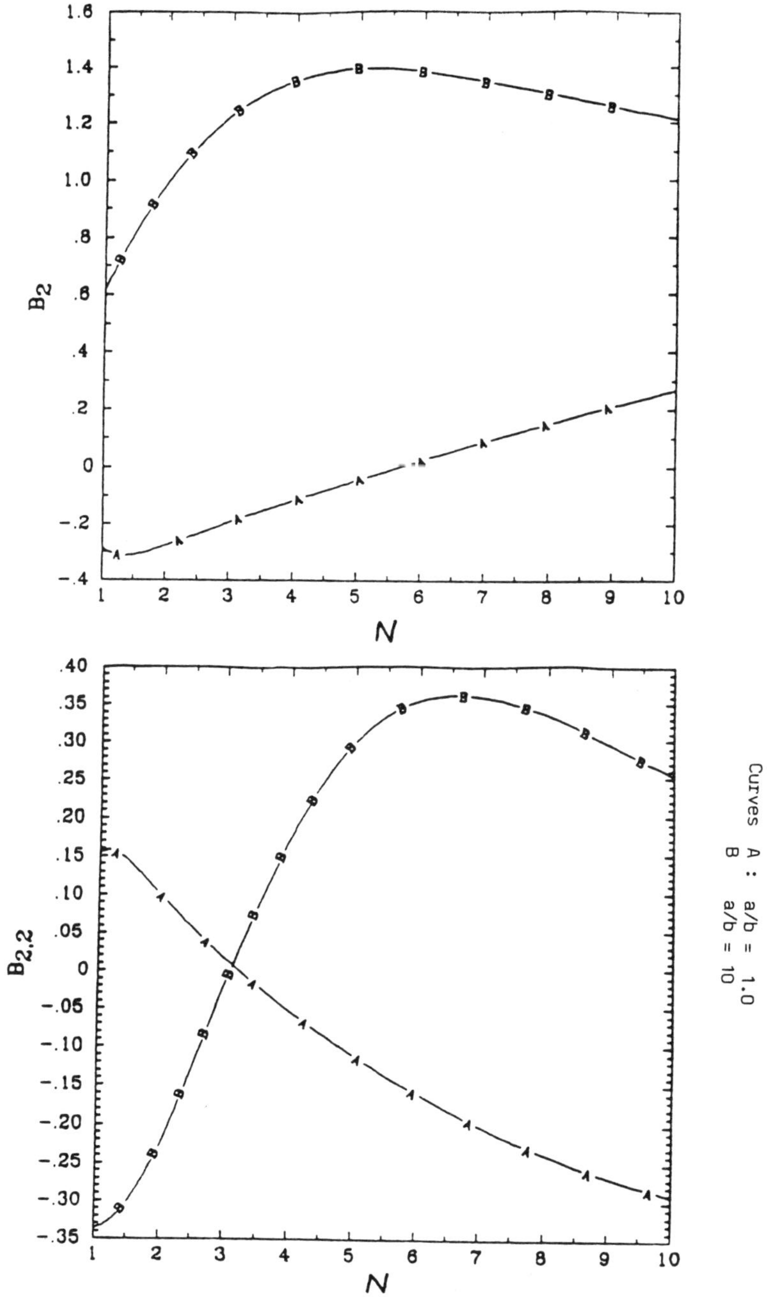

Figure 1.2.17 Dilution and distortion factors as a function of N for various a/b for $\rho_2/\rho_1 = 50$.

We present graphical results for B_2 and B_{22} for this borehole configuration using a similar format to the case above for the two layer planar model. The curves are shown in Figs. 1.2.13 to 1.2.17 where the parameters are the resistivity ratio ρ_2/ρ_1, spacing/hole radius ratio a/b, and spacing number N. Here, it is of interest to note that the dilution factor B_2 can become negative (e.g. as in Fig. 1.2.14). This property is consistent with the observed property that the slope of $\ln \rho_a$ vs. $\ln \rho_2$ of the classical resistivity departures are indeed negative in some cases.

References

[1] Daniel, V., *Dielectric Relaxation*, Academic Press 1967.

[2] Gruszka, T. P. and J. R. Wait, "Dilution and distortion effects in the IP response of a two layer earth," *IEEE Trans.*, **GE-23**, 606-609, 1985.

[3] Gruszka, T. P., "PhD Thesis, Applied Math.Prog.," Univ. of Arizona, Dec. 1987.

[4] Guptasarma, 3 D., "True and apparent spectra of buried polarizable targets," *Geophysics*, **49**, 171–176, 1984

[5] Seigel, H. O., "Mathematical formulation and type curves for IP," *Geophysics*, **24**, 547–565, 1985.

[6] Sokolnikoff, I., *Advanced Calculus*, McGraw-Hill, 1939.

[7] Song, L. and K. Vozoff, "The complex resistivity spectra of models consisting of two polarizable media of different intrinsic properties," *Geophysical Prospecting*, **33**, 1029–1062, 1985.

[8] Wait, J. R., "Towards a general theory of IP in geophysical exploration," *IEEE Trans.*, **GE-19**, 231–234, 1981.

[9] Wait, J. R., *GeoElectromagnetism*, Academic Press, New York, 1982.

[10] Wait, J. R., "Relaxation Phenomena and IP, Geoexploration," **22**, 107–127, 1984.

[11] Wait, J. R., *Electromagnetic Wave Theory*, Harper and Row/ Wi-

ley, New York 1985.

[12] Wait, J. R., *Introduction to Antennas and Propagation*, Peter Peregrinus Ltd.,UK/IEEE Service Center, New York, 1986a.

[13] Wait, J. R., "Extensions to the phenomenological theory of IP," *IEEE Trans.*, **GE-24**, 409–414, 1986b.

[14] Wait, J. R. and T. P. Gruszka, "Resistivity and IP response for a borehole model," *IEEE Trans.*, **GE-25**, 368–372, 1987.

1.3 Resistivity and I.P. Response for an Electrode Near an Interface

a. Introduction

There are a number of instances where it is desirable to measure the complex resistivity of a geological structure near a plane interface or fault plane. A notable example is when the current and potential electrodes are located in a bore hole where the latter intersects the plane of the formation at an angle. Another example is when an in-line surface-based electrode array crosses an outcropping vertical contact at an oblique angle.

Our purpose here is to present explicit results for both the resistivity and the induced polarization response for a two-electrode or normal array in the vicinity of a plane interface between two homogeneous regions. We will describe the problem in the context of a borehole geometry but later we indicate the equivalence to the surface based scheme. The relevant frequencies are assumed to be sufficiently low that all electromagnetic propagation and coupling effects can be neglected. Also the influence of the borehole, whether empty or fluid filled, is also ignored.

b. Basic Formulation

The generic problem is really a standard one in potential theory which is usually phrased in the context of an electrostatic point charge located over a plane interface $(z = 0)$ between two dielectric half-spaces. We begin with the case where a point source C of current I is located at a height h over a plane interface separating two homogeneous conducting half-spaces of complex resistivities ρ_1 and ρ_2. The situation is illustrated in Fig. 1.3.1a and 1.3.1b. The potential Ψ at P can be measured either in the upper region $(z > 0)$ or in the lower region $(z < 0)$ as indicated in Fig. 1.3.2a or 1.3.2b, respectively. There is an obvious symmetry about the z axis so the potential $\Psi(r, z)$ is only a function of r and z.

For the region $z > 0$, we deduce directly from image theory that

$$\Psi = \frac{I\rho_1}{4\pi}\left[\frac{1}{R} + K\frac{1}{R'}\right] \tag{1}$$

where

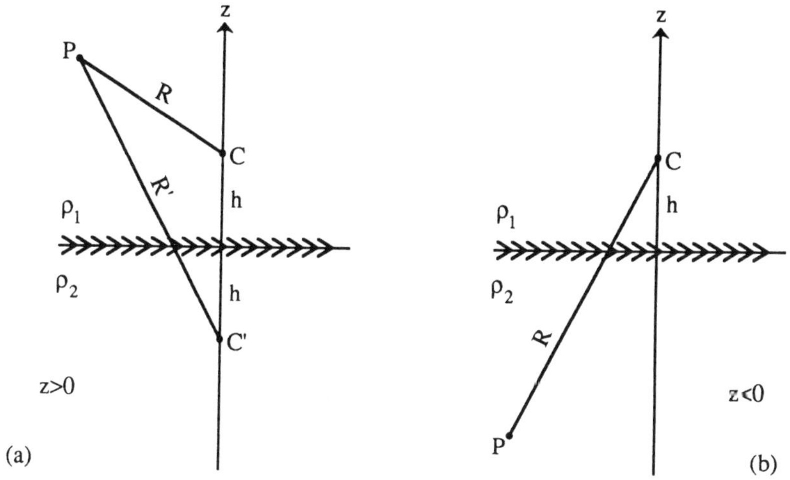

Figure 1.3.1 Generic problem of current electrode at C over a plane interface between two homogeneous regions.

$$R = \left[r^2 + (z-h)^2\right]^{\frac{1}{2}} \tag{2}$$

$$R' = \left[r^2 + (z+h)^2\right]^{\frac{1}{2}} \tag{3}$$

$$K = (\rho_2 - \rho_1)/(\rho_2 + \rho_1) \tag{4}$$

Clearly R is the distance from the source (at $r = 0, z = h$) to the observer at (r, z). On the other hand, R' is the distance from the image at $(0, -h)$ to the observer. For the region $z < 0$,

$$\Psi = \frac{I\rho_1\rho_2}{2\pi(\rho_2 + \rho_1)R} \tag{5}$$

where R is given by (2).

We can readily verify that the potential Ψ and the normal current density J_z are indeed continuous at $z = 0$. In the latter case we should note that, for $z > 0$,

$$J_z = -(1/\rho_1)\partial\Psi/\partial z \tag{6}$$

while, for $z < 0$,

$$J_z = -(1/\rho_2)\partial\Psi/\partial z \tag{7}$$

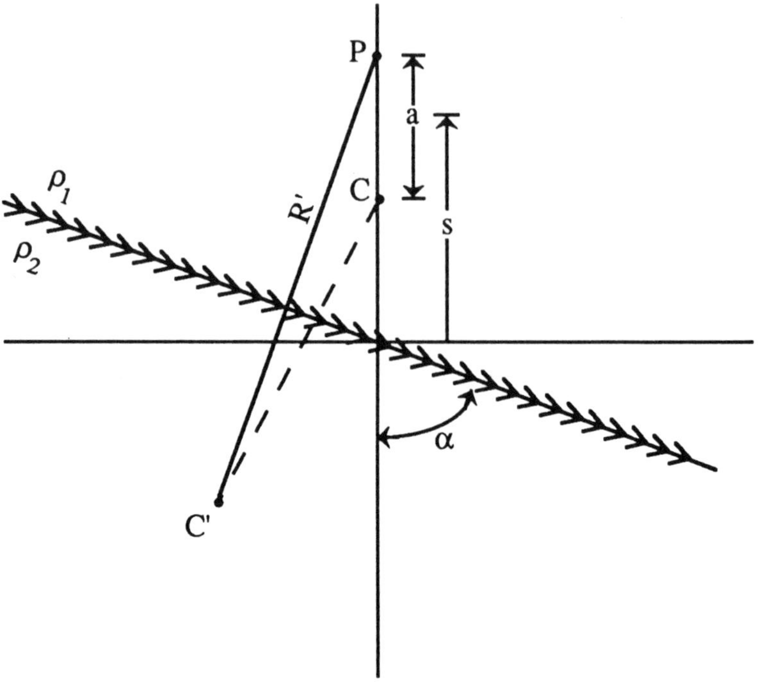

Figure 1.3.2 Geometry for borehole axis intersecting interface at an angle α.

Subject to the low frequencies, the above expressions for the potential are valid functions of frequency $\omega/(2\pi)$ when we insert the appropriate forms for the complex resistivities $\rho_1(i\omega)$ and $\rho_2(i\omega)$ for the appropriate time factor $\exp(i\omega t)$.

We now allow the borehole axis to intersect the planar interface at an angle α where $0° < \alpha < 90°$. Furthermore, the current electrode C and the potential electrode P are located on the axis of the borehole. The situation is illustrated in Fig. 1.3.2 for the case where the electrodes are both located in region 1 where the complex resistivity is $\rho_1(i\omega)$. The distance R between C and P is now denoted by a, the inter-electrode spacing. The distance R' from P to the image source C' is obtained from trigonometry and given by

$$R' = \left[2s^2(1 - \cos 2\alpha) + \frac{a^2}{2}(1 + \cos 2\alpha)\right]^{1/2} \tag{8}$$

where s is the distance from the center of the array to the interface shown in Fig. 1.3.2.

In the case where the electrodes straddle the interface, (5) is still valid where we merely replace R by a. When the electrodes are both in the lower region, the expression for the potential at P, for a current source at C, are symmetric to the case where the electrodes are in the upper region (i.e. we interchange subscripts 1 and 2). In summary, we may write

$$\Psi = \begin{cases} \dfrac{I\rho_1}{4\pi}\left[\dfrac{1}{a} + \dfrac{\rho_2 - \rho_1}{\rho_2 + \rho_1}\dfrac{1}{R'}\right] & ; s > \dfrac{a}{2} \\ \dfrac{I\rho_1\rho_2}{2\pi(\rho_1 + \rho_2)a} & ; -\dfrac{a}{2} < s < \dfrac{a}{2} \\ \dfrac{I\rho_2}{4\pi}\left[\dfrac{1}{a} + \dfrac{\rho_1 - \rho_2}{\rho_1 + \rho_2}\dfrac{1}{R'}\right] & ; s < -\dfrac{a}{2} \end{cases} \tag{9}$$

where R' is given by (8) and it is now valid for $(-\infty < s < \infty)$.

c. Apparent Resistivity

Following tradition, it is convenient and desirable to define an apparent complex resistivity $\rho_a(i\omega)$ for all values of s such that

$$\Psi = I\rho_a(i\omega)/(4\pi a) \tag{10}$$

In other words, if the region were effectively homogeneous and of infinite extent, ρ_a would be the actual resistivity.

Now we are specifically interested in knowing how the ratio ρ_a/ρ_1 varies with the normalized distance s/a for various values of the ratio ρ_1/ρ_2. The explicit functional forms are

$$\frac{\rho_a}{\rho_1} = \begin{cases} 1 + \dfrac{(\rho_2 - \rho_1)/(\rho_2 + \rho_1)}{2(s/a)^2(1 - \cos 2\alpha) + (1/2)(1 + \cos 2\alpha)} & ; \dfrac{s}{a} > \dfrac{1}{2} \\ 2\rho_2/(\rho_2 + \rho_1) & ; -\dfrac{1}{2} < \dfrac{s}{a} < \dfrac{1}{2} \\ \dfrac{\rho_2}{\rho_1}\left[1 + \dfrac{(\rho_1 - \rho_2)/(\rho_1 + \rho_2)}{2(s/a)^2(1 - \cos 2\alpha) + (1/2)(1 + \cos 2\alpha)}\right] & ; \dfrac{s}{a} < -\dfrac{1}{2} \end{cases} \tag{11}$$

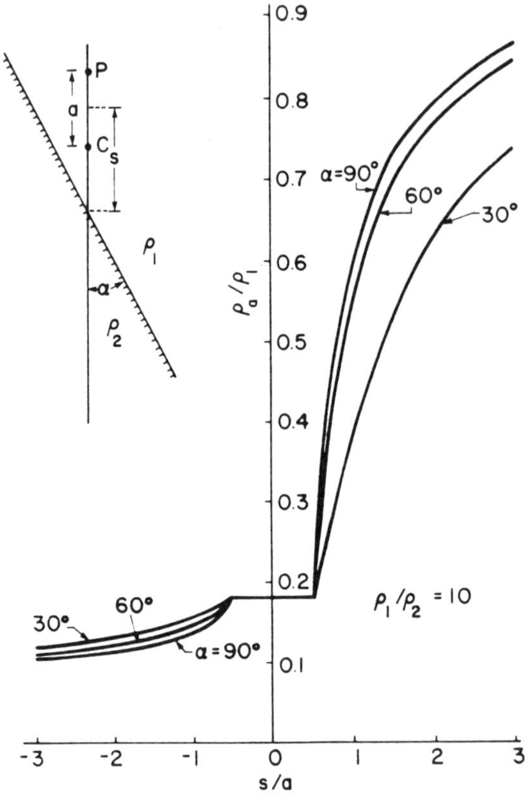

Figure 1.3.3 Apparent resistivity as a function of the vertical distance above or below the tilted interface for $\rho_1/\rho_2 = 10$.

To illustrate the results, we regard ρ_1/ρ_2 as real. In other words any relative phase shifts are ignored. Thus the normalized apparent resistivity ρ_a/ρ_1 is also real. We later show some results relevant to the case where ρ_1/ρ_2 is complex.

Using the tripartite expression (11), ρ_a/ρ_1 is plotted as a function of the normalized spacing s/a in Figs. 1.3.3 to 1.3.8 for various angles α and real values of ρ_1/ρ_2. In Fig. 1.3.3, for $\rho_1/\rho_2 = 10$, three angles $\alpha = 90°, 60°$, and $30°$ are selected. As indicated, ρ_a/ρ_1 asymptotically approaches 1.0 for large values of s/a and approaches 0.1 for large negative values of s/a. The limiting forms are approached most rapidly for $\alpha = 90°$ when the borehole intersects the interface at right angles. Very similar curves are shown in Fig. 1.3.4 where $\rho_1/\rho_2 = 0.1$. Now the limiting values of ρ_a/ρ_1 are 1.0 for large positive s/a and 10 for

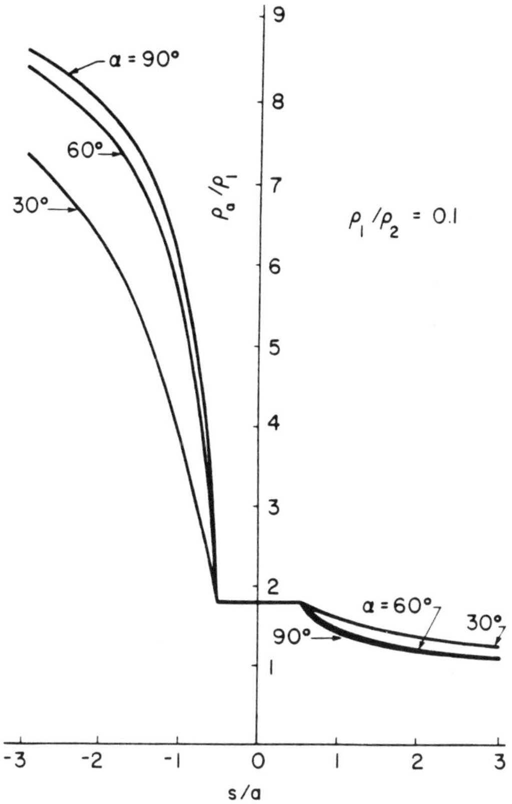

Figure 1.3.4 Apparent resistivity as a function of the vertical distance above or below the tilted interface for $\rho_2/\rho_1 = 0.1$.

large negative values of s/a. Actually the curves in Fig. 1.3.4 have a mirror symmetry with those in Fig. 1.3.3.

The influence of changing the resistivity ratio ρ_1/ρ_2 is illustrated in Fig. 1.3.5 for the case $\alpha = 90°$ only. As indicated, ρ_a/ρ_1 approaches 1.0 for large positive s/a but ρ_a/ρ_1 approaches 3, 5 and 10 for large negative s/a for $\rho_1/\rho_2 = 1/3$, $1/5$ and $1/10$, respectively.

An attempt to illustrate the symmetry of the curves when the ratio ρ_1/ρ_2 is inverted is shown in Fig. 1.3.6. In the conventional manner, we plot ρ_a/ρ_1 as a function of s/a with $\rho_1/\rho_2 = 3$ for various α values. The same curves apply to the case $\rho_1/\rho_2 = 1/3$ if the abscissa $[s/a]$ at the top and the ordinate $[\rho_a/\rho_1]$ on the right of the figure are used. Many other results could also be presented in this format at the slight risk that the reader would be confused. The curves in Fig. 1.3.6 are

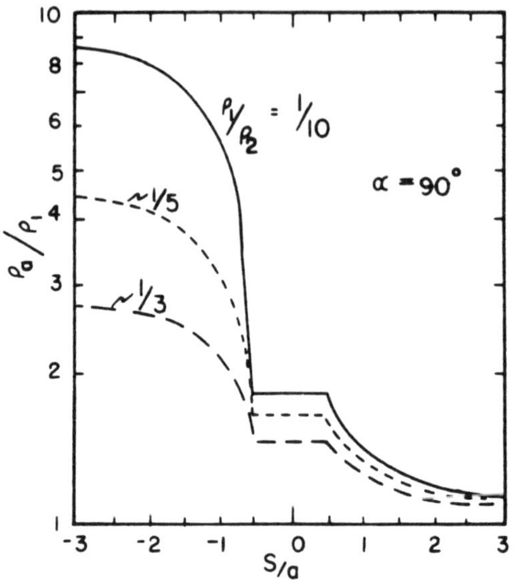

Figure 1.3.5 Apparent resistivity as a function of the distance above or below the horizontal interface for several resistivity ratios.

also intended to show the effect of a highly oblique intersection of the borehole axis with the interface. For example, at $\alpha = 5°$, the change in ρ_a/ρ_1 is within 10% over the range of s/a from -3 to $+3$. Of course, in the limit $\alpha \to 0°$, the ratio ρ_a/ρ_1 degenerates to a straight line given by

$$\rho_a/\rho_1 = 2\rho_2/(\rho_1 + \rho_2) \tag{12}$$

corresponding to the unlikely case when the borehole is coincident with the interface.

To illustrate the effect of a high resistivity contrast on each side of the interface, we choose $\rho_1/\rho_2 = 100$ and $\rho_1/\rho_2 = 0.01$. These results are shown in Figs. 1.3.7 and 1.3.8, respectively.

d. Induced Polarization and Dilution

We now wish to touch briefly on the case where one or both of the half-space regions are polarizable in the sense that the frequency dependencies of the resistivities are to be accounted for. Here, we will employ a first order theory. To this end, we write

$$\rho_a(i\omega) = \rho_a(0)[1 + \delta_a(i\omega)] \tag{13}$$

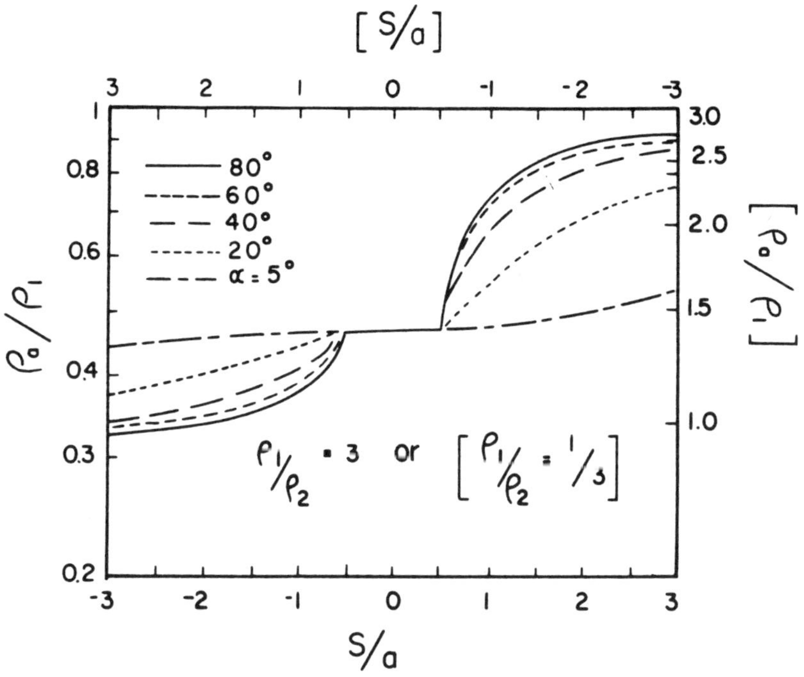

Figure 1.3.6 Apparent resistivity as a function of the distance *s* for a range of tilt angles for $\rho_1/\rho_2 = 3$; the results are also applicable to the case $\rho_1/\rho_2 = 1/3$ if the top and right hand scales are employed.

$$\rho_1(i\omega) = \rho_1(0)[1 + \delta_1(i\omega)] \tag{14}$$

$$\rho_2(i\omega) = \rho_2(0)[1 + \delta_2(i\omega)] \tag{15}$$

where $\delta_a(i\omega)$, $\delta_1(i\omega)$ and $\delta_2(i\omega)$ are the departures of the complex resistivities from their *DC* or zero-frequency values. By making an appropriate Taylor expansion, it is evident that

$$\delta_a(i\omega) \simeq B_1\delta_1(i\omega) + B_2\delta_2(i\omega) \tag{16}$$

where

$$B_1 = \left[\frac{\partial \ln \rho_a(i\omega)}{\partial \ln \rho_1(i\omega)}\right]_{\omega\to 0} \tag{17}$$

and

$$B_2 = \left[\frac{\partial \ln \rho_a(i\omega)}{\partial \ln \rho_2(i\omega)}\right]_{\omega\to 0} \tag{18}$$

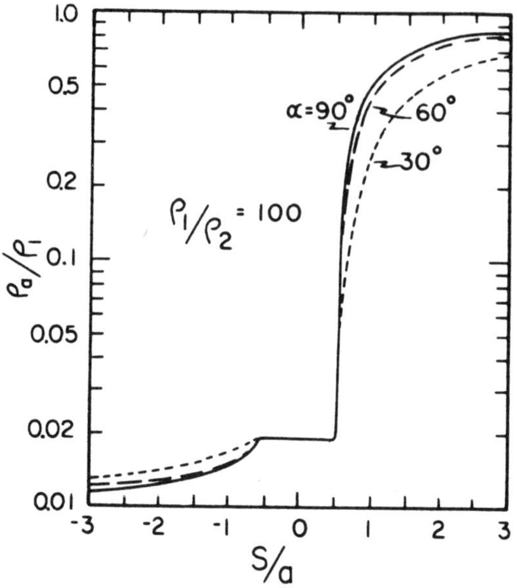

Figure 1.3.7 Apparent resistivity as a function of the vertical distance above or below the tilted interface for $\rho_1/\rho_2 = 100$.

Equivalent definitions of these dilution factors are

$$B_1 = \frac{\partial \ln \rho_a}{\partial \ln \rho_1} = \frac{\rho_1}{\rho_a}\frac{\partial \rho_a}{\partial \rho_1} \tag{19}$$

and

$$B_2 = \frac{\partial \ln \rho_a}{\partial \ln \rho_2} = \frac{\rho_2}{\rho_a}\frac{\partial \rho_a}{\partial \rho_2} \tag{20}$$

where we employ *DC* resistivities. It can also be shown that

$$B_1 + B_2 = 1 \tag{21}$$

so it is only necessary to compute one of these coefficients.

It should be realized that the simple representation given by (16) is only valid in a first order sense; higher order terms containing higher powers of δ_1 and δ_2 are neglected. We will not pursue this matter here. Suffice it to say that we are dealing with weakly dispersive media where $|\delta_1|$ and $|\delta_2| \ll 1$.

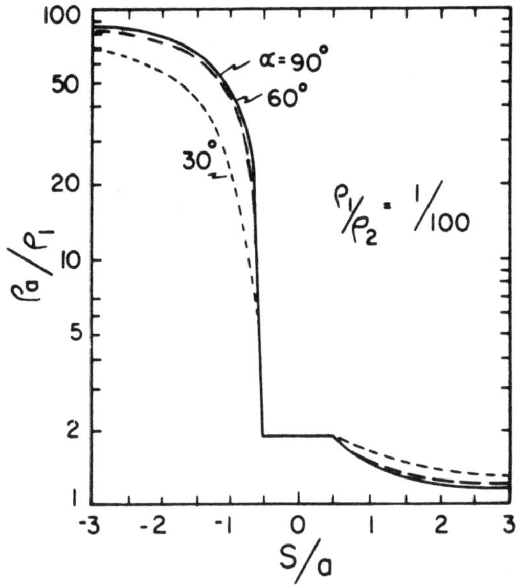

Figure 1.3.8 Apparent resistivity as a function of the vertical distance above or below the tilted interface for $\rho_1/\rho_2 = 0.01$.

Using (20) in conjunction with (12), it is easy to show that

$$B_2 = \begin{cases} \dfrac{2\rho_1\rho_2}{(R'/a)(\rho_1+\rho_2)^2 + {\rho_2}^2 - {\rho_1}^2} & ; \dfrac{s}{a} > \dfrac{1}{2} \\ \rho_1/(\rho_1+\rho_2) & ; -\dfrac{1}{2} > \dfrac{s}{a} > \dfrac{1}{2} \\ \dfrac{(R'/a)(\rho_1+\rho_2) + \rho_1 - \rho_2 - 2\rho_1\rho_2/(\rho_1+\rho_2)}{(R'/a)(\rho_1+\rho_2)+\rho_1-\rho_2} & ; \dfrac{s}{a} < -\dfrac{1}{2} \end{cases} \tag{22}$$

where $R' = R'(s)$ is given by (8).

Some examples of the dilution factor B_2 are shown in Fig 1.3.9a, 1.3.9b, 1.3.10a and 1.3.10b. To facilitate the presentation, we choose a linear-log format. Also positive and negative values of s/a are displayed separately. In Fig. 1.3.9a, we plot B_2 as a function of s/a from $+0.3$ to $+10$ where it is evident that B_2 is approaching zero as the array moves away from medium (2) into medium (1). Of course, the corresponding value of $B_1 = (1 - B_2)$ would then approach 1.0. The values of B_2 for negative s/a are shown in Fig. 1.3.9b for the same conditions. As expected, B_2 approaches 1.0 as the array moves away

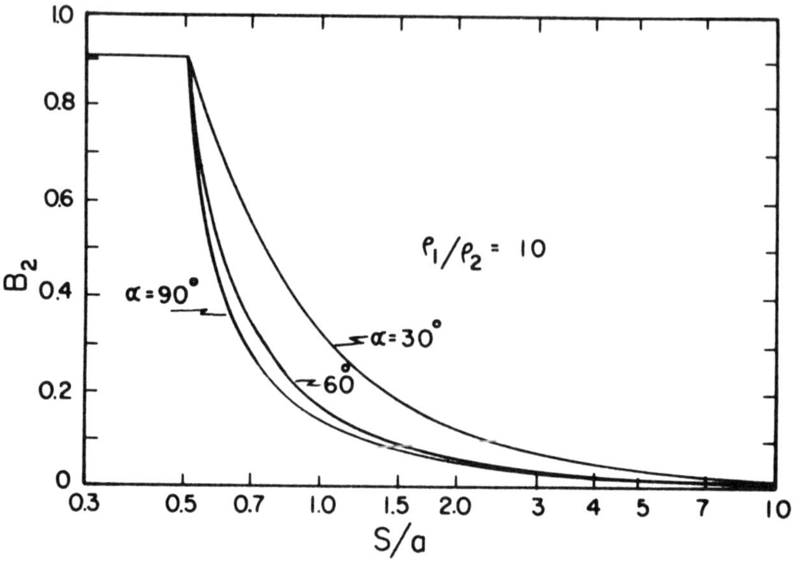

Figure 1.3.9a The dilution factor (relevant to the induced polarization response) for several tilt angles, $\rho_1/\rho_2 = 10$, and positive values of the distance s.

from the interface into medium (2). For the region from $s/a = +0.5$ to -0.5 where the electrodes are straddling the interface, B_2 has a constant value 0.909. The corresponding dilution factor B_2 for $\rho_1/\rho_2 = 0.1$ is shown in Figs. 1.3.10a and 1.3.10b for positive and negative values, respectively. Similar limiting behavior is noted. Also the constant value of B_2, in the range $s/a = +0.5$ to -0.5, is now 0.091.

e. Concluding Remarks

We have adopted a very simple model here in order to illustrate the interesting phenomena when the resistivity and the induced polarization are measured near an interface between two homogeneous regions. The problem is formulated in the context of a drill hole intersecting the plane interface. Actually, the problem is fully equivalent to the case where the two electrodes C and P are located on the surface of plane earth where there is vertical contact between two homogeneous regions (i.e. quarter spaces). For example, in Fig. 1.3.2, we can visualize the

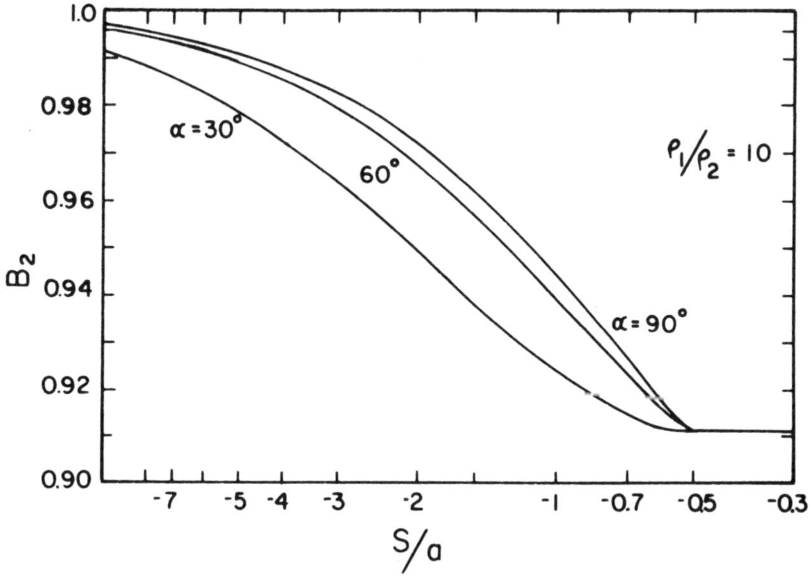

Figure 1.3.9b The dilution factor (relevant to the induced polarization response) for several tilt angles, $\rho_1/\rho_2 = 10$, and negative values of the distance s.

plane of the paper as the surface of the earth and where the array is to be moved along a line making an angle α with the trace of the contact. All the apparent resistivity curves and the dilution factor results are valid for this case under the initial quasi-static assumptions.

f. Exercises

Exercise 1: Consider a two-electrode array on the surface of a two-layer earth. The spacing between the electrodes C and P is denoted by a. The thickness of the upper layer is d and its conductivity is σ_1. The conductivity of the lower homogeneous half-space is σ. Beginning with the exact expression for the potential at P, derive an expression for the apparent resistivity ρ_a in the limiting case where the upper layer thickness is vanishingly thin and the conductivity is indefinitely large in such a manner that their product g, in mhos, is finite. Plot the apparent resistivity as function of the electrode spacing in normalized form.

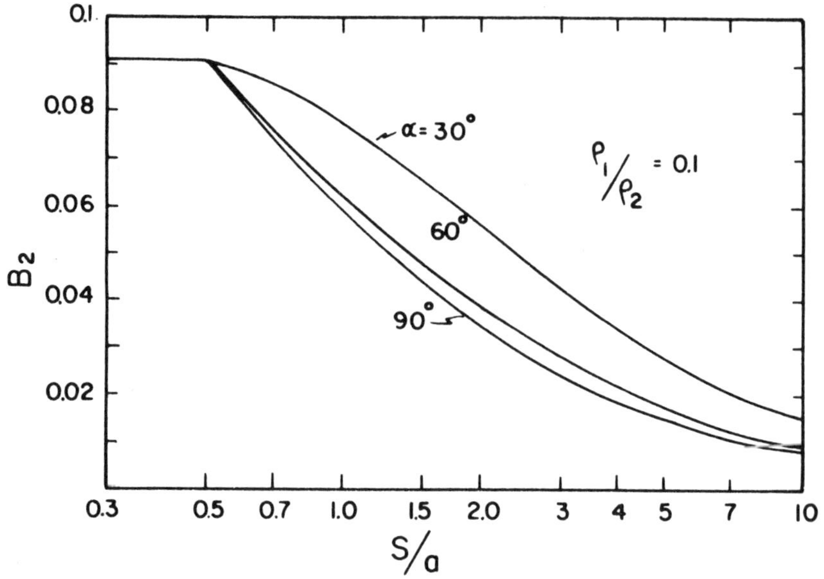

Figure 1.3.10a The dilution factor (relevant to the induced polarization response) for several tilt angles, $\rho_1/\rho_2 = 0.1$, and positive values of the distance s.

Solution 1: The exact expression for the potential at P, assuming DC conditions, is

$$\Psi = \frac{1}{2\pi\sigma_1}\left[\int_0^\infty J_0(\lambda a)d\lambda + 2\int_0^\infty \frac{(\sigma_1-\sigma)e^{-2d\lambda}J_0(\lambda a)}{(\sigma_1+\sigma)-(\sigma_1-\sigma)e^{-2d\lambda}}d\lambda\right] \tag{1}$$

Now noting that $\sigma/\sigma_1 \to 0, \sigma_1 d \to g$, and $e^{-2d\lambda} \to 1-2d\lambda$, we see that

$$\Psi = \frac{1}{2\pi g}\int_0^\infty \frac{J_0(\lambda a)}{\lambda+(\sigma/g)}d\lambda \tag{2}$$

This integral will now be evaluated by first inserting the identity

$$\frac{1}{\lambda+(\sigma/g)} = \int_0^\infty e^{-(\lambda+(\sigma/g))x}dx \tag{3}$$

into the integrand. Then the potential is given by

$$\Psi = \frac{1}{2\pi g}\int_0^\infty\int_0^\infty e^{-(\sigma/g)x}e^{-\lambda x}J_0(\lambda a)d\lambda dx \tag{4}$$

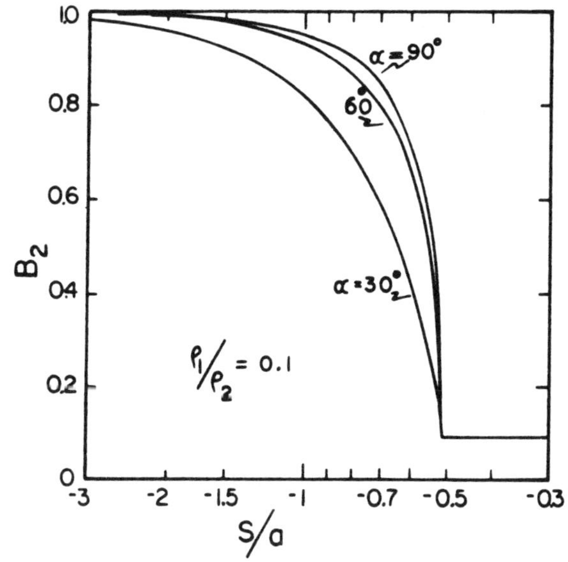

Figure 1.3.10b The dilution factor (relevant to the induced polarization response) for several tilt angles, $\rho_1/\rho_2 = 0.1$, and negative values of the distance s.

The integration over λ can be readily performed to give

$$\Psi = \frac{1}{2\pi g}\int_0^\infty \frac{e^{-(\sigma/g)x}}{(x^2+a^2)^{1/2}}dx$$
$$= \frac{1}{4g}\left[\frac{2}{\pi}\int_0^\infty \frac{\exp\left[-\frac{\sigma a}{g}\left(\frac{x}{a}\right)\right]}{\left[1+\left(\frac{x}{a}\right)^2\right]^{1/2}}d\left(\frac{x}{a}\right)\right] \qquad (5)$$

The integral can now be expressed in terms of Struve's function H_0 and the Bessel function Y_0, both of order 0, by using an identity given by Watson [pg. 331, #3 with $\omega = 0$ and $\nu = 0$]. Thus

$$\Psi = \frac{1}{4g}\left[H_0\left(\frac{a\sigma}{g}\right) - Y_0\left(\frac{a\sigma}{g}\right)\right] \qquad (6)$$

We now define an apparent resistivity ρ_a or an apparent conductivity σ_a according to

$$\Psi = 1/2\pi\sigma_a a = \rho_a/2\pi a \qquad (7)$$

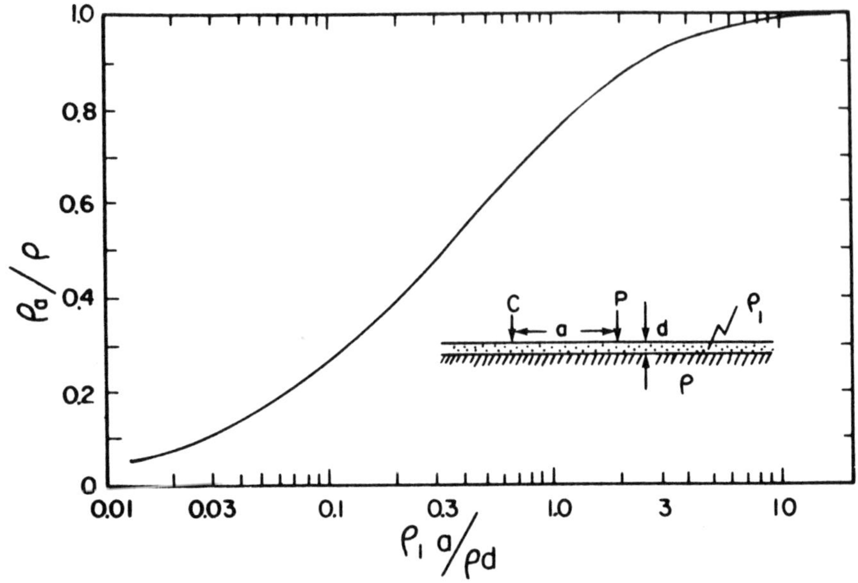

Figure 1.3.11 Normalized apparent resistivity as a function of the normalized electrode spacing for a two electrode array and a thin conductive sheet on the surface of a homogeneous half space.

Then, using dimensionless forms, we write

$$\frac{\rho_a}{\rho} = \frac{\sigma}{\sigma_a} = \pi \frac{\sigma a}{2g} \left[H_0 \left(\frac{a\sigma}{g} \right) - Y_0 \left(\frac{a\sigma}{g} \right) \right] \tag{8a}$$

$$= \pi \frac{a\rho_1}{2d\rho} \left[H_0 \left(\frac{a\rho_1}{d\rho} \right) - Y_0 \left(\frac{a\rho_1}{d\rho} \right) \right] \tag{8b}$$

Using tabulated values of H_0 and Y_0, we plot the normalized resistivity ratio ρ_a/ρ as a function of the normalized spacing parameter $\rho_1 a/\rho d$ for a range of values from 0.01 to 20. As it should, the ordinate approaches 1.0 as the spacing parameter becomes sufficiently large. Thus, we have a convenient single universal curve to estimate the importance of thin conductive surface layer on resistivity measurements.

Exercise 2: Using the same model as in the previous exercise, deduce the sensitivity functions B_1 and B_2 for the same two-electrode array C and P. The relevant quantities are

$$B_1 = \frac{\partial \ln \rho_a}{\partial \ln \rho_1} \tag{1}$$

and

$$B = \frac{\partial \ln \rho_a}{\partial \ln \rho} \tag{2}$$

Plot these parameters as a function of the normalized spacing parameter used in the preceding exercise.

Solution 2: To simplify the development, we write

$$\frac{\rho_a}{\rho} = F(y) = \frac{\pi}{2} y \left[H_0(y) - Y_0(y)\right] \tag{3}$$

where $y = a\rho_1/d\rho$ is the spacing parameter. Now clearly

$$B_1 = \frac{\partial \ln F(y)}{\partial \ln y} = \frac{y}{F(y)} \frac{\partial F(y)}{\partial y} \tag{4}$$

where

$$\frac{\partial F(y)}{\partial y} = \frac{\pi}{2} \left\{ H_0(y) - Y_0(y) - y \left[\frac{2}{\pi} - H_1(y) + Y_1(y) \right] \right\} \tag{5}$$

Here, we have used the identities

$$\frac{dH_0(y)}{dy} = \frac{2}{\pi} - H_1(y) \tag{6}$$

and

$$\frac{dY_0(y)}{dy} = -Y_1(y) \tag{7}$$

where H_1 and Y_1 are Struve and Bessel functions of order one, respectively. Then, the explicit expression to calculate B_1 is

$$B_1 = 1 + y \frac{(2/\pi) - H_1(y) + Y_1(y)}{H_0(y) - Y_0(y)} \tag{8}$$

and

$$B = 1 - B_1 \tag{9}$$

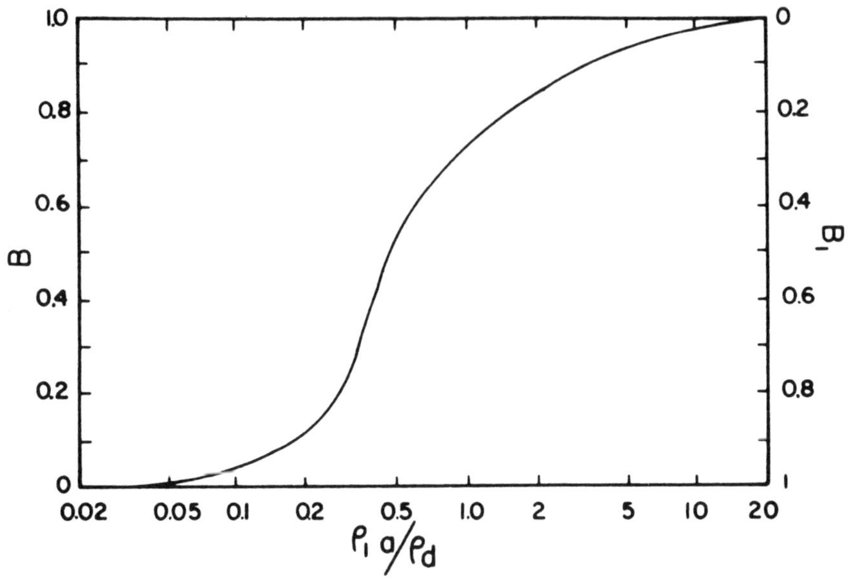

Figure 1.3.12 Sensitivity or dilution factors B and B_1.

Using tabulated values of the Struve and Bessel functions, the functions B and B_1 are plotted as a function of the spacing parameter in Fig 1.3.12. As indicated, for large values of the spacing parameter, the sensitivity factor (i.e. B) for the lower half space tends to 1 meaning that the masking effect of the conductive surface layer is negligible. On the other hand, for small values of the spacing parameter, the surface layer is dominating the response.

Acknowledgments

Most of the numerical results presented are due to Alan Peters in connection with an assignment for the Graduate Course ECE 586. Some results were also included from Gerardo Aguirre's assignment. Alan Peters also contributed to the analysis.

I would like to thank Hulya Akman for her assistance with the calculations in the exercises in subsection 1.3.f.

References

[1] Wait, J. R., *GeoElectromagnetism*, Academic Press, chapters 1 and 2, 1982.

[2] Watson, G.N., *Theory of Bessel Functions, Cambridge University Press*, 2nd. Ed., 1944.

1.4 Low Frequency Electromagnetic Response

a. Introduction

A basic problem in electrical prospecting is how the currents are injected into the earth from a horizontal current-carrying cable grounded at its end points. The basic building block for this analysis is the horizontal electric dipole of current moment $I\,ds$ located in the air-earth interface. Such a solution was given by Sommerfeld [1909-1926] many years ago for the case where the earth was idealized as a homogeneous isotropic half-space. An exhaustive treatment of dipoles in the presence of a conducting half-space was published by Banos [1966] in a textbook. It is our purpose here to outline the solution of the problem in the context of induced polarization. We also consider cylindrical geometry which is relevant to borehole measurements.

b. Basis Anisotropic Half-Space Model

We choose a model of an anisotropic but homogeneous half-space as indicated in Fig. 1.4.1. To facilitate the solution, the source dipole is located at height h over the interface at $z = 0$ and oriented in the x direction. We later allow h to tend to zero corresponding to a grounded electric circuit. The upper region ($z > 0$) is taken to be free space with electrical properties ϵ_o and μ_o (i.e. permittivity and permeability). The lower region, $z < 0$, has a complex resistivity $\rho_h(\omega)$ in the horizontal direction and a complex resistivity $\rho_v(\omega)$ in the vertical direction. As indicated these complex resistivities are functions of the angular frequency ω for an implied time factor $\exp(j\omega t)$. However to abbreviate the subsequent discussion, we shall designate these frequency dependent complex resistivities by ρ_h and ρ_v .

Ohm's law for the lower region can be succinctly stated as follows

$$\overline{J} = \overline{\overline{\sigma}} \cdot \overline{E} \tag{1}$$

where $\overline{J}$ is the current density vector, $\overline{E}$ is the electric field vector, and $\overline{\overline{\sigma}}$ is the (complex) conductivity tensor. The latter quantity takes the form

$$\overline{\overline{\sigma}} = \begin{pmatrix} \rho_h^{-1} & 0 & 0 \\ 0 & \rho_h^{-1} & 0 \\ 0 & 0 & \rho_v^{-1} \end{pmatrix} \tag{2}$$

where the off-diagonal elements are zero. To be specific, we are saying that

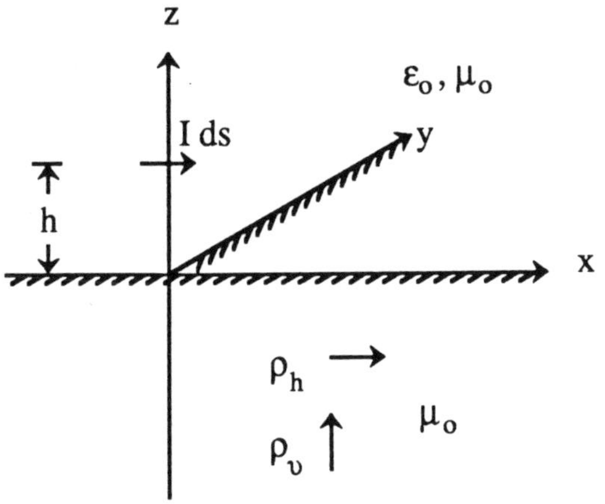

Figure 1.4.1 Horizontal electric dipole over an anisotropic half-space.

$$J_x = (\rho_h)^{-1} E_x \tag{3a}$$

$$J_y = (\rho_h)^{-1} E_y \tag{3b}$$

and

$$J_z = (\rho_z)^{-1} E_z \tag{3c}$$

Then Maxwell's equations for $z < 0$ are

$$\text{curl } \overline{H} = \overline{\overline{\sigma}}\overline{E} \tag{4}$$

and

$$\text{curl } \overline{E} = -j\mu_o\omega\overline{H} \tag{5}$$

excluding any source regions and the corresponding forms for $z > 0$ are identical if $\overline{\overline{\sigma}}$ is replaced by $j\epsilon_o\omega$.

c. Vector Potential Formulation

To solve the boundary value problem, we find it convenient to introduce the vector potential $\overline{A}$ and a scalar potential Ψ such that

$$\overline{H} = \text{curl } \overline{A} \tag{6}$$

and

$$\overline{E} = -j\mu_o\omega\overline{A} - \text{grad } \Psi \tag{7}$$

which clearly are consistent with (4) and (5).

We now choose the gauge condition [Wait 1966a]

$$\mathrm{div}\overline{A} + (\rho_h)^{-1}\Psi = 0 \tag{8}$$

for $z < 0$ and

$$\mathrm{div}\overline{A} + j\epsilon_o\omega\Psi = 0 \tag{9}$$

for $z > 0$. Then it is not difficult to show that for $z > 0$

$$\nabla^2 A_x - \gamma^2 A_x = 0 \tag{10}$$

$$\nabla^2 A_y - \gamma^2 A_y = 0 \tag{11}$$

and

$$\nabla^2 A_z - \left(\frac{1}{\rho_h} - \frac{1}{\rho_v}\right)\rho_h \frac{\partial}{\partial z}(\mathrm{div}\overline{A}) - \frac{\rho_h}{\rho_v}\gamma^2 A_z = 0 \tag{12}$$

where

$$\gamma^2 = (j\mu_o\omega/\rho_h) \tag{13}$$

On the other hand, for $z > 0$, we simply have

$$(\nabla^2 - \gamma_o^2)A_x = 0 \tag{14}$$

$$(\nabla^2 - \gamma_o^2)A_y = 0 \tag{15}$$

and

$$(\nabla^2 - \gamma_o^2)A_z = 0 \tag{16}$$

where $\gamma_o^2 = (j\epsilon_o\omega)(j\mu_o\omega) = -\epsilon_o\mu_o\omega^2$.

In order to solve the problem, we need to choose the vector potential to have both a component A_x in the direction of the dipole and component A_z normal to the interface. However, we can set $A_y = 0$ and still be able to satisfy all the boundary conditions (i.e. tangential field components E_x, E_y, H_x, and H_y must be continuous at the interface $z = 0$). Another condition is that A_x, for the region $z > 0$, must have the proper singularity at the source; that is

$$A_x \to (I\,ds/4\pi r_o)e^{-\gamma_o r_o} \tag{17}$$

as $\rho = (x^2 + y^2)^{1/2} \to 0$ and $z \to h$ where $r_o = [\rho^2 + (z-h)^2]^{1/2}$. This fact is incorporated neatly into the analysis by noting that

$$\frac{e^{-\gamma_o r_o}}{r_o} = \int_0^\infty \frac{\lambda}{u_o} e^{-u_o|z-h|} J_o(\lambda\rho)\,d\lambda \tag{18}$$

where $u_o = (\lambda^2 + \gamma_o^2)^{1/2}$. This "Sommerfeld Integral" is the key for solving many such problems involving localized sources in the presence of layered media [Wait, 1966a; Kong, 1976].

For $z < 0$, we write

$$A_x = \frac{I\,ds}{4\pi} \int_0^\infty F(\lambda, z) J_o(\lambda\rho) d\lambda \tag{19}$$

and

$$A_z = \frac{I\,ds}{4\pi} \frac{\partial}{\partial x} \int_0^\infty G(\lambda, z) J_o(\lambda\rho) d\lambda \tag{20}$$

where $\rho = (x^2 + y^2)^{1/2}$ and where F and G satisfy

$$\frac{\partial^2 F}{\partial z^2} - (\lambda^2 + \gamma^2) F = 0 \tag{21}$$

and

$$\frac{\partial^2 G}{\partial z^2} - (\lambda^2 K^2 + \gamma^2) G = (K - 1) \frac{\partial F}{\partial z} \tag{22}$$

where $K = \rho_v/\rho_h$. Solutions of (21) and (22) have the form

$$F(\lambda, z) = A(\lambda) e^{uz} \tag{23}$$

and

$$G(\lambda, z) = B(\lambda) e^{vz} - (u/\lambda^2) A(\lambda) e^{uz} \tag{24}$$

where $u = (\lambda^2 + \gamma^2)^{1/2}$ and $v = (\lambda^2 K + \gamma^2)^{1/2}$ and where $A(\lambda)$ and $B(\lambda)$ are yet to be determined.

d. Statement of the Formal Solution

It is important to observe that (21) and (22) are coupled differential equations which, in effect, mean that the solutions for A_x and A_z are coupled by virtue of the anisotropy (i.e. $K \neq 1$) even within the homogeneous region.[+] Of course, in the upper anisotropic region ($z > 0$), this complication does not arise. In fact, for $z > 0$, we write the solution in the somewhat more familiar form

$$A_x = \frac{I\,ds}{4\pi} \int_0^\infty \frac{\lambda}{u_o} \left[e^{-u_o|z-h|} + R_\perp(\lambda) e^{-u_o(z+h)} \right] J_o(\lambda\rho) d\lambda \tag{25}$$

+ An alternative approach to the problem is to follow Kong [1976] and use Debye potentials or z directed electric and magnetic hertz vectors.

and

$$A_z = \frac{I\,ds}{4\pi}\frac{\partial}{\partial x}\int_0^\infty S(\lambda)e^{-u_o(z+h)}J_o(\lambda\rho)d\lambda \tag{26}$$

where $R_\perp(\lambda)$ and $S(\lambda)$ are also to be determined by the boundary conditions at $z = 0$.

As it turns out [Wait, 1982]:

$$A(\lambda) = \frac{\lambda}{u_o}\left[1 + R_\perp(\lambda)\right]e^{-u_oh} \tag{27}$$

$$R_\perp(\lambda) = \frac{u_o - u}{u_o + u} \tag{28}$$

$$S(\lambda) = \left[B(\lambda) - \frac{u}{\lambda^2}A(\lambda)\right]e^{u_oh} \tag{29}$$

$$B(\lambda) = \frac{1}{\lambda^2}\left[\frac{\gamma_o^2 + u_ou + \lambda^2}{u_o + (\gamma_o^2/\gamma^2)v}\right]A(\lambda) \tag{30}$$

which constitute the complete formal solution of the problem.

e. Quasi-Static Limiting Forms

We now do two things. First of all, we let $h \to 0$ corresponding to having the dipole lying in the interface. Then we proceed to the quasi-static limit and set $\gamma_o = 0$. The latter approximation or idealization is valid when all significant distances ℓ satisfy $|\gamma_o\ell| \ll 1$. In the limit, where $z = h = 0$, we then find that

$$E_x = -j\mu_o\omega A_x - \frac{\partial\Psi}{\partial x} \tag{31}$$

and

$$E_y = -\frac{\partial\Psi}{\partial y} \tag{32}$$

where

$$A_x = \frac{I\,ds}{2\pi}\int_0^\infty \frac{\lambda}{\lambda + u}J_o(\lambda\rho)d\lambda \tag{33}$$

$$= \frac{I\,ds}{2\pi\gamma^2\rho^3}\left[1 - (1 + \gamma\rho)e^{-\gamma\rho}\right] \tag{34}$$

and

$$\Psi = -\rho_h\left(\frac{\partial A_x}{\partial x} + \frac{\partial A_z}{\partial z}\right)_{z=0} \tag{35}$$

$$= -\frac{I\,ds}{2\pi}\rho_h\frac{\partial}{\partial x}\int_0^\infty\left[\frac{v - u}{\lambda} + 1\right]J_o(\lambda\rho)d\lambda \tag{36}$$

The expression above for A_x is the same as that for an isotropic half-space [Wait, 1961] and it depends only on ρ_h the complex resistivity in the horizontal direction. On the other hand, the surface potential Ψ depends on both ρ_h and ρ_v. To clarify this point, we write

$$\Psi = \Psi_1 + \Psi_2 \tag{37}$$

where

$$\Psi_1 = -\frac{I\,ds}{2\pi}\rho_h \frac{\partial}{\partial x}\frac{1}{\rho} = \frac{I\,ds}{2\pi}\rho_h \frac{x}{\rho^3} \tag{38}$$

and

$$\Psi_2 = \frac{I\,ds}{2\pi}\rho_h \int_0^\infty (v-u)J_1(\lambda\rho)d\lambda\frac{x}{\rho} \tag{39}$$

where we have employed the identities

$$\int_0^\infty J_o(\lambda\rho)d\lambda = \frac{1}{\rho} \tag{40}$$

$$\frac{\partial}{\partial x}J_o(\lambda\rho) = -J_1(\lambda\rho)\lambda\frac{x}{\rho} \tag{41}$$

In the isotropic limit, Ψ_2 vanishes and the limiting form $\Psi = \Psi_1$ corresponds to the expected result.

Actually, Ψ_2 can also be evaluated in closed form [Wait, 1982] and the result is given by

$$\Psi_2 = -\frac{I\,ds}{2\pi}\rho_h\frac{x}{\rho^3}\left[\exp\left(-\gamma\rho\right) - K^{1/2}\exp\left(-\gamma\rho K^{-1/2}\right)\right] \tag{42}$$

In the static limit (i.e. $|\gamma\rho| \to 0$), we see that

$$\Psi = \Psi_1 + \Psi_2 \to \frac{I\,ds}{2\pi}(\rho_h\rho_v)^{1/2}\frac{x}{\rho^3} \tag{43}$$

which is the expected DC solution. Another interesting observation is that $\Psi_2 \to 0$ if $|\gamma\rho| \gg 1$. Then, in this asymptotic limit

$$\Psi \to \Psi_1 = \frac{I\,ds}{2\pi}\rho_h\frac{x}{\rho^3} \tag{44}$$

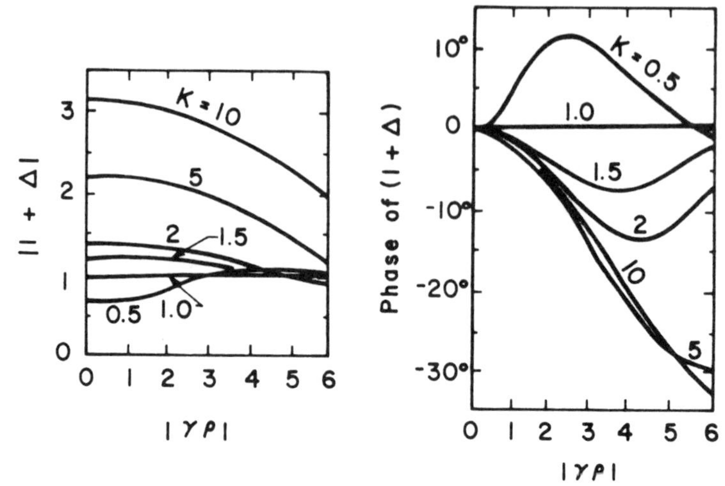

Figure 1.4.2 Amplitude and phase of the E_y field component for an x directed electric dipole on the surface ($z = 0$) of an anisotropic homogeneous half-space model of the earth.

which depends only on the horizontal resistivity.

The key point in the above development is that the potential function Ψ is dependent on frequency as $|\gamma\rho|$ varies from 0 to ∞ unless the half-space is isotropic. An interesting demonstration of this point is seen when we look at E_y (for $z = h = 0$) for the x directed electric dipole source. Using (32), (37), and (42), we deduce that

$$E_y = \frac{3(I\,ds)xy}{2\pi}\rho_h(1+\Delta) \tag{45}$$

where

$$\Delta = \frac{1}{3}\left[\left(3K^{1/2} + \gamma\rho\right)\,\exp\,(-\gamma\rho K^{-1/2}) - (3+\gamma\rho)\,\exp\,(-\gamma\rho)\right] \tag{46}$$

Clearly, in the isotropic limit, Δ vanishes. Also, if $|\gamma\rho| \gg 1$, Δ vanishes.

To illustrate the frequency dependence of E_y , as given by (45), we plot the amplitude and phase of $(1+\Delta)$ in Figs. 1.4.2a and 1.4.2b, as a function of $|\gamma\rho|$ for the cases where $K = 0.5, 1.0, 2, 5,$ and 10 . In this example, all displacement currents have been neglected. Thus

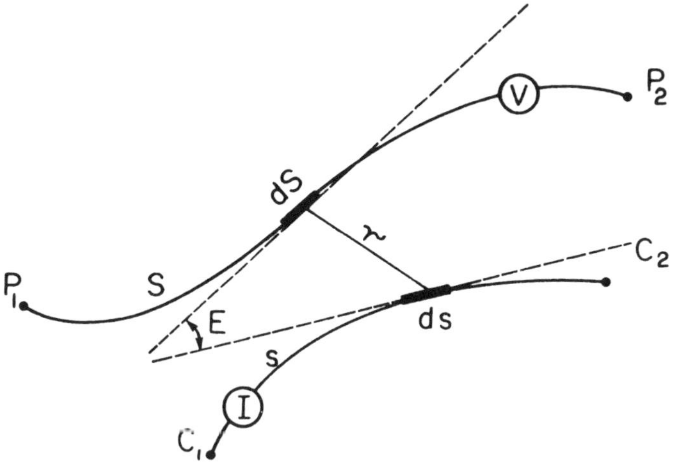

Figure 1.4.3 Plan view of the general non-intersecting current and potential cable configuration.

$|\gamma\rho| = (\mu_o\omega/\rho_h)^{1/2}$ is proportional to $(\text{frequency})^{\frac{1}{2}}$ where ρ_h is real and frequency independent. Clearly the frequency dependence of the normalized field $1 + \Delta$ is quite significant except in the isotropic limit where $K = 1$.

f. General Coupling Theory

Conceptually, it is evident that the tangential electric fields E_x and E_y of a horizontal electric dipole in the interface, can be used to deduce the mutual impedance between two grounded circuits. Such a configuration is illustrated in Fig. 1.4.3. For example, C_1 and C_2 could be the current electrodes which are connected by insulated wires to the generator which supplies the current I. The voltage V, induced in the potential circuit, is then the integrated electric field along the contour of the insulated wires connecting the "voltmeter" to the potential electrodes P_1 and P_2.

The first step is to regard the element $I\,ds$ in the current circuit as a source dipole at a distance s from C_1. The voltage V induced in the element dS is $-E_t dS$ where E_t is the tangential electric field at a distant r from ds. Here we designate ϵ as the angle subtended by ds and dS. The resulting induced voltage V is then obtained by integrating over both s and S. This procedure was illustrated by Sunde [1949] and generalized later [Wait, 1982]. In the present context, we find that

$$\frac{V}{I} = Z_m = \int_{C_1}^{C_2} \int_{P_1}^{P_2} \left[P(r) \cos \epsilon + \frac{\partial^2 Q(r)}{\partial s \partial S} \right] ds\, dS \tag{47}$$

where

$$P(r) = \frac{j \mu_o \omega}{2\pi \gamma^2 r^3} \left[1 - (1 + \gamma r) e^{-\gamma r} \right] \tag{48}$$

and

$$Q(r) = \frac{\rho_h}{2\pi} \left[\frac{1}{r} + \int_0^\infty \left(\frac{v-u}{\lambda} \right) J_o(\lambda r) d\lambda \right] \tag{49}$$

At sufficiently low frequencies (i.e. where $|\gamma r| \ll 1$) we see that

$$P(r) \cong j \mu_0 \omega / 2\pi r \tag{50}$$

and

$$Q(r) \cong (\rho_h \rho_v)^{1/2} / 2\pi r \tag{51}$$

In this case $P(r)$ and the resulting integral in (47) has a purely inductive character while $Q(r)$ leads to a purely resistive contribution. In general, γr is arbitrary and the frequency dependence is more complicated and simple closed-form expressions are not possible. Nevertheless, it is useful to note that $Q(r)$ can be expressed in terms of exponential integrals. To show this, we observe that

$$\begin{aligned} \frac{\partial}{\partial r} \int_0^\infty \frac{v-u}{\lambda} J_0(\lambda r) d\lambda &= - \int_0^\infty (v-u) J_1(\lambda r) d\lambda \\ &= \left[K^{1/2} \exp\left(-\gamma \rho K^{-1/2}\right) - \exp\left(-\gamma \rho\right) \right] / \rho^2 \end{aligned} \tag{52}$$

Then we can write

$$\begin{aligned} Q(r) &= \frac{1}{2\pi} \left[\frac{1}{r} + \int_0^\infty \left[K^{1/2} e^{(-\gamma r K^{-1/2})} - e^{(-\gamma r)} \right] \frac{1}{r^2} dr \right] \\ &= \frac{1}{2\pi} \left[\frac{1}{r} + K^{1/2} \frac{1}{r^2} E_2 \left(\gamma r K^{-1/2} \right) - \frac{1}{r^2} E_2(\gamma r) \right] \end{aligned} \tag{53}$$

where

$$E_2(\alpha) = \int_{t=1}^\infty \frac{e^{-\alpha t}}{t^2} dt \tag{54}$$

is an exponential integral of order 2 as defined by Abramowitz and Stegun [1964].

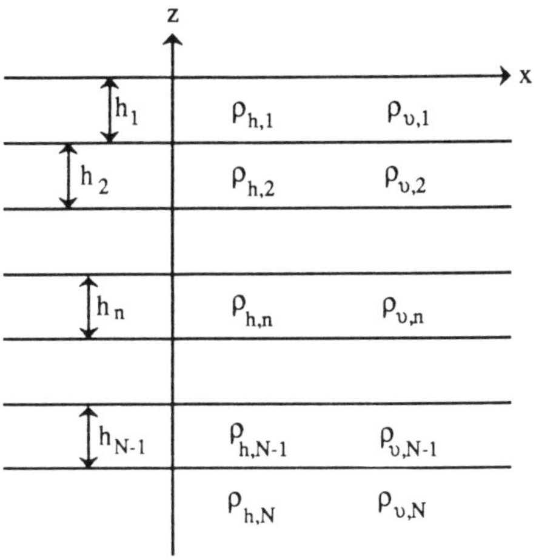

Figure 1.4.4 The N layered anisotropic half-space model of the earth.

g. Extensions to Layered Anisotropic Earth

There is some merit in extending the formulation of the electrocoupling to the case where the earth is represented as a layered half-space. Also it is desirable to let one or more of the layers be anisotropic. The situation is illustrated in Fig. 1.4.4. The solution of N-layer problem is carried through in a manner very similar to the homogenous half-space model. The details are given elsewhere [Wait 1966b, 1982]. As noted in Fig. 1.4.4, the n'th layer has a complex resistivity $\rho_{h,n}$ in the horizontal direction, and $\rho_{v,n}$ in the vertical direction. Here $n = 1, 2, 3, \ldots N$.

The mutual impedance formula given by (47) is still applicable but $P(r)$ and $Q(r)$ are now given by

$$P(r) = \frac{j\mu_o\omega}{2\pi}\int_o^\infty \frac{\lambda}{\lambda + j\mu_o\omega Y_1} J_0(\lambda r)d\lambda \tag{55}$$

and

$$Q(r) = \frac{1}{2\pi}\int_o^\infty \left[Z_1 - \frac{j\mu_o\omega}{\lambda + j\mu_o\omega Y_1}\right]\lambda J_0(\lambda r)d\lambda \tag{56}$$

where Y_1 and Z_1 are admittance and impedance functions defined as follows

$$Y_1 = N_1\frac{Y_2 + N_1\tanh u_1h_1}{N_1 + Y_2\tanh u_1h_1} \tag{57}$$

$$Y_2 = N_2\frac{Y_3 + N_2\tanh u_2h_2}{N_2 + Y_3\tanh u_2h_2} \tag{58}$$

$$\ldots$$

$$Y_n = N_n \frac{Y_{n+1} + N_n \tanh u_n h_n}{N_n + Y_{n+1} \tanh u_n h_n} \tag{59}$$

$$\ldots$$

$$Y_{N-1} = N_{N-1} \frac{N_N + N_{N-1} \tanh u_{N-1} h_{N-1}}{N_{N-1} + N_N \tanh u_{N-1} h_{N-1}} \tag{60}$$

where $N_n = u_n/(j\mu_o\omega)$ and $u_n = (\lambda^2 + \gamma_n^2)^{1/2}$ and $\gamma_n = (j\mu_o\omega/\rho_{h,n})^{1/2}$ for $n = 1, 2, 3, \ldots N$. The iterative formula for Z_1 has precisely the same form as (57) to (60) if the Y 's are replaced by Z 's, the N 's are replaced by K 's and the u 's by v 's, where

$$K_n = v_n \rho_{h,n} \quad \text{and} \quad v_n = \left(\lambda^2 \frac{\rho_{v,n}}{\rho_{h,n}} + \gamma_n^2\right)^{1/2}$$

In the limiting case of a homogeneous half-space (i.e. $h_1 \to \infty$ and $u_1 \to u$), $Y_1 \to u/j\mu_o\omega$ and $Z_1 \to v\rho_h$. Then $P(r)$ and $Q(r)$ reduce to the forms (48) and (49), respectively. We also recover the results derived by Hohmann [1973] for the 2-layer isotropic case.

Any general configuration of grounded wires connecting the electrodes can be handled with the above general forms for $P(r)$ and $Q(r)$ used in conjunction with (47). A rather extreme special case is the homogeneous half-space. It is often used as the basis for an electromagnetic coupling estimate in connection with induced polarization surveys. But, perhaps one should bear in mind here that the resistivity is complex and matters can become complicated; we illustrate the point below for a half-space of isotropic complex resistivity $\rho(\omega)$.

h. An Illustrative Example

The so called dipole-dipole array is considered as illustrated in the inset in Fig. 1.4.5a. The electrodes are on a common line and the spacing is such that $C_1C_2 = P_1P_2 = a$ and $P_2C_1 = na$. As shown by Wait and Gruszka [1986a], the mutual impedance for this case can be expressed in analytical form:

$$Z_m = \frac{\rho(\omega)}{4\pi}\gamma \left\{ \frac{2}{an(n+1)(n+2)\gamma} - 2G[\gamma a(n+1)] + G(\gamma an) + G[\gamma a(n+2)] \right\} \tag{61}$$

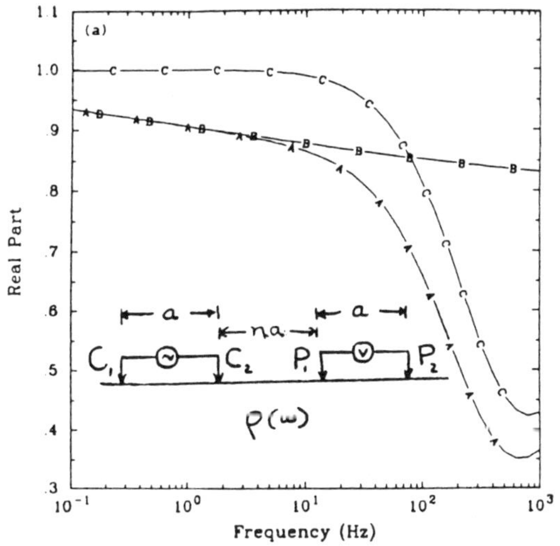

Figure 1.4.5a The real part of the complex resistivity for a dipole-dipole array as a function of frequency. Curve A – apparent resistivity, Curve B – pure IP, Curve C – pure EM.

where

$$\gamma = [j\mu_o\omega/\rho(\omega)]^{1/2} \tag{62}$$

and

$$G(z) = e^{-z}\left(\frac{1}{z} - 1\right) + zE_1(z) \tag{63}$$

in terms of the exponential integral [Abramowitz and Stegen, 1964] defined by

$$E_1(z) = \int_z^\infty \frac{e^{-v}}{v} dv = \int_1^\infty \frac{e^{-zt}}{t} dt \tag{64}$$

In the DC limit (i.e. $\omega \rightarrow 0$), we recover the expected form

$$Z_m \rightarrow R_0 = \frac{\rho(0)}{\pi an(n+1)(n+2)} \tag{65}$$

which is well known in resistivity formulations (e.g. [Chap. 1, Wait, 1982]).

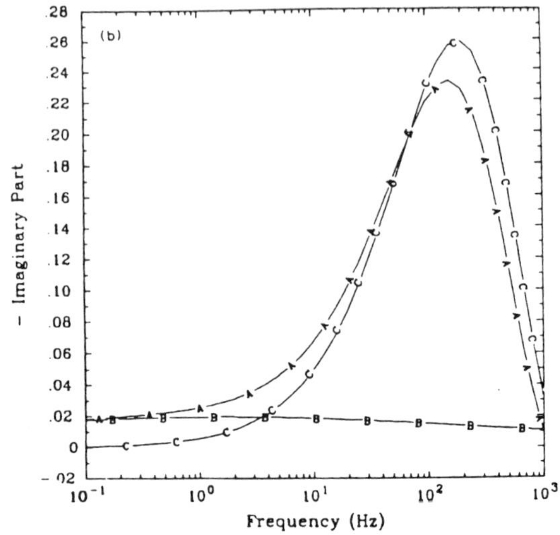

Figure 1.4.5b The imaginary part of the complex resistivity for a dipole-dipole array as a function of frequency.

We are now led to define an apparent (complex) resistivity as follows, for any frequency:

$$Z_m = \frac{\rho_a(\omega)}{\pi an(n+1)(n+2)} \tag{66}$$

Then clearly

$$\frac{\rho_a(\omega)}{\rho(0)} = \frac{Z_m}{R_0} = R(\omega) + jX(\omega) \tag{67}$$

where $R(\omega)$ and $X(\omega)$ are the real and imaginary parts of the apparent complex resistivity.

Actually, Z_m is valid for any frequency provided the appropriate frequency dependent form for $\rho(\omega)$ is employed. For purposes of illustration, we choose the Cole-Cole form [Pelton, 1978] given by

$$\rho(\omega) = \rho(0)\left[1 - m\left(1 - \frac{1}{1+(j\omega\tau)^k}\right)\right] \tag{68}$$

where $\rho(0)$ is the DC value, m is the chargeability, τ is a time constant, and k is a dispersion index. One should note that at a sufficiently high frequency $\rho(\omega) \to \rho(0)(1-m)$. Thus m is a good measure of the total dispersion over the whole significant frequency range (i.e. 10^{-1} to 10^3 Hz). The adjustable parameters τ and k are chosen to represent a particular rock or mineral type. For our numerical example, we choose $m = 0.2$, $n = 2$, $k = 1/4$, $\tau = 0.1$ sec., $\rho(0) = 10^2$ ohm-m and $a = 200$ m.

Using (61) and (68), we plot $R(\omega)$ and $X(\omega)$, as defined by (67), as a function of frequency in Hz in Figs. 1.4.5a and 1.4.5b respectively. The resulting curves are designated A. Then we plot what might be called pure induced polarization (IP) which, in the present context, would be defined by

$$R(\omega) + jX(\omega) = \frac{\rho(\omega)}{\pi a n (n+1)(n+2) R_0} = \frac{\rho(\omega)}{\rho(0)} \tag{69}$$

In other words, electromagnetic (EM) coupling is ignored. These curves are designated B. It is significant that the A and B curves begin to diverge even for frequencies as low as 1 Hz. This fact is not surprising when the pure EM coupling curves are also plotted. In this case we use (61) but $\rho(\omega)$ is replaced everywhere by $\rho(0)$. Clearly the EM coupling is dominating the response for frequencies above about 10 Hz. This fact should be borne in mind in interpreting such measurements. Wait and Gruszka [1986a,b] discuss a possible decoupling procedure where the actual or composite response function is "corrected" by subtracting out the pure EM coupling results. The method is not particularly effective at the higher frequencies.

In the applied geophysical literature [Wait 1959a, Dey and Morrison 1973, Brown 1985, Song 1985, Wynn and Zonge 1975, 1977, Wynn 1979], one finds various proposals to mitigate the annoying effects of electromagnetic coupling in induced polarization surveys. Some rather ingenious attempts to "remove" the EM coupling by special data processing techniques have been made. No attempt will be made here to evaluate these procedures which may be more "art" than science. Our purpose is merely to identify the nature of the problem and to illustrate the inherent complexity of the situation that confronts the exploration geophysicists.

Rather than attempting to remove or correct the resultant complex resistivity data for EM coupling, it might be better to carry out the interpretation using a dynamic formulation such as proposed earlier [Wait 1981]. Such an approach has been developed by Mahmoud et al [1987] which requires the use of two receiving grounded dipoles

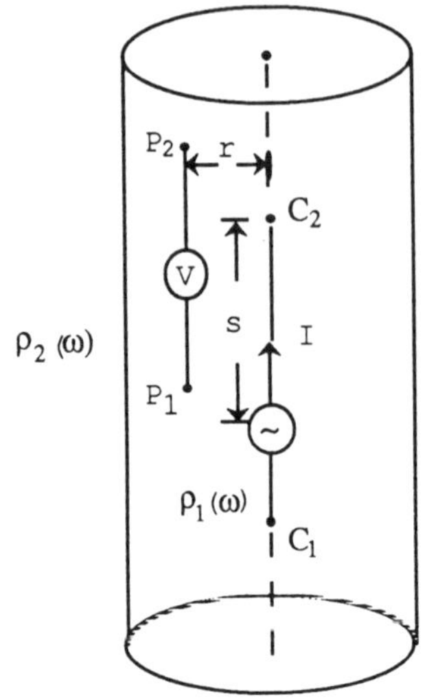

Figure 1.4.6 Four electrode array located in a bore-hole where potential circuit is offset by a distance r from the axis.

with special orientations relative to the source dipole. The procedure was applied to computer simulated data on a two layer earth with encouraging results.

i. Borehole Configuration

Here we wish to discuss the EM coupling problem when the electrodes are located in a borehole. The starting point is to deal with the electromagnetic field of an oscillating electric dipole located on the axis of the borehole. Such a solution, for a homogeneous cylinder of radius b of complex resistivity ρ_1 and an external homogeneous region of complex resistivity ρ_2, is straight forward. It is the special case of a general analysis for dipoles in the presence of cylindrical structures [Wait 1959b, Hill and Wait 1979]. A similar model was considered recently by Freedman and Vogiatzis [1986].

The special geometry is shown in Fig. 1.4.6. The current electrodes C_1 and C_2 are fed effectively by a generator on the axis of the borehole. The potential electrodes P_1 and P_2 are displaced from the axis by a distance r. An expression for the mutual impedance Z is given in explicit form by Gruszka [1987] as follows:

$$Z = \frac{\rho_1}{4\pi} \sum_{i,j=1}^{2} (-1)^{i+j+1} \left\{ \frac{e^{-\gamma_1 R_{ij}}}{R_{ij}} - \gamma_1^2 \int_0^\infty \frac{\lambda}{u_1^2} e^{-u_1 z_{ij}} J_0(\lambda r) d\lambda \right.$$
$$\left. + \frac{2}{\pi} \int_0^\infty \frac{u_1^2 A(\lambda) I_0(u_1 r) - \gamma_1^2 A(0) I_0(\gamma_1 r)}{\lambda^2} \cos(\lambda z_{ij}) d\lambda \right\}$$
$$- \frac{\gamma_1^2 \rho_1 s}{2\pi} \left\{ \int_0^\infty \frac{\lambda}{u_1^2} J_0(\lambda r) d\lambda + A(0) I_0(\gamma_1 r) \right\} \quad (70)$$

where

$$A(\lambda) = - \left[\frac{u_1 \rho_1 K_0(u_1 b) - u_2 \rho_2 \frac{K_0(u_2 b)}{K_1(u_2 b)} K_1(u_1 b)}{u_1 \rho_1 I_0(u_1 b) + u_2 \rho_2 \frac{K_0(u_2 b)}{K_1(u_2 b)} I_1(u_1 b)} \right] \quad (71)$$

$$u_1 = (\lambda^2 + \gamma_1^2)^{1/2} \quad , \quad Re\, u_1 > 0$$
$$u_2 = (\lambda^2 + \gamma_1^2)^{1/2} \quad , \quad Re\, u_2 > 0$$
$$R_{ij} = (r^2 + z_{ij}^2)^{1/2} \quad , \quad z_{ij} = |P_j - C_i|$$
$$\gamma_1 = (j\mu_1\omega/\rho_1)^{1/2} \quad , \quad \gamma_2 = (j\mu_2\omega/\rho_2)^{1/2}$$

and s is the overlap distance as indicated in Fig. 1.4.6.

A special case of the formulation is when we locate a dipole-dipole array on the axis of the bore-hole (i.e. $r = 0$). To illustrate this case, we let the borehole fluid have a dispersionless or non-polarizable resistivity ρ_1 while the external region of complex resistivity $\rho_2(\omega)$ is characterized by the Cole-Cole form given by (68). The particular parameter values are indicated in the caption for Fig. 1.4.7a.

The real and imaginary parts of the apparent complex resistivity are plotted as a function of frequency in Figs. 1.4.7a and 1.4.7b and identified as curve A in each figure. Curves B correspond to "pure IP" in the sense that the propagation constants γ_1 and γ_2 are set equal to zero. Then curves C are what we might call "pure EM" in the sense that $\rho_2(\omega)$ is replaced by its DC value $\rho(0)$. Curves D are obtained by subtracting the EM anomaly from the actual complex resistivity so the quantity plotted is

$$\frac{\rho_a(\omega)}{\rho_a(0)} - \left(\frac{\rho_a^{EM}}{\rho_a(0)} - 1 \right)$$

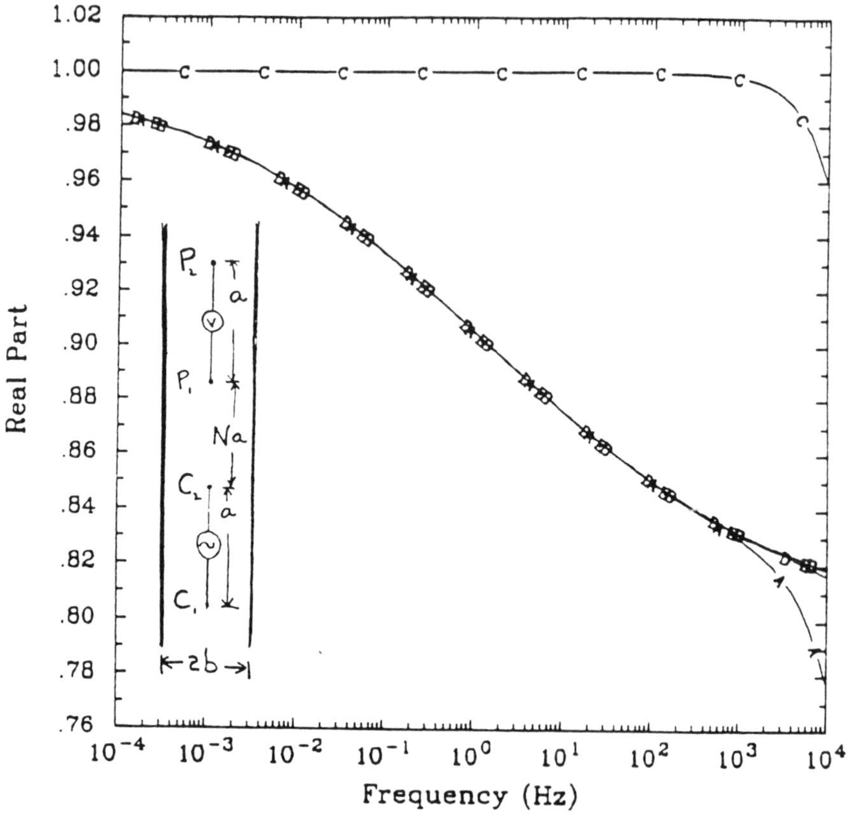

Figure 1.4.7a Dipole-dipole array located on the axis of a bore-hole with the following parameters: $\rho_1 = 1\ \Omega\text{m}$, $\rho_2/\rho_1 = 50$, $m_2 = 0.2$, $\tau = 0.1$ $k_2 = 0.25$, $N = 1.0$, $b = 0.1$ m, and $a = 10$ m. The real part of the normalized complex resistivity is shown for the five cases listed below. Curve A – $\rho_a(\omega)/\rho_a(0)$, Curve B – $\rho_a(\omega)/\rho_a(0)$ for $\gamma_1 = \gamma_2 = 0$, Curve C – $\rho_a(\omega)/\rho_a(0)$ for $m_2 = 0$, Curve D – EM corrected, see text, Curve E – $\rho_2(\omega)/\rho_2(0)$.

Finally, in Figs. 1.4.7a and 1.4.7b, curves E correspond to the complex resistivity ratio $\rho_2(\omega)/\rho_2(0)$ or "true IP" of the formation or external region.

It is evident that the electromagnetic coupling (curves C in Figs. 1.4.7a and 1.4.7b) is important at frequencies above about 100 Hz for the parameters chosen. Not surprisingly, the coupling is most noticeable in the imaginary part (i.e. the reactive part) of the apparent complex resistivity where, to the first order, we would expect the EM coupling to be proportional to $j\omega L$ where L is an inductance. In the

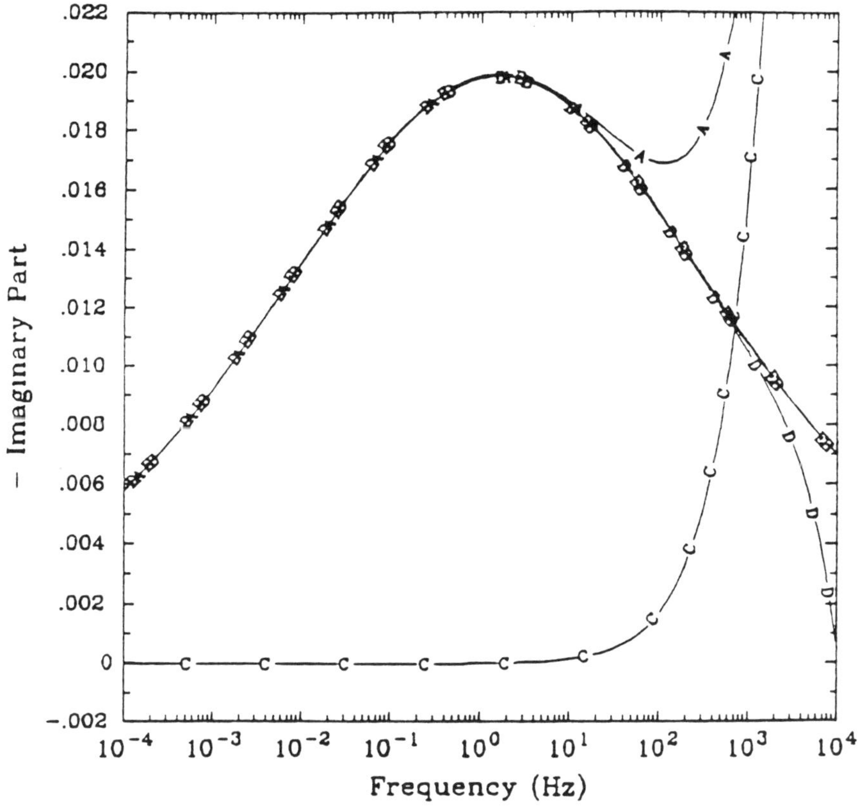

Figure 1.4.7b Dipole-dipole array located in a bore-hole with parameters indicated in caption for Fig. 1.4.7a. Here the imaginary part of the normalized complex resistivity is shown for the same five cases.

hypothetical case where the electromagnetic effects were absent (i.e. $\gamma_1 = \gamma_2 = 0$), we can see from the closeness of curves B and E that the resistivity of the formulation or external region would be almost identical to the measured or apparent resistivity.

To "correct" the apparent complex resistivity in the presence of EM coupling, we perform the single subtraction process indicated above. For the real part, we see from Fig. 1.4.7a that the resultant corrected resistivity coincides, within the graphical accuracy, with the actual resistivity of the external medium. However, the correction process is not so effective (i.e. curves A and D diverge at frequencies above about 200 Hz).

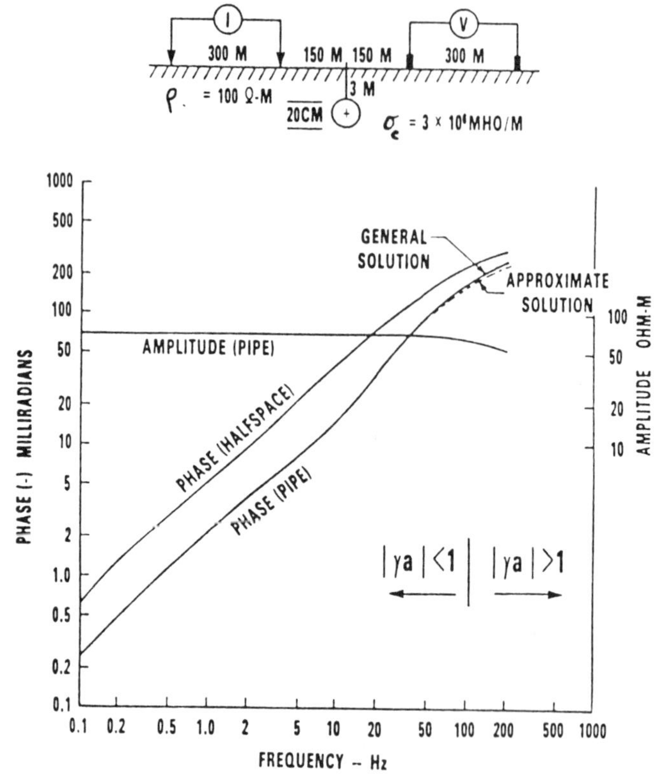

Figure 1.4.8a Amplitude and phase spectra, dipole-dipole array perpendicular to the pipe (spread $N = 1$).

j. More Complicated Cylindrical/Half-Space Problems

Often in dealing with complex resistivity data in the field, the so-called cultural problem arises. The best example is buried conductor such as a metallic pipe or cable that may be in the vicinity of the measuring electrodes. A case in point occurs when we employ a dipole-dipole array on the surface of a half space of the earth which is homogeneous except for a relatively thin conductor located at a fixed depth h. General formulations [Wait 1977a,b, 1978] for this type of problem have been published. The solutions are tractable under the following conditions: the buried conductor or cable is of infinite length, the conductor radius is small compared with other typical dimensions, and the conductor can be described by an axial (spatially dispersive) impedance. Parra [1984] has applied such a formulation to the case

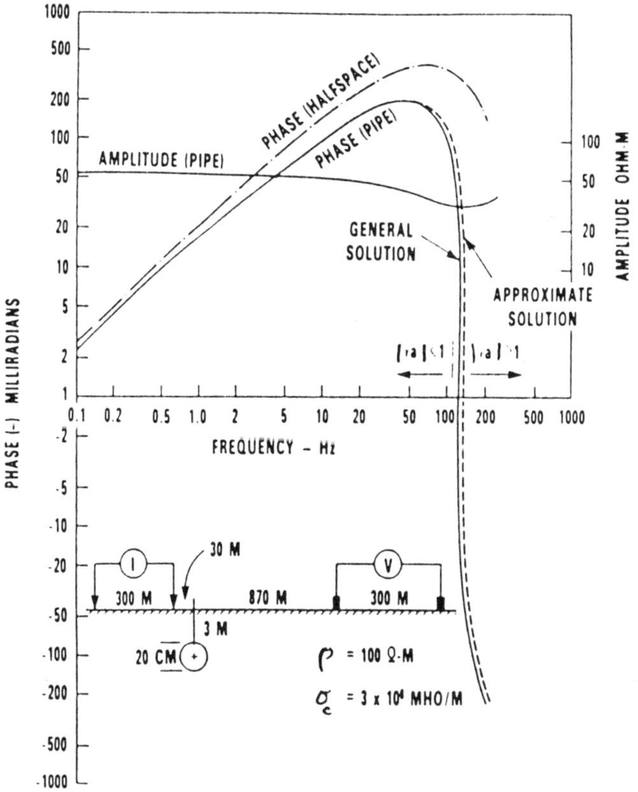

Figure 1.4.8b Amplitude and phase spectra, dipole-dipole array perpendicular to the pipe (spread $N = 3$).

of solid metal pipe of radius 10 cm buried at a depth of $h = 3\,\text{m}$. The half-space has a resistivity of $100\,\Omega\,\text{m}$ while the pipe has a real conductivity $\sigma_c = 3 \times 10^6$ mhos/m. Parra shows various results for a dipole-dipole array that may be oriented at any angle relative to the pipe. His results for two cases where the pipe is at right angles to the dipole-dipole array are shown in Figs. 1.4.8a and 1.4.8b. The amplitude and phase of the effectives complex resistivity are shown as a function of frequency from 0.1 to 10^3 Hz. For comparison, the phase is also shown for the half space in the absence of the pipe. In Fig. 1.4.8a, the pipe is located centrally between the dipoles and buried a distance of 3m. The electrode spacing $a = 300\,\text{m}$ and the spread factor $N = 1$. In Fig. 1.4.8b the pipe is located nearer the transmitting dipole and the spread factor $N = 3$. In the latter case, there is a drastic change of the phase curve for frequencies greater than 100 Hz. This rapid change of phase seems to occur when $|\gamma a| > 1$ where γ is the propagation constant of the half-space. In Parra's calculations, the magnetic permeability of the pipe is assigned the free space value μ_o. Also, in the curves shown in Figs. 1.4.8a and 1.4.8b, the induced polarization at the pipe surface is neglected. Similar results were presented by Wait and Williams [1985] for the closely related problem of a vertical metal well casing in the vicinity of a surface-based dipole-dipole array. They allowed for the large magnetic permeability of the steel (e.g. $\mu/\mu_o \simeq 500$) and also the appropriate value of the interface impedance η at the steel/electrolyte boundary was incorporated into the calculations. In such a situation, both the eddy currents in the pipe or casing and the induced polarization contribute to the apparent complex resistivity.

Concluding Remarks

As we have demonstrated, the mutual impedance or apparent complex resistivity is a complicated function of the intrinsic complex resistivity of the medium. The influence of direct EM coupling, of the lead wires and effects of buried pipes and wires need to be estimated for a given situation. While it does not seem possible to mathematically justify claims to "remove" EM coupling, it is certainly not inconceivable that special data processing such as devised by Wynn and Zonge [1975, 1977] will emphasize the relative contribution of the induced polarization.

References

[1] Abramowitz, M., and I. A. Stegun (ed.), *Handbook of Mathematical Functions*, Dover, 1964.

[2] Brown, R. J., "EM coupling in multi-frequency IP and a generalization of the Cole-Cole impedance model," *Geophysical Prospection,* **33**, 282–302, 1985.

[3] Dey, A. and H.F. Morrison, "EM coupling in frequency and time domain IP surveys over a multi-layered earth," *Geophysics,* **38**,

380–405, 1973.

[4] Freedman, R. and J. P. Vogiatzis, "Theory of IP logging in a borehole," *Geophysics,* **51,** 1830–1849, 1986.

[5] Hill, D. A. and J. R. Wait, "Comparison of loop and dipole antennas in leaky feeder communication systems," *Int'l. Jour. of Electronics,* **47,** 155–166, 1979.

[6] Gruszka, T. P., "Induced polarization and electromagnetic coupling and their interaction in low frequency geophysical exploration," PhD Thesis, Univ. of Arizona, 1987.

[7] Hohmann, G. W., "EM coupling between grounded wires at the surface of a two-layer earth," *Geophysics,* **38,** 854–863, 1973.

[8] Mahmoud, S. F., S. G. Tantawi, and J. R. Wait, "Interpretation of multi-frequency complex resistivity data for a layered earth model," *IEEE Trans.,* **GE-26,** 399–408, 1988.

[9] Parra, J. O., "Effects of pipelines on spectral IP surveys," *Geophysics,* **49,** 1979–1992, 1984.

[10] Kong, J. A., *Theory of Electromagnetic Waves,* Wiley, 1976.

[11] Pelton, W. H., S. H. Ward, P. G. Hallof, W. R. Sill, and P. H. Nelson, "Mineral discrimination and the removal of inductive coupling with multi-frequency IP," *Geophysics,* 588–609, 1978.

[12] Sommerfeld, A. N., "The propagation of waves in wireless telegraphy," *Annalen der Physik,* Series 4, **28,** pp. 665 and 81, pp. 1135, 1909 and 1926, (in German).

[13] Song, L., "A new decoupling scheme," *Exploration Geophysics,* 1599–112, 1984.

[14] Sunde, E. D., *Earth Conduction Effects in Transmission Systems,* Van Nostrand, 1949.

[15] Wait, J. R., *Electromagnetic Radiation from Cylindrical Structures,* Pergamon, 1959b (reprinted with corrections, Peter Peregrinus Ltd. Stevenage UK, 1988).

[16] Wait, J. R., "EM fields of a horizontal dipole in the presence of a conducting half-space," *Can. Jour. Phys.,* **39,** 1017–1028, 1961.

[17] Wait, J. R., "The variable frequency method," *Overvoltage Research and Geophysical Applications* (edited by J. R. Wait), Perg-

amon Press, 1959a.

[18] Wait, J. R., "EM fields of a dipole over a stratified anisotropic half-space," *IEEE Trans. on Antennas and Propagat.*, **AP-14**, 790–792, 1966b.

[19] Banos, A., *Dipole Radiation in the Presence of a Conducting Half-space*, Pergamon, 1966.

[20] Wait, J. R., "Excitation of a coaxial cable or wire conductor located over the ground by a dipole radiator," *Archiv fur Elektronik und Electragungtechnik*, **31**, 121–127, 1977a.

[21] Wait, J. R., "Excitation of an ensemble of parallel cables by an external dipole over a layered ground," *A.E.U.*, **31**, 489–493, 1977b.

[22] Wait, J. R., "Towards a general theory of induced electric polarization in geophysical exploration," *IEEE Trans. on Geoscience and Remote Sensing,* **GE–19**, 231–234, 1981.

[23] Wait, J. R., "Excitation of currents on a buried insulated cable," *J. Appl. Phys.,* **49**, 876–880, 1978.

[24] Wait, J. R., *GeoElectromagnetism*, Academic Press, 1982.

[25] Wait, J. R. and T. P. Gruszka, "Interaction of EM and electrochemical effects in geophysical probing," *Electronics Letters,* **22**, 393–394, 1986b.

[26] Wait, J. R. and T.P. Gruszka, "On EM coupling removal from IP surveys," *Geoexploration,* **24**, 21–27, 1968a.

[27] Wait, J. R. and J. T. Williams, "EM and IP response of a steel well casing for a four electrode surface array," *Geophysical Prospecting,* **33**, 723–745, 1985.

[28] Wynn, J. C. and K. L. Zonge, "EM coupling," *Geophysical Prospecting,* **25**, 31–51, 1977.

[29] Wynn, J. C. and K. L. Zonge, "EM coupling, its intrinsic value, its removal and the cultural coupling problem," *Geophysics*, **40**, 831–850, 1975.

[30] Wynn, J. C., "EM coupling with a collinear array on a two layer earth," *US Geological Professional Paper 1077*, US Gov't Printing Office, 1979.

1.5 I.P. Response of Prolate Spheroidal Ore Grains

a. Introduction

In the induced polarization method of geophysical prospecting, one exploits the interfacial polarization at the surfaces of dissimilar material. The best example is the interface between a metallic conductor such as an ore grain and the adjacent electrolytic region. Here we will consider the alternating current response of an idealized spheroidal model of the ore grain which has a specified interface impedance $Z(j\omega)$ for a time-harmonic factor $\exp(j\omega t)$. We then consider an ensemble of such particles and obtain an expression for the apparent or effective complex resistivity $\rho_e(j\omega)$ as a function of volume loading and particle shape. We shall restrict attention to frequencies that are low enough to allow potential theory to be used.

The following theoretical development is based partly on previous publications [Wait 1982, 1983, Flanagan 1983, Flanagan and Wait 1985]. Essentially, it is a generalization of the spherical particle model which also utilized the interface impedance concept. We contrast our approach with the electrochemical formulation by Wong and Strangway [1981]. Not withstanding their diligence in fitting their theory to published experimental data, one could question their method in handling the infinite system of coupled mode equations for ionic perturbations. We discuss this point below.

b. Basic Formulation

To conform with the shape of the basic particle, we choose prolate spheroidal coordinates as indicated in Fig. 1.5.1. The relationships between the spheroidal system (η, δ, ϕ) and the cylindrical system (r, ϕ, z) are

$$r = c\left[(1-\delta^2)(\eta^2-1)\right]^{1/2} \tag{1}$$

$$z = c\eta\delta \tag{2}$$

$$\phi = \phi \tag{3}$$

where c is the semi-focal distance. The angular coordinate ranges from -1 to $+1$ while the radial coordinate ranges from 1 to ∞. As indicated in Fig. 1.5.2, the surface of the particle is defined by $\eta = \eta_0$ or, in terms

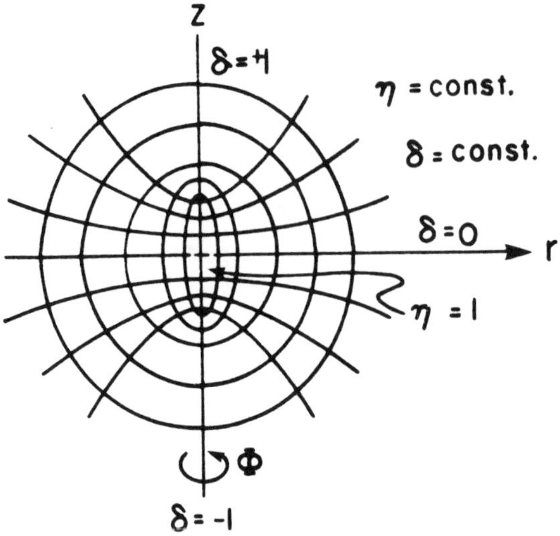

Figure 1.5.1 The prolate spheroidal coordinate system.

of cylindrical coordinates

$$\frac{r^2}{a^2} + \frac{z^2}{b^2} = 1 \tag{4}$$

where a and b are the semi-minor and semi-major axes, respectively. Confocal spheroidal coordinates are then

$$\frac{r^2}{c^2(\eta^2 - 1)} + \frac{z^2}{c^2\eta^2} = 1 \tag{5}$$

provided $a = c({\eta_0}^2 - 1)^{1/2}$ and $b = c\eta_0$.

Initially, we assume $b > a$ corresponding to the prolate spheroidal geometry. Later, we indicate the extension to the oblate spheroidal case. In terms of a and b, we have $c = (b^2 - a^2)^{1/2}$ and $\eta_0 = b(b^2 - a^2)^{-1/2}$.

The particle is assigned a resistivity ρ_1 (for $\eta < \eta_0$) and the external region is homogeneous with resistivity ρ. The surface of the particle is characterized by an interface impedance $Z(j\omega)$. In this way, we account for the discontinuity of the potential across the surface.

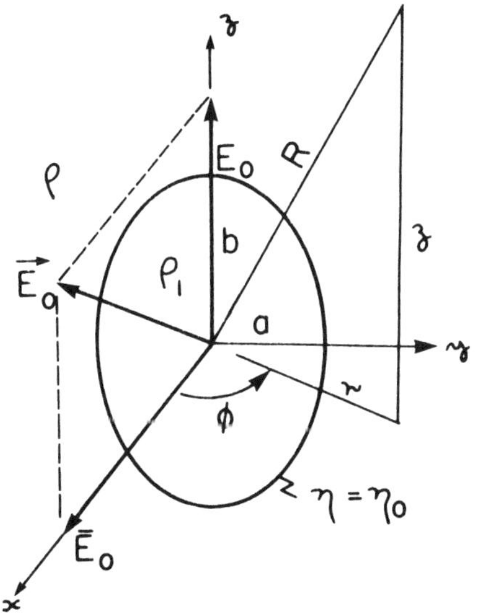

Figure 1.5.2 The spheroidal particle in a locally uniform applied field.

c. Spheroidal Harmonic Solution

Our objective is to determine the response of the spheroid when it is immersed in a uniform electric field $\vec{E}_0$. As indicated in Fig. 1.5.2, $\vec{E}_0$ is in the plane $\phi = 0$ or the (x, z) plane in the cartesian system. Thus, it is convenient to write

$$\vec{E}_0 = \hat{x}\overline{E}_0 + \hat{z}E_0 \tag{6}$$

where $\overline{E}_0$ and E_0 are the transverse and longitudinal components, respectively. Basically, we need to derive the response of the sphere to both transverse and longitudinal field excitations. The general case is then handled by superposition.

The electric fields are obtained from

$$\overline{E} = -\text{ grad } \Phi \tag{7}$$

where Φ, the resultant potential, satisfies

$$\nabla^2\Phi = 0 \tag{8}$$

Solutions are of the form

$$\Phi = \begin{matrix} P_n{}^m(\eta)P_n{}^m(\delta)\cos m\phi \\ Q_n{}^m(\eta)Q_n{}^m(\delta)\sin m\phi \end{matrix} \tag{9}$$

where, in general, any linear combination is allowed. Here $P_n{}^m$ and $Q_n{}^m$ are Legendre polynomials of order m and degree n. The arguments are η for the radial functions and δ for the angular functions. To assure single-valueness, m is an integer. Also, because the field is finite at $\delta = \pm 1$, we can reject the $Q_n{}^m(\delta)$ functions. Furthermore, we recognize that for the interior (i.e. $\eta < \eta_0$), $Q_n{}^m(\eta)$ is singular as $\eta \to 1$ so it is not acceptable. Also, in the external region (i.e. $\eta > \eta_0$), we note that $P_n{}^m(\eta)$ does not vanish as $\eta \to \infty$.

In the case of purely axial excitation (i.e. $\overline{E}_0 = 0$), we write the resultant potentials in the following forms:

For $\eta > \eta_0$

$$\Phi = A^0 P_1(\eta)P_1(\delta) + \sum_{n=1,3,5\cdots}^{\infty} B_n{}^0 Q_n(\eta)P_n(\delta) \tag{10}$$

For $\eta < \eta_0$

$$\Phi = \sum_{n=1,3,5\cdots}^{\infty} C_n{}^0 P_n(\eta)P_n(\delta) \tag{11}$$

Here, A^0 is known while $B_n{}^0$ and $C_n{}^0$ are coefficients to be determined. We note that $m = 0$ because the problem has obvious axial symmetry. In fact, $A^0 = -cE_0$.

In the case of purely transverse excitation (i.e. $E_0 = 0$), the solution forms are as follows:

For $\eta > \eta_0$

$$\Phi = A^1 P_1{}^1(\eta)P_1{}^1(\delta)\cos\phi + \sum_{n=1,3,5\cdots}^{\infty} B_n{}^1 Q_n{}^1(\eta)P_n{}^1(\delta)\cos\phi \tag{12}$$

For $\eta < \eta_0$

$$\Phi = \sum_{n=1,3,5\cdots}^{\infty} C_n{}^1 P_n{}^1(\eta)P_n{}^1(\delta)\cos\phi \tag{13}$$

Here, A^1 is known while B_n^1 and C_n^1 are to be determined. Here, we have taken $m = 1$ and noted that Φ is an even function of ϕ. In fact, $A^1 = -c\overline{E}_0$.

In writing out (10) and (12), we have located the primary potential as the first expression on the R.H.S.(right hand side). In both cases, the summation term represents the secondary potential which vanishes as $\eta \to \infty$. More explicit details are given elsewhere [Wait, 1983].

The first boundary condition is that the normal current density J_η is continuous at the particle surface. To be specific,

$$\frac{1}{\rho_1}\frac{\partial\Phi}{\partial\eta}\bigg|_{\eta=\eta_0-0} = \frac{1}{\rho}\frac{\partial\Phi}{\partial\eta}\bigg|_{\eta=\eta_0+0} \tag{14}$$

The second boundary condition is that the potential drop across the interface is proportional to the product of the interface impedance $Z(j\omega)$ and the normal current density. This is equivalent to the statement

$$\Phi\bigg|_{\eta=\eta_0-0} = \Phi\bigg|_{\eta=\eta_0+0} - \frac{Z(j\omega)(\eta_0{}^2-1)^{1/2}}{\rho c(\eta_0{}^2-\delta^2)^{1/2}}\frac{\partial\Phi}{\partial\eta}\bigg|_{\eta=\eta_0+0} \tag{15}$$

which must hold for $-1 \le \delta \le +1$ [Wait 1982, 1983].

On applying the boundary condition (14) to either (10) and (11) or to (12) and (13), we get

$$\begin{aligned}\frac{1}{\rho_1}\sum_n C_n{}^m \dot{P}_n^m(\eta_0)P_n{}^m(\delta) =& \frac{1}{\rho}A^m P_1(\eta_0)P_1{}^m(\delta)\\ &+\frac{1}{\rho}\sum_n B_n{}^m \dot{Q}_n^m(\eta_0)P_n{}^m(\delta)\end{aligned} \tag{16}$$

where

$$\dot{P}_n^m(\eta_0) = \frac{\partial}{\partial\eta}P_n{}^m(\eta)\bigg|_{\eta=\eta_0} \tag{17}$$

and

$$\dot{Q}_n^m(\eta_0) = \frac{\partial}{\partial\eta}Q_n{}^m(\eta)\bigg|_{\eta=\eta_0} \tag{18}$$

Here $m = 0$ or 1 corresponding to longitudinal or transverse excitation, respectively.

We now multiply both sides of (16) by $P_k{}^m(\delta)$ where k is an integer and then integrate over δ from -1 to $+1$. Making use of the orthogonality relation

$$\int_{-1}^{+1} P_n{}^m(\delta)P_k{}^m(\delta)d\delta = \begin{cases} \dfrac{2}{2k+1}\dfrac{(k+m)!}{(k-m)!} & \text{for } k = n \\ 0 & \text{for } k \neq n \end{cases} \tag{19}$$

we deduce that

$$C_k = A^m \frac{\rho_1}{\rho}\frac{P_1{}^m(\eta_0)}{\dot{P}_k^m(\eta_0)}\delta_{1,k} + B_k \frac{\rho_1}{\rho}\frac{\dot{Q}_k^m(\eta_0)}{\dot{P}_k^m(\eta_0)} \tag{20}$$

which holds for any value of k. Here

$$\delta_{1,k} = \begin{cases} 1 & \text{for } k = 1 \\ 0 & \text{for } k = 3, 5, 7\cdots \end{cases}$$

Now, we apply the boundary condition (15) which leads to

$$\begin{aligned} &\sum_n C_n{}^m P_n{}^m(\eta_0)P_n{}^m(\delta) = \\ &\qquad A^m P_1{}^m(\eta_0)P_1(\delta) + \sum_n B_n{}^m Q_n{}^m(\eta_0)P_n{}^m(\delta) \\ &\quad - \frac{Z(j\omega)(\eta_0{}^2-1)^{1/2}}{\rho c(\eta_0{}^2-\delta^2)^{1/2}}\left[A^m \dot{P}_1^m(\eta_0)P_1{}^m(\delta) + \sum B^m \dot{Q}_n^m(\eta_0)P_n{}^m(\delta)\right] \end{aligned} \tag{21}$$

which holds for $-1 < \delta < +1$. We multiply both sides by $P_k{}^m(\delta)$ and integrate over δ from -1 to $+1$. With (19), this process yields

$$\begin{aligned} C_k{}^m P_k{}^m(\eta_0) &= A^m P_1{}^m(\eta_0)\delta_{1,k} + B_k Q_k(\eta_0) \\ &- \frac{Z(j\omega)(\eta_0{}^2-1)^{1/2}}{\rho c}\left[A^m \dot{P}_1^m(\eta_0)a_{1,k} + \sum_{n=1,3\cdots}^{\infty} B_n{}^m \dot{Q}_n^m(\eta_0)a_{n,k}\right] \end{aligned} \tag{22}$$

where

$$a_{n,k} = \frac{2k+1}{2}\frac{(k-m)!}{(k+m)!}\int_{-1}^{+1}\frac{P_n{}^m(\delta)P_k{}^m(\delta)}{(\eta_0{}^2-\delta^2)^{1/2}}d\delta \tag{23}$$

Combining (20) and (22), we have the single equation set to determine B_k :

$$\frac{1}{A^m}\left[\sum_{n=1,3,\cdots}^{\infty} B_n{}^m \dot{Q}_n^m(\eta_0)a_{n,k} + B_k{}^m \Lambda F_k\right]$$
$$= -\dot{P}_1^m(\eta_0)a_{1,k} + \Lambda f_k \quad (\text{for } k = 1,3,5\cdots) \tag{24}$$

where

$$\Lambda = \frac{\rho c}{Z(j\omega)(\eta_0{}^2-1)^{1/2}} = \frac{a(b^2/a^2-1)\rho}{Z(j\omega)} \tag{25}$$

$$F_k = \frac{\rho_1}{\rho} P_k{}^m(\eta_0)\frac{\dot{Q}_k^m(\eta_0)}{\dot{P}_k^m(\eta_0)} - Q_k{}^m(\eta_0) \tag{26}$$

and

$$f_k = \left(1 - \frac{\rho_1}{\rho}\right) P_1{}^m(\eta_0)\delta_{1,k} \tag{27}$$

The reader, if you are still with us, is reminded that for the purely axial excitation, $m = 0$ while for the purely transverse excitation, $m = 1$.

d. The Confocal Model

Before discussing concrete results, it is useful to consider the nature of the equation set given by (24). Actually this system is complicated by the presence of the coupling integral given by (23). In our earlier work [Wait 1982, 1983], we have avoided the non-orthogonality by choosing an angle dependent form for the interface impedance. To illustrate this point, we allow the interface impedance to be $Z(\delta, j\omega) = Z_0(j\omega)f(\delta)$ where $f(0) = 1$. Thus, in the boundary condition (15), we replace $Z(j\omega)$ by $Z(\delta, j\omega)$ as defined. The equation set (24) has the same form but now

$$a_{n,k} = \frac{2k+1}{2}\frac{(k-m)!}{(k+m)!}\int_{-1}^{+1}\frac{P_n{}^m(\delta)P_k{}^m(\delta)}{(\eta_0{}^2-\delta^2)^{1/2}}f(\delta)d\delta \tag{28}$$

We now choose the functional form of $f(\delta)$ such that

$$f(\delta) = (\eta_0{}^2 - \delta^2)^{1/2}/\eta_0 \tag{29}$$

over the range $-1 \le \delta \le +1$. In view of (19), we see that

$$a_{n,k} = \begin{cases} 1/\eta_0 & \text{if } n = k \\ 0 & \text{if } n \neq k \end{cases}$$

Then, (24) reduces to

$$\frac{1}{A^m} B_k{}^m \left[\dot{Q}_k^m(\eta_0) + \Lambda_0 F_k \right] = -P_1{}^m(\eta_0) + \Lambda_0 f_k \tag{30}$$

where Λ_0 is now a constant given by $\Lambda_0 = \rho c / Z_0(j\omega)$. Equation (30) can now be used to determine an explicit expression for B_k^m. In this case, the interface impedance can be regarded as a confocal sheath. Indeed, in accordance with (29)

$$f(\delta) = \left[1 - \delta^2 \left(1 - \frac{a^2}{b^2} \right) \right]^{1/2} \tag{29'}$$

which varies from $f(0) = 1$ at the waist to $f(\pm 1) = a/b$ at the tips of the particle. Essentially, this is the model considered by Wait [1983]. While this special angular variation may not be too realistic, it does lead to a simpler solution. Also, it is worth noting that other non-uniform models can be handled by modifying the form of $f(\delta)$.

e. Reduction of General Solution

The extension of the general solution given above for the prolate spheroidal geometry (i.e. $b > a$) to the oblate spheroidal case (i.e. $b < a$) is straight-forward. Analytically, the solutions are identical but we need to allow various parameters to pass from real to imaginary values in a consistent manner [Wait 1983]. We omit further discussion of this matter here.

To apply the general solution for the potential outside the particle, it is desirable to express the results in terms of spherical coordinates (R, θ, ϕ) as indicated in Fig. 1.5.2. Here, we will also restrict attention to the case where $R \gg a$ or b. Then, it is not difficult to show that for $n = 1$

$$Q_1(\eta) \simeq \frac{1}{3\eta^2} \simeq \frac{c^2}{3R^2} \tag{31}$$

$$Q_1{}^1(\eta) \simeq -\frac{2}{3\eta^2} \simeq -\frac{2c^2}{3R^2} \tag{32}$$

and, for $n = 3$

$$Q_3(\eta) \simeq \frac{2}{35\eta^4} \simeq \frac{2c^4}{35R^4} \tag{33}$$

$$Q_3{}^1(\eta) \simeq -\frac{8}{35\eta^4} \simeq -\frac{8c^4}{35R^4} \tag{34}$$

In general, we may note that for $\eta \gg 1$

$$\begin{aligned} Q_n{}^m(\eta) &\simeq (-1)^m \frac{n!(n+m)!2^n}{(2n+1)!\eta^{n+1}} \\ &\simeq (-1)^m \frac{n!(n+m)!2^n}{(2n+1)!R^{n+1}} c^{n+1} \end{aligned} \tag{35}$$

In this sense, only the dipole term is significant. Also, in this same limit, we note that

$$P_1(\eta) \simeq R/c \tag{36}$$
$$P_1{}^1(\eta) \simeq R/c \tag{37}$$
$$P_1(\delta) \simeq z/R \tag{38}$$
$$P_1{}^1(\delta) \simeq z/R \tag{39}$$

We can now express (10) and (12) in convenient forms using the preceding approximations. For axial excitation

$$\Phi_z = -E_0 z + E_0 a^2 b M_z / R^3 \tag{40}$$

and for transverse excitation,

$$\Phi_x = -\overline{E}_0 r \cos\phi + \overline{E}_0 a^2 b (M_x / R^3) r \cos\phi \tag{41}$$

Here M and $\overline{M}$ are normalized dipole moments which are given by

$$M = \left(\frac{c^3}{a^2 b}\right) B_1{}^0 / A^0 \tag{42}$$

and

$$\overline{M} = -\left(\frac{2c^3}{3a^2 b}\right) B_1{}^1 / A^1 \tag{43}$$

This normalization, in effect, means that M and $\overline{M}$ both reduce to 1.0 in the limiting case of a perfectly conducting sphere with zero interface impedance and with the same volume.

f. The Normalized Induced Dipoles

At this stage, it is desirable to show some numerical results for the simpler confocal model formulated earlier [Wait 1982, 1983]. We restrict attention to the case of axial excitation. Thus we need to deduce

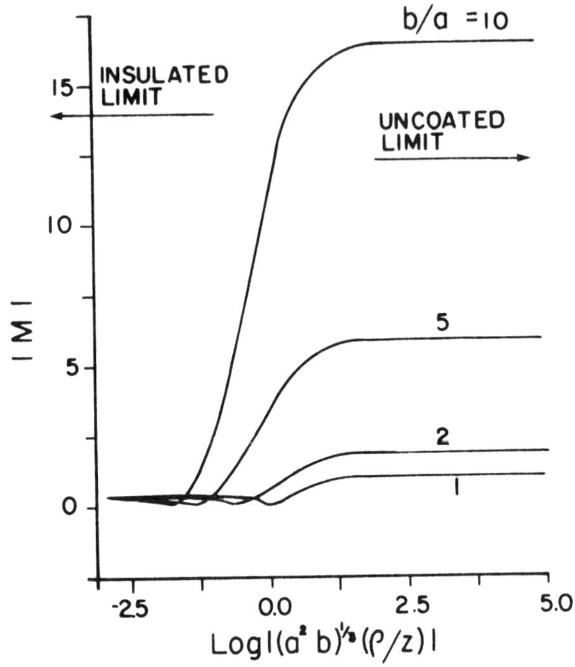

Figure 1.5.3a The magnitude of the normalized axial dipole moment for various spheroidicities calculated from the confocal model.

${B_1}^m/A^m$ from (30) for $m = 0$, or use the relatively simpler but equivalent formula for M_z given by Wait [1982, 1983]. A useful formula for plotting purposes is the dimensionless quantity $(a^2b)^{1/3}\rho/Z(j\omega)$ which is proportional to $(\text{particle volume})^{1/3}$ and inversely proportional to the interface impedance for a fixed value of the surrounding medium resistivity ρ. Here, we will let ρ be real but $Z(j\omega)$ is intrinsically complex. For purposes of illustration, the phase of $Z(j\omega)$ is taken to be 0.1 radian (i.e. 5.7°). Because we are thinking in terms of metallic particles, it is also permissible to set the particle resistivity $\rho_1 = 0$.

The results for $|M|$ and the phase of M for this confocal model are shown in Figs. 1.5.3a and 1.5.3b for values of $b/a = 1.1, 2, 5,$ and 10. As the coating parameter $Z(j\omega)$ tends to zero, the amplitude of the dipole moment becomes very large for the elongated particle (i.e.

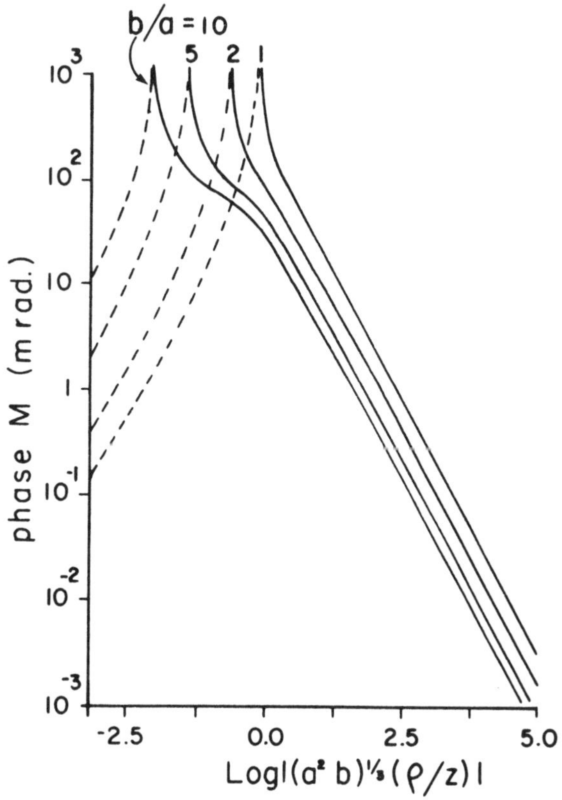

Figure 1.5.3b The phase of the normalized axial dipole moment for various spheroidicities calculated from the confocal model.

$b/a = 10$). This is a current channeling effect which is pronounced for axial excitation. However, the values of $|M|$ are greatly reduced as $Z(j\omega)$ is increased. In the limit of sufficiently large values of $Z(j\omega)$, the particle is acting as a prolate spheroidal void. In the intermediate range, interesting things happen. In fact, there is a point where $|M|$ becomes very small but not quite zero because $Z(j\omega)$ is complex. Here, there is also a rapid change in the phase as indicated in Fig. 1.5.3b.

To deal with the general case of a uniform coating (i.e. $f(\delta) = 1$), we must solve the system of equations given by (24). It is also necessary to evaluate the integral for $a_{n,k}$ given by (23). Both tasks have been implemented by Flanagan [1983] and described by Flanagan and Wait [1985] in detail. The main objective is to get a satisfactory numerical

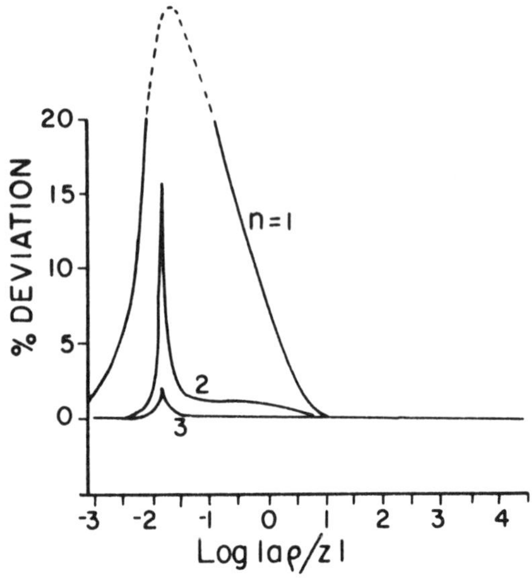

Figure 1.5.4a The convergence of the normal-axial dipole moment for various dimensions of the linear system involved.

solution for B_1/A. To this end, (24) is truncated at a sufficiently large value of n to simulate the infinite system. Not surprisingly, the highly elongated case requires a larger value of n. To illustrate the point, we choose $b/a = 10$ and successively increase n until there is no appreciable change in the solution for B_1/A and the corresponding value of the normalized dipole moment M. For $n = 4$, the error for $|M|$ was less than 1% (relative to the $n = 5$ case). To show the nature of the convergence of the solution process, we plot both the amplitude and the phase of M in Figs. 1.5.4a and 1.5.4b for this example, with $n = 1, 2,$ and 3 relative to the $n = 4$ case. The vertical scale is the percentage deviation so obtained. The convergence is very rapid. Of course, the $n = 1$ case corresponds to neglecting the higher order terms. This is akin to dealing with a confocal model where there is no such coupling.

The values of $|M|$ and the phase of M, are plotted in Figs. 1.5.5a and 1.5.5b using the (non-confocal) interface impedance which does not vary with δ. These results are compared with those in Figs. 1.5.3a and 1.5.3b where $f(\delta)$ is defined by (29). The general shapes and trends in the curves are remarkably similar in spite of the physically different

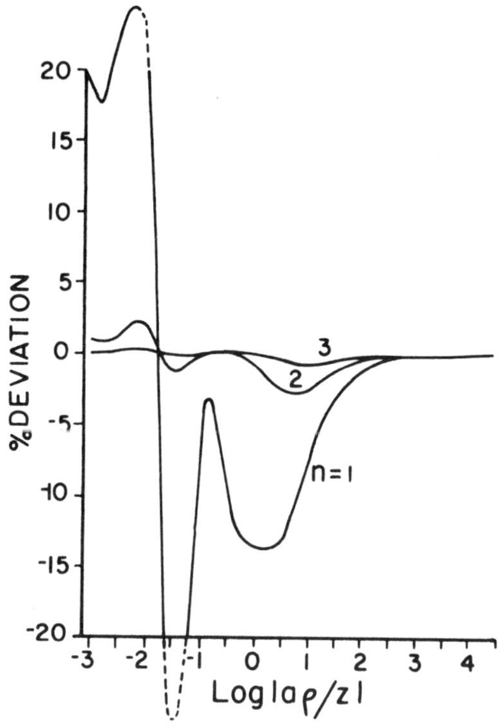

Figure 1.5.4b The convergence of the phase of the normalized axial dipole moment for various dimensions of the linear system involved.

models. In both cases, the phase of the interface impedance was taken to be 0.1 radians.

g. Effective Resistivity of Disseminated Particles

We now consider a uniform distribution of spheroidal particles in a bounded region. The objective is to determine the effective complex resistivity of the mixture. To simplify the problem, we will make a number of assumptions that will become clear in what follows.

As indicated in Fig. 1.5.6, we consider a spherical region of radius R_0. Within this volume there are N identical prolate spheroidal particles all aligned in the axial direction parallel to the applied electric field E_0. We assume that the particle locations are random.

At an external point $P(R, \theta)$ where $R \gg R_0$, we follow Maxwell

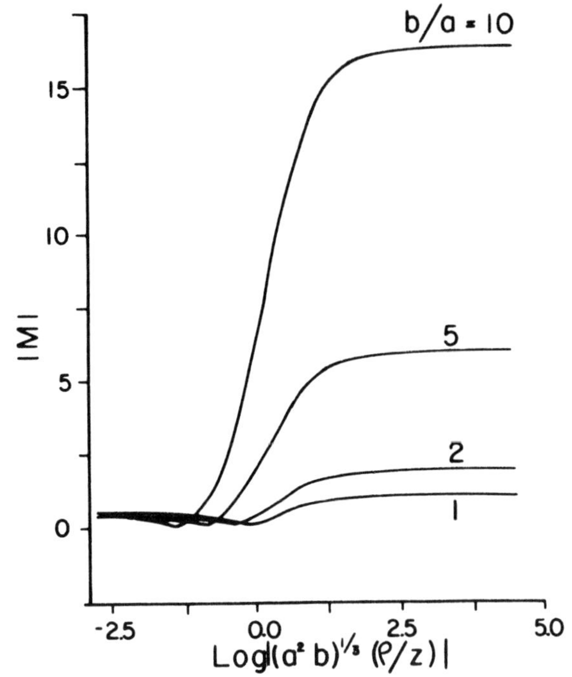

Figure 1.5.5a The magnitude of the normalized axial dipole moment for various spheroidicities.

[1891] and express the potential Φ as the total contribution of all the induced dipoles. Thus,

$$\Phi \simeq -E_0 R \cos\theta + N \frac{M(a^2 b)}{R^2} E_0 R \cos\theta \tag{44}$$

An equivalent expression is

$$\Phi \simeq -E_0 R \cos\theta + \frac{\rho - \rho_e}{\rho + 2\rho_e} {R_0}^3 E_0 R^{-2} \cos\theta \tag{45}$$

in terms of the effective resistivity ρ_e of the spherical volume and the resistivity ρ of the background. Equating these two expressions, we have

$$\frac{\rho - \rho_e}{\rho + 2\rho_e} = vM \tag{46}$$

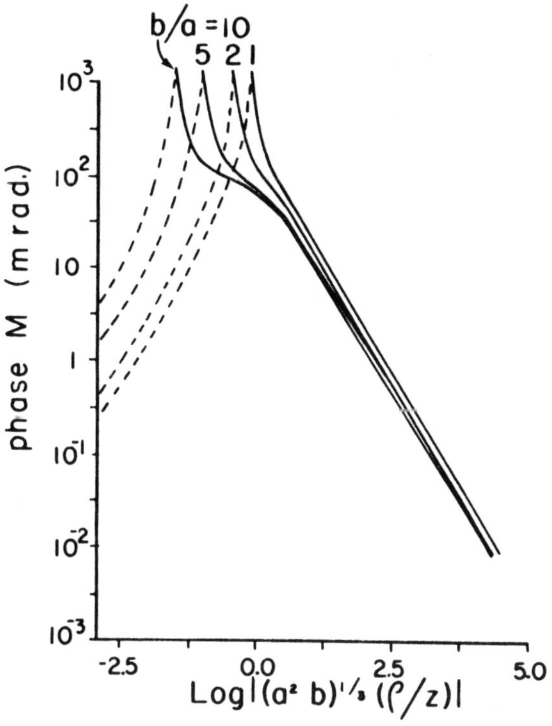

Figure 1.5.5b The phase of the normalized axial dipole moment for various spheroidicities.

where $v = Na^2b/R_0{}^3$ is the fractional volume loading of the particles within the spherical volume. From (46), we see that

$$\frac{\rho_e}{\rho} = \frac{1 - vM}{1 + 2vM} \simeq 1 - 3vM \tag{47}$$

Following the Maxwell prescription, we have assumed $v \ll 1$ which justifies the superposition of the individual induced dipoles.

Equation (47) can be employed to deduce the effective complex resistivity for axial excitation of a dissemination of aligned prolate spheroidal particles in terms of their normalized dipole moment M. The same reasoning applies to the case where the electric field is applied in the direction transverse to the axes of the particles. Then, in (47), we

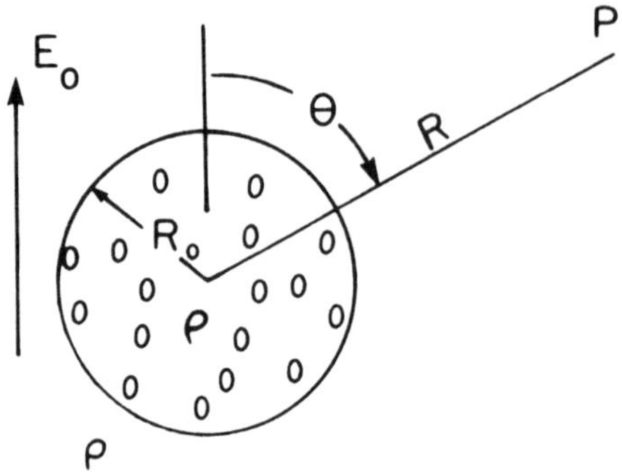

Figure 1.5.6 Ensemble of spheroidal particles.

merely replace M by $\overline{M}$ which is the appropriate normalized dipole moment for the transverse excitation.

A more sophisticated discussion of the effective medium approach is given elsewhere [Flanagan 1983, Wait 1983, Flanagan and Wait 1985]. In the present context, alternative formulas for the effective resistivity of the disseminated particles would not differ from (47) provided v were less than about 5%.

h. Results for the Apparent Complex Resistivity

To present meaningful results for the effective or apparent complex resistivity of the loaded medium, it is necessary to specify the interface impedance as a function of frequency. There are many choices; here, we will adopt a form recommended by Olhoeft [1982]. It has the form

$$Z(j\omega) = A + B/(j\omega)^{\nu} \tag{48}$$

where ν, A and B are constants to be fitted to observed data. For a pyrite/electrolyte interface, Olhoeft found that $\nu \cong 1/2$, $A = 0.9536\ \Omega\text{m}^2$ and $B = 1.7931\ \Omega\text{m}^2(\text{sec})^{-1/2}$ were consistent with experimental data. Leaving aside the physical or electrochemical implications, we will adopt this "Warburg" model. A plot of the amplitude and phase of $Z(j\omega)$ is shown in Fig. 1.5.7.

We now show the corresponding results for the effective or apparent complex resistivity $\rho_e(j\omega)$ of the disseminated model for various

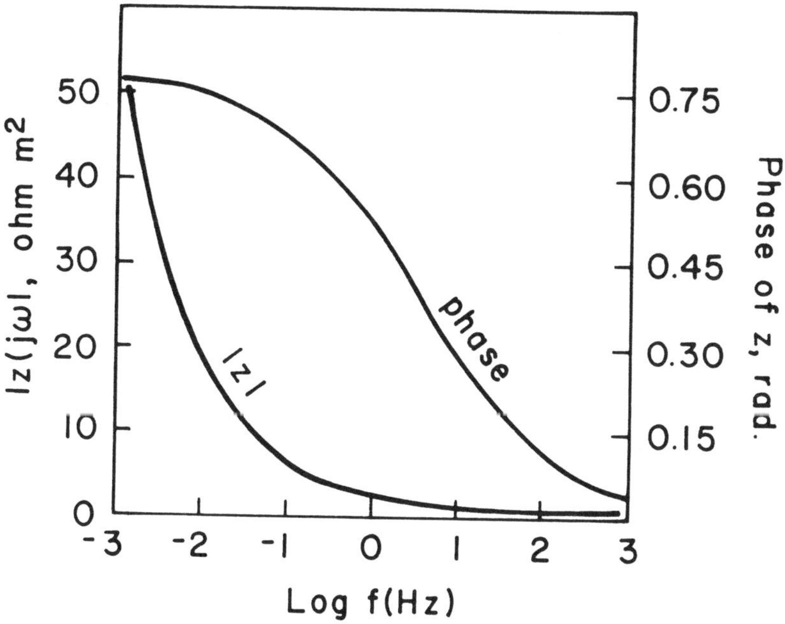

Figure 1.5.7 Interface impedance as a function of frequency for a pyrite/ electrolyte boundary according to Olhoeft [1982].

assigned parameters. We use the approximate form given by (47) for the case of axial excitation. For transverse excitation we replace M by $\overline{M}$. The particle resistivity ρ_1 is assigned the value zero and the ambient medium resistivity ρ is set equal to 200 Ωm. The interface impedance is the plotted in Fig. 1.5.7 as described above.

The amplitude $|\rho_e/\rho|$ and the negative of the phase of ρ_e are plotted as a function of frequency in Figs. 1.5.8a and 1.5.8b for the case where $v = 0.05$ and $b/a = 5$ and for four particles (i.e. $a =$ 0.33, 1, 3.3 and 10 mm). Both axial and transverse excitations are shown. The results illustrate that the small particles have a maximum in the phase at higher frequencies as indicated in an early study using a coated sphere model [Wait 1958]. This behaviour is also present in the results computed by Wong and Strangway [1981] with their electrochemical model. In Fig. 1.5.8a, we can also note that $|\rho_e/\rho|$ changes most rapidly at frequencies near where the phase shift is maximum.

In Figs. 1.5.9a and 1.5.9b, we show the influence of the spheroidic-

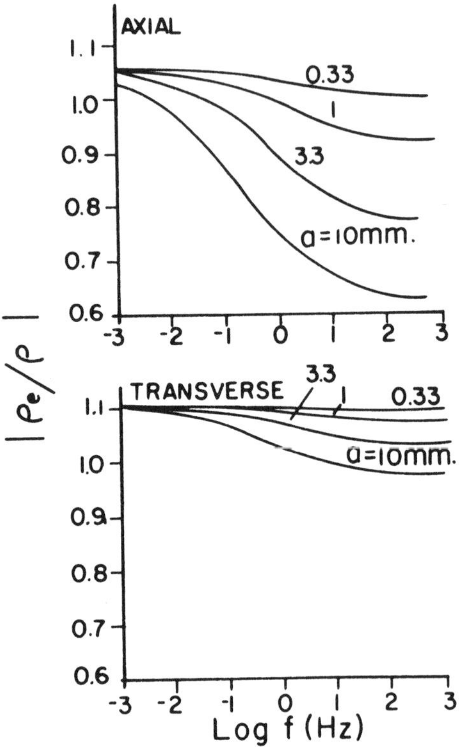

Figure 1.5.8a The magnitudes of the effective axial and transverse resistivities for various particle sizes.

ity ratio b/a on the frequency response curves using the same format as the curves in Figs. 1.5.8a and 1.5.8b. The axial and transverse excitations are designated by A and T, respectively. For the curves in Figs. 1.5.8a and 1.5.8b, $a = 1.0$ mm and $v = 0.05$. Clearly, the elongated particles have an enhanced IP response for axial excitation but the response for purely transverse excitation is only slightly dependent on b/a. In fact, it can be shown that for b/a greater than about 5, the response would be the same as a grating of infinitely long cylinders with the same diameter as the minor axis of the elongated spheroidal particles and with the same volume loading. We might add that the

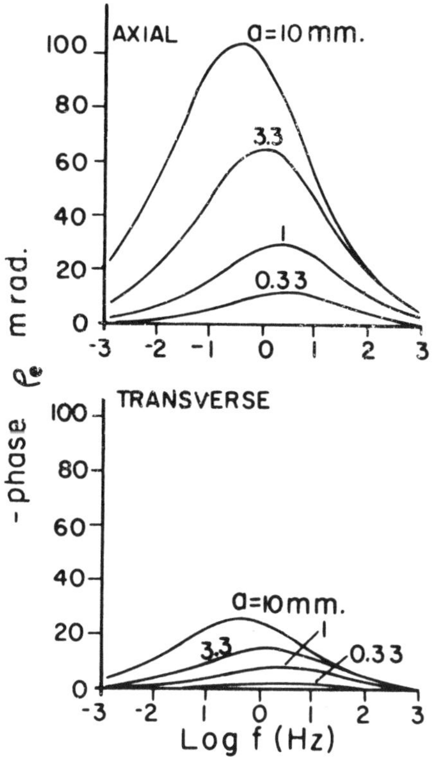

Figure 1.5.8b The phase-shifts of the effective axial and transverse resistivities for various particle sizes.

case designated $b/a = 1.1$ is essentially the same as a spherical particle insofar as the plots are concerned.

Finally, in Figs. 1.5.10a and 1.5.10b, we illustrate the dependence on volume loading v again using the same format as in Figs. 1.5.8a and 1.5.8b. In this case, v varies from 0.01 to 0.2 (i.e. 1 to 20%). The fixed parameters are $a = 1$ mm and $b/a = 5$. As indicated in the phase response, there is almost a linear increase with the volume loading v over this range. According to our simple adaptation of the Maxwell loading law, such a behavior is not surprising. For v greater than 0.05, the simple form given by (47) is in error. As discussed by Flanagan and Wait [1985], a modified mixture formula such as the Brueggman-Hanai form should be preferred but the differences in the

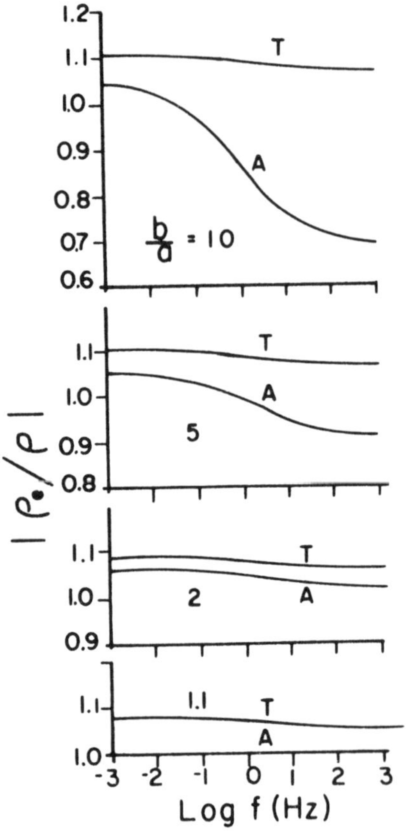

Figure 1.5.9a The magnitudes of the effective axial and transverse resistivities for particles of various spheroidicities.

plots in Figs. 1.5.10a and 1.5.10b, even for $v = 0.20$, would not be significant.

Final Remarks

There are many similarities between our results and those of Wong and Strangway [1981]. In their electrochemical model, an interface impedance descriptor was not used. They dealt explicitly with the active cations which reacted at the metal interface to yield a net charge

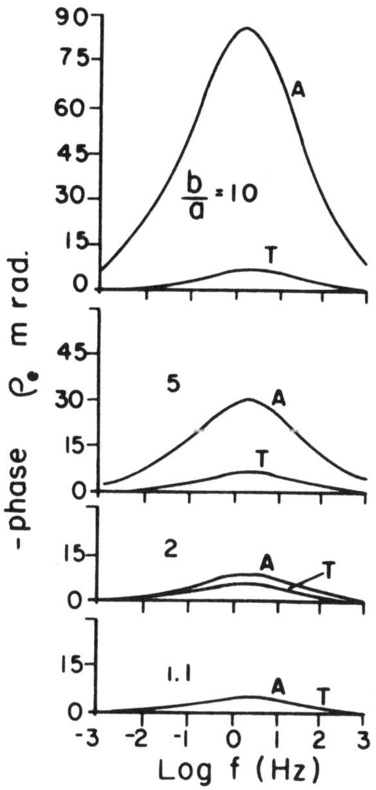

Figure 1.5.9b The phase-shifts of the effective axial and transverse resistivities for particles of various spheroidicities.

transfer. They also needed to solve the coupled time-dependent wave equation to describe the concentration perturbations of both the active and the inactive ions. They neglected the coupling of higher modes for the perturbation solutions. It is difficult to see how this step is justified although the qualitative behavior of the results should not be changed appreciably for the slightly elongated particles. Of course the coupling disappears in the spherical limit [Wong 1979]. But any serious attempt to model elongated particles should address the question of mode coupling which becomes most significant when the particle becomes highly elongated such as a needle. In our interface impedance model, we have attempted to show how to cope with this problem by dealing with the coupled equation set for the modal coefficients.

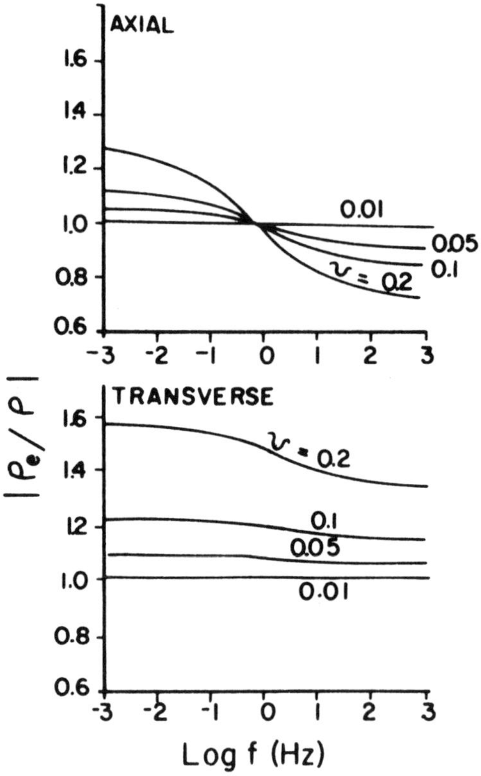

Figure 1.5.10a The magnitudes of the effective axial and transverse resistivities for various volume loadings.

Our results are limited by the assumption that the electrochemical processes can be modelled by an interface impedance. When the radius of curvature of the particle becomes small, such a descriptor will clearly be in doubt. In this case, the local current density at the tips of the particle may be high enough to cause non-linearities in the response. Of course, this limitation also appears in the so called electrochemical formulations.

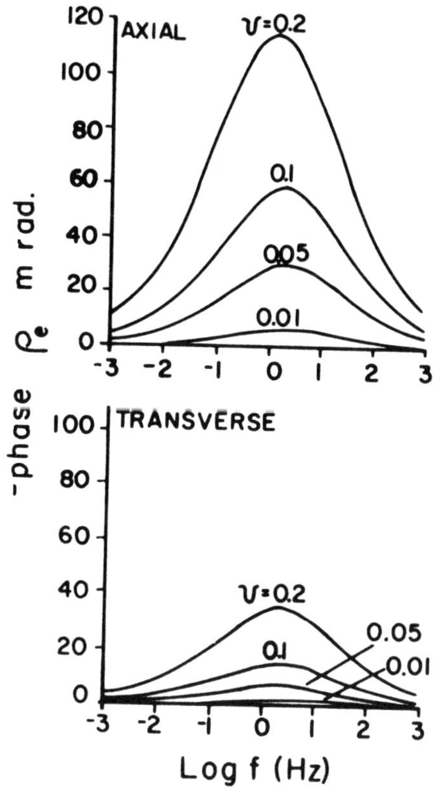

Figure 1.5.10b The phase-shifts of the effective axial and transverse resistivities for various volume loadings.

References

[1] Flanagan, P. W., *M.S. Thesis*, Dept. of Geosciences, University of Arizona, 1983.

[2] Flanagan, P. W. and J. R. Wait, "Induced polarization response of disseminated mineralization for spheroidal geometries," *Radio Science*, **30**, 147–148, 1985.

[3] Maxwell, J. C., *A Treatise on Electricity and Magnetism*, **3rd. ed. part 1**, 314–315, Clarendon Press, Oxford, 1891.

[4] Olhoeft, G. R., "Warburg impedances at single water/mineral interfaces," *Extended Abstracts, Society of Exploration Geophysicists Meeting, Dallas*, 17–21, Oct. 1982.

[5] Wait, J. R., "A phenomenological theory of induced electrical polarization," *Can. Jour. Phys.*, **36**, 1634–1644, 1958.

[6] Wait, J. R., "EM response of a medium loaded with conductive particles," *IEEE Trans. on Geoscience and Remote Sensing*, **GE-20**, 500–503, 1983.

[7] Wait, J. R., "Complex conductivity of disseminated spheroidal ore grains," *Gerlands Beitr. Geophysik (Leipzig)*, **92**, 49–69, 1983.

[8] Wong, J., "An electrochemical model of the induced polarization phenomena in disseminated sulphide ores," *Geophysics*, **44**, 1245–1265, 1979.

[9] Wong, J. and D. W. Strangway, "Induced polarization in disseminated sulphide ores containing elongated mineralization," *Geophysics*, **46**, 1258–1268, 1981.

1.6 Response of Disperse Systems for Simple Particles

a. *Introduction*

A common feature of geologic media is the lumpy nature of the structure. Thus, at least in an idealized sense, we can represent such a region as a homogeneous host with various kinds of suspended particles. Central to this viewpoint is the characterization of the electromagnetic response of a single particle when it is placed in a uniform field. Also it is appropriate to allow for any boundary layer effects at the interface between the host and the particle. To treat this problem fully, we should really build in the requisite electrochemistry to properly describe the ionic reactions and the transfer of charges across the interface. This aspect of the problem is described in a separate chapter. Here, we argue that the interface can be characterized, at least in a phenomenological sense, by a surface admittance and/or an interface impedance. The meaning of this statement should become clear as we proceed.

b. *Formulation*

We begin with a very simple model of a solid conducting spherical particle of radius a and of resistivity ρ_1 immersed in a homogeneous host medium of resistivity ρ. To allow for a possible discontinuity of potential across the surface of the particle, we endow it with an interface impedance $Z(j\omega)$ that may be frequency dependent. The applied or primary field E_0 is uniform and directed along the polar axis of a spherical coordinate system (r, θ) centered at the particle. The time factor is $\exp(j\omega t)$ and the frequency is sufficiently low that Laplace's equation governs the fields both inside and outside the particle.

The imposed condition at the boundary $r = a$ amounts to saying that voltage drop $v = ZJ_r$ where J_r is the radial current density. It is understood that Z is both complex and frequency dependent. However, we will avoid non-linear effects and say that Z does not depend on J_r.

c. *Potential Theory Solutions (Spherical Model)*

Suitable forms for the potentials are

$$\Psi_1 = A_0 r \cos\theta \qquad \text{for} \qquad 0 < r < a \qquad (1)$$

and

$$\Psi = -E_0 r \cos\theta + A r^{-2} \cos\theta \qquad \text{for} \qquad r > a \tag{2}$$

Here A_0 and A are constants to be determined from the boundary conditions. It is evident that $\Psi = -E_0 r \cos\theta$ if $\rho = \rho_1$.

Continuity of radial current density requires that

$$\frac{1}{\rho_1}\frac{\partial \Psi_1}{\partial r} = \frac{1}{\rho}\frac{\partial \Psi}{\partial r} \qquad \text{for} \qquad r = a \tag{3}$$

Also the required potential discontinuity means that

$$\Psi_1 = \Psi - \frac{Z}{\rho}\frac{\partial \Psi}{\partial r} \qquad \text{for} \qquad r = a \tag{4}$$

Applying (3) and (4) to (1) and (2) leads immediately to the solution

$$A_0 = \frac{-3E_0\rho_1}{2\rho_1 + \rho + (2Z/a)} \tag{5}$$

and

$$A = E_0 a^3 \frac{\rho - \rho_1 - (Z/a)}{2\rho_1 + \rho + (2Z/a)} \tag{6}$$

Thus, for the case $r > a$, we can write

$$\Psi = -E_0 r \cos\theta + a^3 E_0 \chi_0 r^{-2} \cos\theta \tag{7}$$

where

$$\chi_0 = \frac{\rho - \rho_0 - (Z/a)}{2\rho_1 + \rho + (2Z/a)} \tag{8}$$

In the case where the double layer is absent, we have $Z = 0$ and the solution reduces to the case where the potential is continuous across the boundary.

The solution given above for the sphere with the interface impedance property is really a special case of that given in the preceding chapter for the spheroid model. In the spherical limit the higher order multipoles are zero.

A slightly more complicated case is to allow the normal current to be discontinuous at the boundary. As a consequence, the tangential electric field just inside the boundary will drive a surface current that is proportional to the surface admittance $Y(j\omega)$. Now the boundary condition given in (3) is to be replaced by

$$\left.\frac{1}{\rho_1}\frac{\partial\Psi}{\partial r}+\frac{Y}{a}\frac{\partial}{\partial r}\left(r^2\frac{\partial\Psi_1}{\partial r}\right)\right]_{r=a-0}=\left.\frac{1}{\rho_1}\frac{\partial\Psi}{\partial r}\right]_{r=a+0} \tag{9}$$

and (4) is written in a slightly amended form

$$\left.\Psi_1\right]_{r=a-0}=\left.\Psi-\frac{Z}{\rho}\frac{\partial\Psi}{\partial r}\right]_{r=a+0} \tag{10}$$

For this double shell model, we can write the exterior potential in the form

$$\Psi=-E_0 r\cos\theta+\frac{a^3E_0\chi}{r^2}\cos\theta \tag{11}$$

where

$$\chi=\frac{1-\delta}{1+2\delta} \tag{12}$$

with

$$\delta=\frac{(1/\rho)}{(1/\rho_1)+(2Y/a)}+\frac{Z}{\rho a} \tag{13}$$

Here χ can be identified as the normalized induced dipole of the particle. If $Y\to 0$, we recover the solution given by (7) and (8).

The mathematics of the double shell model has been discussed in more detail elsewhere [Wait 1985]. Physically, at zero frequency, the situation corresponds to a solid core of radius a encased by a thin conductive shell which, in turn, is covered by a thin resistive coating. For non-zero frequency, the displacement currents or similar phase shifted currents cause both Y and Z to be complex. Thus χ is complex in general. The electrochemical theorists [e.g. Wong 1979, Chew and Sen 1982] obtain representations for χ involving interactions within the double layer. As we have already indicated, χ, at least in a phenomenological sense, is a characteristic of the particle and its surface layers. Similar suggestions were put forth by Schwan et al. [1962] and Schwartz [1962] who may not have been aware of earlier work by the author and his colleagues [Wait et al. 1956, Wait 1958].

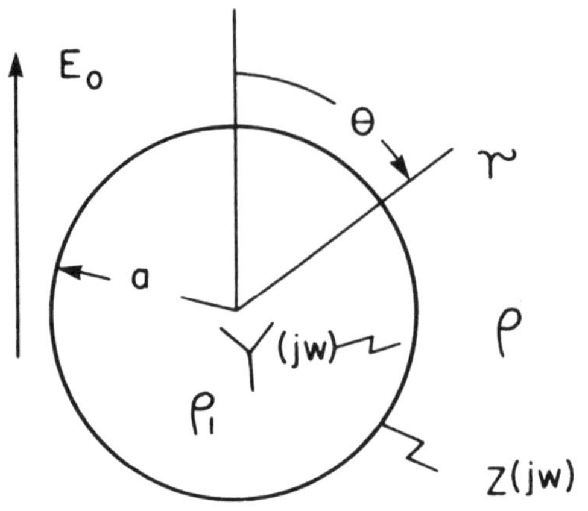

Figure 1.6.1 Basic spherical model as characterized by an interface impedance and/or admittance.

d. Cylindrical Rod Model

While it may seem artificial, another useful model is an infinitely long circular cylinder of radius b which has a resistivity ρ_1. The host medium is homogeneous with resistivity ρ. The cylinder is oriented with its axis at right angle to the applied primary field E_0. We can still refer to Fig. 1.6.1 if this is viewed as a cross section. Again the surface of the *particle* is characterized by an interface impedance $Z(j\omega)$ and a surface admittance $Y(j\omega)$ which, in general, are both complex and frequency dependent.

We now employ cylindrical coordinates (r, θ) not to be confused with the spherical coordinates used above. The appropriate forms for the potentials are now

$$\Psi_1 = B_0 r \cos\theta \qquad \text{for} \qquad 0 < r < a \tag{14}$$

and

$$\Psi = -E_0 r \cos\theta + B r^{-1} \cos\theta \qquad \text{for} \qquad r > a \tag{15}$$

Here B_0 and B are constants to be determined. The boundary conditions, in analogy to the spherical counterpart, are now written

$$\frac{1}{\rho_1}\frac{\partial \Psi_1}{\partial r} + \frac{Y}{a} r \frac{\partial}{\partial r}\left(r \frac{\partial \Psi_1}{\partial r}\right)\Bigg]_{r=a-0} = \frac{1}{\rho_1}\frac{\partial \Psi}{\partial r}\Bigg]_{r=a+0} \tag{16}$$

and

$$\Psi_1\Big]_{r=a-0} = \Psi - \frac{z}{\rho}\frac{\partial \Psi}{\partial r}\Big]_{r=a+0} \tag{17}$$

which, apart from the term involving Y, have the same form as (9) and (10). Using the obtained value for B, we can write (for $r > a$)

$$\Psi = -E_0 r \cos\theta + E_0 a^2 r^{-1} \chi \cos\theta \tag{18}$$

where

$$\chi = \frac{1-\delta}{1+\delta} \tag{19}$$

with

$$\delta = \frac{(1/\rho)}{(1/\rho_1) + (Y/a)} + \frac{Z}{\rho a} \tag{20}$$

This solution for the cylindrical model is almost the same for the spherical model (i.e. compare (18), (19), and (20) with (11), (12), and (13) respectively). An independent justification for the boundary conditions (16) and (17), for the cylindrical model, is given in the Appendix.

e. Ensemble of Spherical Particles

To deal with an ensemble of basic particles, it is desirable to ignore the interaction between the particles. At least, such an idealization is a first step as we have indicated in the preceding section. Thus, we consider the configuration shown in Fig. 1.6.2 where N spherical particles of the same radius a are contained within a spherical volume of radius r_0. The resistivity of the host region both inside and outside the spherical volume is ρ. A uniform field E_0 is applied in the axial direction.

We now follow the original concept put forth by Maxwell that we have alluded to in the preceding section. We begin by writing down an expression for the total potential P at (r, θ) as a superposition of the potentials of each of the particles. Neglecting the interaction between the particles, we have

$$\Psi = -E_0 r \cos\theta + \sum_i^N \frac{E_0 a^3 \chi}{r_i^2} \cos\theta_i \tag{21}$$

where the spherical coordinate system (r_i, θ_i) is chosen for the ith particle with its polar axis (i.e. $\theta_i = 0$) aligned with the applied field.

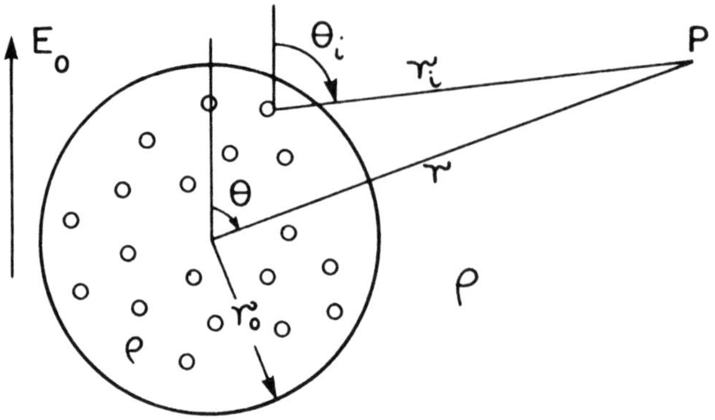

Figure 1.6.2 Ensemble of identical spherical particles enclosed in a spherical volume.

At a sufficiently large distance such that $r \gg r_0$, we can argue that r_i and θ_i in (21) can be replaced by r and θ, respectively. Within this approximation

$$\Psi \simeq -E_0 r \cos\theta + \frac{Na^3\chi}{r^2} E_0 \cos\theta \tag{22}$$

Then, if we regard the spherical region of radius r_0 as a continuum with an effective resistivity ρ_e, we write

$$\Psi = -E_0 r \cos\theta + \frac{r_0^3}{r^2}\frac{\rho - \rho_e}{\rho + 2\rho_e} E_0 \cos\theta \tag{23}$$

We now equate the right hand sides of (22) and (23) to give

$$v\chi = \frac{\rho - \rho_e}{\rho + 2\rho_e} \tag{24}$$

where $v = Na^3/r_0^3$ is the volume of particles per cubic meter; v is called the fractional volume loading. From (24), we deduce the important result

$$\frac{\rho_e}{\rho} = \frac{1 - v\chi}{1 + 2v\chi} \tag{25}$$

which is a generalization of Maxwell's formula. The limitation is that the volume loading v is small compared with 1 but there is no restriction on the particle's properties. In the following section, we review various attempts to extend the theory to allow for higher volume loading.

It is useful to outline the corresponding solution for a cylindrical volume of radius r_0 loaded with N cylindrical rods of radius a. The sketch shown in Fig. 1.6.2 is still applicable if we view this as the cross section of the cylindrical model. The applied electric field E_0 is taken here to be transverse or perpendicular to the axes of the cylindrical rods. We employ the same procedure as used for the spherical model with essentially the same assumptions. Then, we deduce that the apparent resistivity normalized by the homogeneous host resistivity is given by

$$\frac{\rho_e}{\rho} = \frac{1 - v\chi}{1 + v\chi} \tag{26}$$

where now $v = Na^2/r_0^2$ which again can be identified as the fractional volume loading. In this case, χ is defined by (20) as appropriate for the cylindrical geometry.

f. Extension to Cole-Cole Form

It is interesting to note that the derived formulas, for both the spherical and cylindrical models, yield expressions for the complex resistivity which are reducible to the Cole-Cole form. To explain what we mean by this statement, we will write (25) for the spherical model in the simplified form

$$\frac{\rho_e}{\rho} = \frac{1 - v\chi}{1 + 2v\chi} \simeq 1 - 3v\chi \tag{27}$$

which is justified because $v \ll 1$. We also note that

$$\chi = \frac{1 - \delta}{1 + 2\delta} = 1 - \frac{3}{2}\frac{1}{1 + (2\delta)^{-1}} \tag{28}$$

After a little algebra, we see that

$$\frac{\rho_e}{\rho} \simeq 1 + \frac{3}{2}v - \frac{9}{2}v\left[1 - \frac{1}{1 + (2\delta)^{-1}}\right] \tag{29}$$

The Cole-Cole form for the complex resistivity, as discussed in section 1.2, is written

$$\rho_e(j\omega) = \rho_0\left\{1 - m_0\left[1 - \frac{1}{1 + (j\omega\tau)^k}\right]\right\} \tag{30}$$

where $\rho_0 = \rho_e(0)$ is the DC or zero-frequency apparent resistivity.

By inspection, it is clear that (29) and (30) can be made equivalent if we set

$$\rho_0 = 1 + \frac{3}{2}v \tag{31}$$

$$m_0 = \frac{9}{2}\frac{\rho}{\rho_0}v = \frac{9}{2}v\frac{1}{1+(3v/2)} \simeq \frac{9}{2}v \tag{32}$$

and

$$(j\omega\tau)^k = (2\delta)^{-1} \tag{33}$$

The chargeability m_0 as obtained here is consistent with the simple relation

$$m_0 = \frac{\rho_e(0) - \rho_e(\infty)}{\rho_e(0)} \tag{34}$$

To discuss the fractional frequency term, we first specialize the particle to a perfectly conducting core (i.e. $\rho_1 = 0$ or $y = \infty$). Then, in accordance with (13), $\delta = Z(j\omega)/\rho a$ and thus

$$Z(j\omega) \simeq \frac{\rho a/2}{(j\omega\tau)^k} \tag{35}$$

In other words, the interface impedance is proportional to $(j\omega)^{-k}$. As we discuss elsewhere, this is a realistic approximation to the behavior at an interface between a metal and an electrolyte where k could vary from say 0.2 to 0.4. If the interface impedance as measured fits the law

$$Z(j\omega) = \frac{\alpha_0}{(j\omega)^k} \tag{36}$$

where α_0 and k are adjusted constants, we deduce from (35) that the corresponding time constant τ is given by

$$\tau = \left(\frac{\rho a}{2\alpha_0}\right)^{1/k} \tag{37}$$

This tells us that τ is proportional to $(\text{particle radius})^{1/k}$ where the power $1/k$ is typically in the range from 2 to 5.

Jumping now to the cylindrical model we would write, in place of (29), that

$$\frac{\rho_e}{\rho} \simeq 1 + 2v - 4v\left[1 - \frac{1}{1+\delta^{-1}}\right] \tag{38}$$

This representation is the same as the Cole-Cole form if in (30) we set

$$\rho_0 = 1 + 2v \tag{39}$$

$$m_0 = 4v(1+2v)^{-1} \simeq 4v \tag{40}$$

and

$$(j\omega\tau)^k = \delta^{-1} \tag{41}$$

Here a^2/r_0^2 is the fractional volume loading of the cylindrical rods and δ is defined by (20). The remarks made above for the spherical model are unchanged for the cylindrical model but we need replace 2 in (35) and (37) by 1.

It is tempting to argue that, for the more general case, the replacement of δ by a fractional frequency dependence [i.e. proportional to $(j\omega)^{-k}$] is a reasonable supposition. In fact, if we are dealing with an insulating particle, then $\rho_1 \to \infty$ and the corresponding assumption is that surface admittance Y is now proportional to $(j\omega)^k$. Some support for this contention is found in a study by Olhoeft [1985] who found that the empirical law, time constant = $(\text{radius})^2/3.4 \times 10^{-6}$ sec., fitted the experimental data for diffusion-limited processes for particle radii varying from 10^{-9} to 10^{-1} m. In this case, the effective value of the dispersion index k is $1/2$.

In the next section, we consider further generalizations of this particle model and take up again the question of non-spherical particles.

References

[1] Chew, W. C., and P. N. Sen, "Dielectric enhancement due to electrochemical double layer; thin double layer approximation," *J. Chem. Phys.*, **77**, 4683–4692, 1982.

[2] Schwan, H. P., G. Schwarz, J. Maczuk, and H. Pauly, "On the low frequency dielectric dispersion of colloidal particles in electrolyte solution," *J. Phys. Chem. Phys.*, **66**, 2626–2635, 1962.

[3] Schwarz, G., "A theory of the low frequency dielectric dispersion of colloidal particles in electrolyte solution," *J. Phys. Chem.*, **66**, 2636–2642, 1962.

[4] Wait, J. R., "A phenomenological theory of induced electric polarization," *Can. J. Phys.*, **36**, 1634–1644, 1958.

[5] Wait, J. R., H. O. Seigel, L. S. Collett, W. E. Bell, and A. A. Brant, *Method and Apparatus for Geophysical Exploration*, US Patent 2,766,421, 1956. (Application No. 273,422, 1952.) [includes brief derivation of coated particle model].

[6] Wait, J. R., *Electromagnetic Wave Theory*, Harper and Row/ Wiley, 1986. (see chap. 2).

[7] Wong, J., "An electrochemical model of the IP phenomenon in disseminated sulphide ores," *Geophysics*, **44**, 1245–1265, 1979.

[8] Olhoeft, G. R., "Low-frequency electrical properties," *Geophysics*, **50**, 2492–2503, 1985.

1.7 Generalizations of the Spheroidal Model

a. Problem Statement

In discussing the response of an ensemble of coated particles, we have assumed spherical geometry. It is useful to indicate how the formulation can be extended to spheroidal shaped particles. Here, we restrict attention to small volume loadings. Our approach is based on the Maxwell/Wagner technique of enclosing the particles in a closed volume and then deducing the macroscopic or complex resistivity of the ensemble. An important question is how we choose the geometry of the reference volume. This point does not seem to have been addressed by previous workers, e.g. van Beek's [1967] extensive review. We present a relevant analysis here without making any claim about whether the last word on the subject has been said.

The situation is illustrated in Fig. 1.7.1. An ensemble of N identical spheroidal particles of resistivity ρ_1 are enclosed in a spheroidally shaped volume whose macroscopic or effective resistivity ρ_e is to be determined. As indicated in the inset, the basic particle has semi-minor axis a and semi-major axis b. The surface of the particle has an interface impedance Z_0. The reference volume has a semi-minor axis a_e and a semi-major axis b_e. The major axes of both the particle spheroid and the reference spheroid are taken to be parallel to each other and to the applied electric field E_0. To facilitate the analysis, we choose the confocal model for the interface impedance [Wait, 1983a].

The total potential at an external point P for the sparse distribution of particles is given approximately by

$$\Psi \cong -E_0 R \cos\theta + E_0 a^2 b M \sum_{i=1}^{N} \frac{1}{R_i^2} \cos\theta_i \tag{1}$$

where M is a normalized dipole moment of the basic ith particle. Here R_i and θ_i are the appropriate spherical coordinates for the individual ith particle. In writing (1), we have assumed that $R_i \gg a$ and b. The next step now is to write down the corresponding expression for the potential at the point P where we regard the spheroidal reference volume, with its enclosed particles, to be an equivalent uniform medium with complex resistivity ρ_e. Thus

$$\Psi \cong -E_0 R \cos\theta + E_0 a_e^2 b_e M_e \frac{1}{R^2} \cos\theta \tag{2}$$

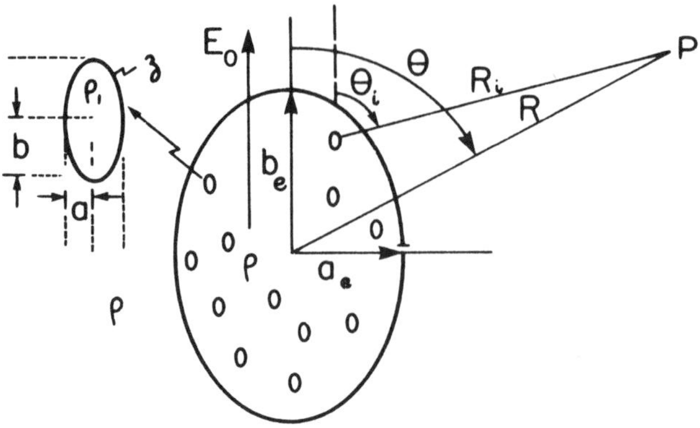

Figure 1.7.1 Ensemble of prolate spheroidal particles within a prolate spheroidal reference volume.

where R is assumed to be large compared with a_e and b_e and where M_e is the effective dipole moment of the volume. In accord with this restriction on R, we can replace R_i and θ_i by R and θ, respectively. Then, we follow the Maxwell/Wagner prescription and equate the righthand sides of (1) and (2) to yield the deceptively simple expression to determine M_e

$$M_e = v\,M \tag{3}$$

where $v = N a^2 b/(a_e^2 b)$ is the fractional volume loading of the particles contained within the reference spheroid.

b. Modified Effective Medium Approach

We now call upon the potential theory solution [Wait 1983] for the dipole moment M for an isolated particle of prolate spheroidal form with an interface impedance under the influence of a uniform applied electric field E_0. The appropriate form is

$$M = M_\infty \left[\frac{1 - (\Omega + \rho_1/\rho)}{1 - (\Omega + \rho_1/\rho)(M_\infty/M_0)} \right] \tag{4}$$

Here M_∞ is the normalized dipole moment for a perfectly conducting spheroidal particle (with axes a and b) *without* any coating (i.e. $Z_0 = 0$). On the other hand, M_0 is the normalized dipole moment of the corresponding spheroidal particle which is perfectly insulating (i.e.

$Z_0 = \infty$). The dimensionless interface parameter Ω is defined by

$$\Omega = \frac{Z_0 a}{\rho b^2}$$

where Z_0 is the actual interface impedance at the waist of the spheroid. In analogy, we have the following expression for the dipole moment parameter for the volume

$$M_e = M_{e\infty} \frac{1 - (\rho_e/\rho)}{1 - (\rho_e/\rho)(M_{e\infty}/M_{e0})} \tag{5}$$

We now combine (3), (4), and (5) to write

$$\frac{\rho - \rho_e}{\rho + \alpha_e \rho_e} M_{e\infty} = vM_\infty \frac{1 - (\Omega + \rho_1/\rho)}{1 + (\Omega + \rho_1/\rho)\alpha} \tag{6}$$

where $\alpha = -M_\infty/M_0$ and $\alpha_e = -M_{e\infty}/M_{e0}$. It is also useful to note that $M_\infty = (1+\alpha)/3$ and $M_{e\infty} = (1+\alpha_e)/3$. In principle, (6) can be employed to deduce the apparent complex resistivity $\rho_e(j\omega)$ in terms of the geometrical parameters and the specified interface impedance function $Z_0(j\omega)$.

While we have been talking about prolate spheroidal particles (i.e. $a < b$), the theory is directly applicable to oblate spheroidal particles (i.e. $a > b$). The results are also applicable to transverse field excitation where the M 's are then replaced by the appropriate $\overline{M}$ functions [Wait 1982, 1983a].

To help understand the problem, we show plots of the normalized dipole moment functions in Fig. 1.7.2. Complete expressions for these functions are given elsewhere [Wait 1982, 1983a]. Here, we just indicate the limiting behavior as follows :

For $b \gg a$ (i.e. needle like particles),

$$\left.\begin{aligned} M_\infty &\simeq \frac{(b/a)^2}{3[\ln(2x) - 1]} \\ M_0 &\simeq -1/3 \end{aligned}\right\} \text{ longitudinal excitation}$$

$$\left.\begin{aligned} \overline{M}_\infty &\simeq 2/3 \\ \overline{M}_0 &\simeq -2/3 \end{aligned}\right\} \text{ transverse excitation}$$

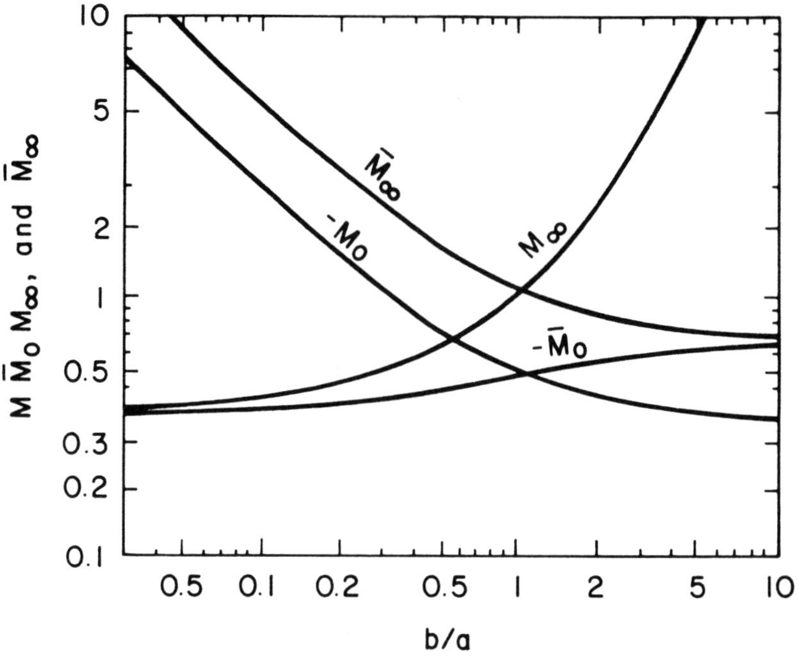

Figure 1.7.2 Normalized dipole moments for perfectly conducting and insulated spheroidal particles for axial and transverse excitations as a function of the ratio of the semi-major axis b to the semi-minor axis a. Prolate spheroidal particles corresponds to $b/a > 1$ while oblate spheroidal particles corresponds to $b/a < 1$.

For $b/a = 1$ (i.e. spherical particles),

$$M_\infty = 1 \qquad \text{and} \qquad M_0 = -1/2$$

For $a \gg b$ (i.e. disc particles),

$$\left. \begin{aligned} M_\infty &\simeq 1/3 \\ M_0 &\simeq -2a/(3\pi b) \end{aligned} \right\} \text{longitudinal excitation}$$

$$\left. \begin{aligned} \overline{M}_\infty &\simeq 4a/(3\pi b) \\ \overline{M}_0 &\simeq -1/3 \end{aligned} \right\} \text{transverse excitation}$$

In the case of the spherical limit for both the particle shape and the reference volume, it is clear that $M = M_e = 1$ and $\alpha = \alpha_e = 2$.

Then (6) reduces to

$$\frac{\rho - \rho_e}{\rho + 2\rho_e} = v\frac{1 - (\Omega_s + \rho_1/\rho)}{1 + 2(\Omega_s + \rho_1/\rho)} \tag{7}$$

where

$$\Omega_s = \frac{Z_0}{\rho a}$$

in terms of the particle radius a and the interface impedance $Z_0(j\omega)$ which is constant over the spherical surface. Equation (7) is the form given by Wait [1958]. If $Z_0 \to 0$, we have

$$\frac{\rho - \rho_e}{\rho + 2\rho_e} = v\frac{\rho - \rho_1}{\rho + 2\rho_1} \tag{8}$$

which is the famous Maxwell [1891] result.

c. *Choice of Reference Volume*

In dealing with non-spherical particles, it is necessary to return to (6). One obvious simplification is to allow both the particle and reference spheroids to have the same spheroidicity. That is $b/a = b_e/a_e$. Then (6) reduces to

$$\frac{\rho - \rho_e}{\rho + \alpha\rho_e} = v\frac{1 - (\Omega + \rho_1/\rho)}{1 + \alpha(\Omega + \rho_1/\rho)} \tag{9}$$

where $\alpha = -M_\infty/M_0 = -M_{e\infty}/M_{e0}$. When $Z_0 = 0$ or $\Omega = 0$, (9) reduces to

$$\frac{\rho - \rho_e}{\rho + \alpha\rho_e} = v\frac{\rho - \rho_1}{\rho + \alpha\rho_1} \tag{10}$$

which is often found in the literature [e.g. van Beek 1967, Sen 1981, Fricke 1924, 1953, Sillars 1937]. Here we might note that $\alpha = -1 + L^{-1}$ where L is the depolarization factor commonly employed.

Another possibility is to choose the reference volume to be spherical even when dealing with elongated particles. This assumption was adopted by Wong and Strangway [1981] and Wait [1982]. Further discussion on the choice of the reference spheroid and other generalizations are to be found in Flanagan [1983], Wait [1983a] and Flanagan and Wait [1985]. An obvious consideration is whether the mean separation of the aligned particles are the same or different in the longitudinal and the transverse directions. The general question is still unresolved.

d. Deduced Frequency Dependence

It is useful to examine the general form given by (6) to see if we can draw some general conclusions about the frequency dependence of the observed conductivity in terms of the specified interface impedance $Z_0(j\omega)$. In particular, we would like to see how the empirical Cole-Cole form [e.g. see Pelton 1977] might be related to this effective medium formulation.

We can easily solve (9) directly for $\rho_e(j\omega)$. If we just retain terms in first order of v, it follows quite readily that

$$\rho_e(j\omega) \simeq \rho\left[1 - v(1+\alpha_e)\frac{M_\infty}{M_{e\infty}}\left(\frac{1-\Omega_e}{1+\alpha\Omega_e}\right)\right]$$
$$\simeq \rho\left[1 - v(1+\alpha)\frac{(1-\Omega_e)}{(1+\alpha\Omega_e)}\right] \tag{11}$$

where

$$\Omega_e = \Omega + \frac{\rho_1}{\rho}$$

We note that by simple algebra

$$\frac{1-\Omega_e}{1+\alpha\Omega_e} = 1 - \left(1+\frac{1}{\alpha}\right)\frac{1}{1+(\alpha\Omega_e)^{-1}} \tag{12}$$

Thus (10) can be cast in the form

$$\rho_e(j\omega) = \rho_0\left\{1 - m_0\left[1 - \frac{1}{1+(\alpha\Omega_e)^{-1}}\right]\right\} \tag{13}$$

where

$$\rho_0 = \rho\left[1 + \frac{1+\alpha_e}{\alpha}\frac{M_\infty}{M_{e\infty}}v\right] = \rho\left[1 + \frac{3M_\infty}{\alpha}v\right]$$
$$= \rho\left[1 + \frac{1+\alpha}{\alpha}v\right] \tag{14}$$

and

$$m_0 = \frac{\rho}{\rho_0}\frac{(1+\alpha_e)(1+\alpha)}{\alpha}\frac{M_\infty}{M_{e\infty}}v = \frac{\rho}{\rho_0}\frac{(1+\alpha)^2}{\alpha}v \tag{15}$$

We now follow previous workers and direct our attention to metallic particles located in an electrolytic host medium. Then $\rho_1/\rho \ll 1$ and furthermore we assume that the frequency dependence of the interface impedance is characterized by

$$Z_0(j\omega) = \frac{\delta_0}{(j\omega)^k} \tag{16}$$

where δ_0 is a constant and where k is an exponent typically in the range from $1/4$ to 1. We can set

$$\frac{1}{\alpha\Omega_e} \simeq \frac{1}{\alpha\Omega} = (j\omega\tau)^k \tag{17}$$

where τ is time constant (in seconds). In fact we see that

$$\tau = \left[\left(\frac{\rho b^2}{a}\right)\frac{1}{\alpha\delta_0}\right]^{1/k} \tag{18}$$

which follows from (5). By using (17), it is evident that (13) becomes

$$\rho_e(j\omega) = \rho_0\left[1 - m_o\left(1 - \frac{1}{1+(j\omega\tau)^k}\right)\right] \tag{19}$$

which has precisely the Cole-Cole form as promulgated by Pelton [1977]. We may identify ρ_0 as the zero frequency or DC response and m_0 is called the "chargeability factor". In fact we may confirm that

$$m_0 = \frac{\rho_e(0) - \rho_e(\infty)}{\rho_e(0)} \tag{20}$$

which is consistent with conventional definitions, at least for small values of m_0 (i.e. m_0 or $v \ll 1$).

It is interesting to note that much of the experimental data on induced polarization in rock/mineral environments shows that $k \simeq 1/2$ as evidenced by the fact that τ varies as the mean particle size [Olhoeft 1981]. This factual statement lends some support for the Warburg diffusion process at metal/electrolyte interfaces.

e. Extension to Asymmetrical Excitation

In our previous discussion we have selected identical spheroidal particles with their axes of symmetry parallel to the applied field. It is a simple matter to extend the formulation to the case when the applied field is transverse or perpendicular to the axes of symmetry of the particles. As we indicated above, we merely replace the functions M_e and α_e by $\overline{M}_e$ and $\overline{\alpha}_e$ in (6). The formula now gives the apparent resistivity $\overline{\rho}_e$ in the transverse direction. Another interesting extension is when we are dealing with a random orientation of the particles. Then, the medium is isotropic at least in a macroscopic sense. In this case, we have

$$\frac{(\rho-\rho_e)(1+\alpha_e)}{\rho+\alpha_e\rho_e} = \frac{v}{3}(1+\alpha)\left[\frac{1-(\Omega+\rho_1/\rho)}{1+(\Omega+\rho_1/\rho)\alpha}\right] + \frac{2v}{3}(1+\overline{\alpha})\left[\frac{1-(\Omega+\rho_1/\rho)}{1+(\Omega+\rho_1/\rho)\overline{\alpha}}\right] \tag{21}$$

where α_e is the polarizability coefficient for the reference spheroid while α and $\overline{\alpha}$ are the polarizability coefficients for the longitudinal and transverse excitations of the particles respectively. In this situation, we should probably select the reference spheroid to be a sphere in which case $\alpha_e = 2$. When the coating parameter Ω is set equal to zero, (21) agrees with Fricke [1953].

The extension of the mixture formulas, to the case where the loading parameter v is not small, is not simple. A sizeable literature on this topic exists at least for particles of spherical shape [e.g. Bruggeman 1931 and Hanai 1968]. Also uncoated spheroidal particles have been given some attention even for fractional volume loadings approaching unity [e.g. Sen 1981 and Wait 1983b]. Here we indicate how this so called Bruggeman-Hanai approach can be applied to the coated spheroid model.

We begin with (6) which is written in equivalent form as

$$\frac{\Lambda_e-\Lambda}{L_e\Lambda_e-(L_0-1)\Lambda} = v\frac{\Lambda_1-\Lambda}{L\Lambda_1-(L-1)\Lambda} \tag{22}$$

where the newly defined admittance functions are defined as follows

$$\frac{1}{\Lambda}=\rho, \quad \frac{1}{\Lambda_e}=\rho_e \qquad \text{and} \qquad \frac{1}{\Lambda_1}=\rho_1+\frac{Z_0}{b^2/a}$$

and where the often used depolarization factors are defined by

$$L_e = 1/(1+\alpha_e) \quad \text{and} \quad L = 1/(1+\alpha)$$

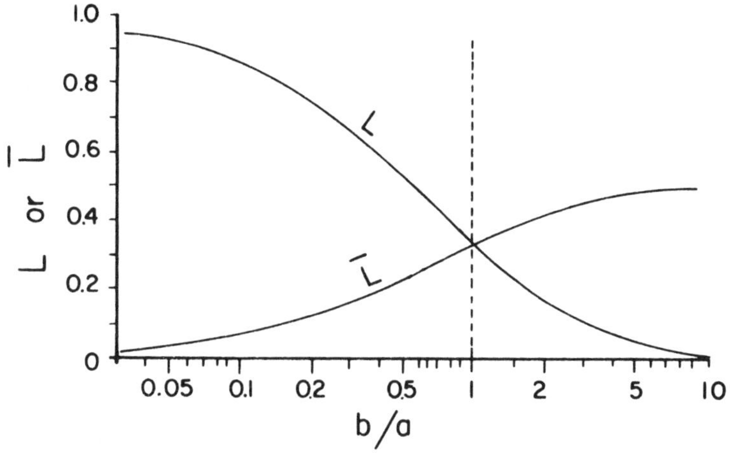

Figure 1.7.3 The axial (L) and the transverse ($\overline{L}$) depolarization factors.

Plots of L and $\overline{L}$ (for transverse excitation) are shown in Fig. 1.7.3.

In (22), we can set $v = 1-\phi$ in terms of the porosity ϕ which is 1 if the particles are absent. We now point out that (11) can be applied in a sequence of small steps to handle any value of v or ϕ. At least that is the claim following the original proposal of Bruggeman [1931]. We return to (22) and we argue that, as a result of filling the composite with grains to the intermediate concentration $1-\phi$, a medium with effective admittivity is obtained. We now augment every unit volume of this intermediate medium with a small volume du of spheroids (with the same common eccentricity and alignment) having an admittivity Λ_1. Then, according to (22), we can assert that the corresponding increment of effective admittivity is

$$\frac{d\Lambda_e}{\Lambda_e} = \frac{\Lambda_1 - \Lambda_e}{L\Lambda_1 - (L-1)\Lambda_e} du \tag{23}$$

which is independent of the parameter L_e for the reference spheroid.

Now some of the augmented grains of volume $v\,du$ have actually replaced grains already present. Thus, the resultant increment dv is related to du by the equation

$$dv = (1-v)du$$

or

$$\frac{d\phi}{\phi} = -du \tag{24}$$

Using (24), we can write (22) in the form

$$\left[\frac{L\Lambda_1 - (L-1)\Lambda_e}{\Lambda_1 - \Lambda_e}\right]\frac{d\Lambda_e}{\Lambda_e} = -\frac{d\phi}{\phi} \tag{25}$$

This can be arranged to

$$\frac{L}{\Lambda_e}d\Lambda_e + \frac{1}{\Lambda_1 - \Lambda_e}d\Lambda_e + \frac{1}{\phi}d\phi = 0 \tag{26}$$

which is equivalent to

$$\frac{d}{d\phi}\left[\Lambda_e^L \frac{1}{\Lambda_e - \Lambda_1}\right] = 0 \tag{27}$$

The solution of this differential equation is trivial if we use the initial condition that $\Lambda_e = \Lambda$ when $\phi = 1$ (or $v = 0$). Then, we obtain the interesting result

$$\left(\frac{\Lambda_e}{\Lambda}\right)^L \left(\frac{\Lambda - \Lambda_1}{\Lambda_e - \Lambda_1}\right)\phi = 1 \tag{28}$$

This formula reduces to the classic formula first derived by Bruggeman [1931] for spherical particles with no interface impedance (i.e. $L = 1/3$ and $Z_0 = 0$). There is now convincing experimental data [e.g. Hanai 1968 and Sen 1981] that the Bruggeman formula, at least for spherical particles or occlusions, is more accurate than the Maxwell-Wagner forms at larger values of v or $1 - \phi$. But the jury is still out on this question and doubts concerning the iterations of the first order formulation to higher volume loadings are without any rigorous foundation.

As we have indicated, (28) satisfies the initial condition $\Lambda_e = \Lambda$ for $\phi = 1$ corresponding to the absence of grains or particles. It is interesting to note that, in the limit $\phi \to 0$, we have $\Lambda_e \to \Lambda$ or $\rho_e \to \rho_1$ in the case of uncoated particles. This limit is highly artificial because the particles would fill all the space and the dipole induced moment concept, for individual particles, would obviously break down. Nevertheless, the trend of this limit does instill some confidence in the usefulness of the Bruggeman/Hanai type formulas for dense packing of the particles (i.e. low porosity).

f. Sen's Geometrical Model

We have tried to be fairly general in our discussion of the mixture or effective medium concept. Both particle shape and interface impedance properties can be included via our definitions of the admittivity concept. However, it is useful to digress here and consider just the influence of particle shape following up some of the cogent comments of Sen [1981]. Thus we dispense with electrochemical characteristics and just focus on the geometrical effects. To this end, we will consider the particles or occlusions as being pure dielectrics characterized by a permittivity ϵ_1. In (28), $\Lambda_1 = j\epsilon_1\omega$ where we have already noted that $Z_0 = 0$. We also assume that the suspending medium is effectively brine with a real conductivity $\Lambda = \sigma$ (i.e. $\sigma \ll \epsilon\omega$). For $L = 1/(1+\alpha)$, (28) can be written

$$\left(\frac{\Lambda_e}{\sigma}\right)\left[\frac{1-(j\epsilon_1\omega/\sigma)}{1-(j\epsilon_1\omega/\Lambda_e)}\right]^{-\frac{1+\alpha}{\alpha}} = \phi^{\frac{1+\alpha}{\alpha}} \tag{29}$$

We further argue that $\epsilon_1\omega/\sigma$ and $\epsilon_1\omega/\Lambda_e$ will be small. To first order, $\Lambda_e \simeq \sigma_e + j\epsilon_e\omega$ where the effective (real) conductivity is given by

$$\sigma_e \simeq \sigma(\phi)^{\frac{1+\alpha}{\alpha}} \tag{30}$$

and the effective (real) permittivity is

$$\epsilon_e \simeq \frac{1+\alpha}{\alpha}\epsilon_1\left(1-\phi^{\frac{1+\alpha}{\alpha}}\right) \tag{31}$$

We note here that (28) has the same form as Archie's law written as $\sigma_e = \sigma\phi^m$ where $m = (1+\alpha)/\alpha$ in our notation. For spheres, we have $m = 3/2$ which indicates that $\sigma_e = \sigma\phi^{3/2}$, a form that has been verified experimentally by Sen, Scala, and Cohen [1981] for polystyrene glass beads in a salt solution. In actual porous media in a geological setting, the observed value of m, called the cementation index, may vary from 1.5 for loosely compacted rocks to 2.5 or higher for tight material [Sumner 1976]. We can also write $m = (1-L)^{-1}$ where L is the depolarization coefficient. For disc shaped occlusions, $L \simeq 1$ and m becomes very large. For example, if $b/a \ll 1$, $m \simeq 2a/\pi b$

which leads to a strong dependence on porosity when the current flow is perpendicular to the axis of such pancake-shaped dielectric particles even when their volume is relatively small.

The effective permittivity ϵ_e, as given by (31) for disc particles with their axes parallel to the field, is given by

$$\epsilon_e \simeq \frac{2a}{\pi b}\epsilon_1 \left(1 - \phi^{\frac{2a}{\pi b}}\right) \tag{32}$$

The relative dielectric constant ϵ_e/ϵ_1, can be very large if $a/b \gg 1$ (i.e. very thin discs) even for moderately large values the porosity ϕ. In this case, the discs are acting as small condensors with very high capacitance. This phenomena is related, of course, to the classical Maxwell-Wagner effect in the context of planar layered media [van Beek 1967].

When we deal with transverse excitation of the aligned spheroids, we replace L by $\overline{L}$ or α by $\overline{\alpha}$. For needle like particles (i.e. $b/a \gg 1$), we see that $\overline{\alpha} = 1$ so that the cementation index $m \simeq 2$. In the case of discs, $\overline{\alpha} \simeq 4a/\pi b \gg 1$ for $m \simeq 1$.

Another possibility is that we work with (21) corresponding to randomly oriented particles. The resulting differential equation for the effective isotropic conductivity is then found to be

$$d\Lambda_e \simeq -\frac{d\phi}{3\phi}(\Lambda_1 - \Lambda_e)\left[\frac{1}{1 + L(\Lambda_1 - \Lambda_e)/\Lambda_e} + \frac{2}{1 + \overline{L}(\Lambda_1 - \Lambda_e)/\Lambda_e}\right] \tag{33}$$

An analogous equation was derived by Veinberg [1967] for magnetic systems such as a ferromagnetic powder cemented with non-magnetic material. Following Veinberg, we may integrate (33) with the initial condition that $\Lambda_e \to \Lambda$ as $\phi \to 1$ but the result is complicated. A simple limiting case is when the occlusions are pure insulators such that $\Lambda_1 \simeq 0$ (which is only strictly possible at DC). In this case, (33) reduces to

$$d\Lambda_e \simeq \frac{d\phi}{3\phi}\Lambda_e\left(\frac{1}{1 - L} + \frac{2}{1 - \overline{L}}\right) \tag{34}$$

The solution now reads

$$\Lambda_e = \Lambda\phi^M \tag{35}$$

where

$$M = \frac{2}{3}\left(\frac{1+\alpha}{2\alpha} + \frac{1+\overline{\alpha}}{\overline{\alpha}}\right) = \frac{1}{3}\left(\frac{1}{1-L} + \frac{2}{1-\overline{L}}\right) \tag{36}$$

In terms of resistivities,

$$\rho_e = \rho\phi^{-M} \tag{37}$$

which again has the same form as Archie's empirically derived law mentioned above.

As expected, M reduces to 3/2 for spherical cavities or occlusions. For needle-shaped cavities $\alpha \to \infty$ and $\overline{\alpha} \to 1$ whence $M \simeq 5/3 = 1.66\ldots$. For disc-shaped cavities, $\alpha \simeq \pi b/2a$ and $\overline{\alpha} \simeq 4a/\pi b$ in which case $M \simeq (1/3)(2a/\pi b)$ can be very large in spite of the weighting factor $1/3$.

References

[1] Archie, G. E., "The electrical resistivity log as an aid in determining some reservoir characteristics," *Trans. AIME*, **146**, 54–62, 1942.

[2] Bruggeman, D. A. G., "Berechnung verschiedner physikalischer Komstanten von heterogeren Substanzen," *Ann. der Physik* (Leipzig), **24**, 636–664, 1931.

[3] Flanagan, P. W., *M.S. Thesis*, Dept. of Geosciences, University of Arizona,1983.

[4] Flanagan, P. W., and J. R. Wait, "Induced polarization response of disseminated mineralization for spheroidal geometries," *Radio Science*, **30**, 147–148, 1985.

[5] Fricke, H., "A mathematical treatment of the electric conductivity and capacity of disperse systems," *Phys. Rev.*, **24**, 575–587, 1924.

[6] Fricke, H., "The Maxwell-Wagner dispersion in a suspension of ellipsoids," *J. Phys. Chem.*, **57**, 934–937, 1953.

[7] Hanai, T., "Electrical properties of emulsions," *Emulsion Science*, (P. Sherman, ed.), Academic Press, 354–478, 1968.

[8] Maxwell, J. C., *A Treatise on Electricity and Magnetism, 3rd. ed. PtII*, Clarendon Press, 1891 (reprinted Dover Publications, New York, 1954).

[9] Olhoeft, G. R., "Electrical properties of rocks," *Physical Proper-*

ties of Rocks and Minerals, (Y. S. Touloukian, ed.), McGraw-Hill, New York, 257–330, 1981.

[10] Pelton, W. H., *Interpretation of Induced Polarization and Resistivity data*, Ph.D. Thesis, Univ. of Utah, 256 pages, 1977.

[11] Sen, P. N., C. Scala, and M. Cohen, "A self similar model for sedimentary rocks with application to the dielectric constant of fused glass beads," *Geophysics*, **46**, 781–795, 1981.

[12] Sen, P. N., "Relation of certain geometrical features to the dielectric anomaly of rocks," *Geophysics*, **46**, 1714–1720, 1981.

[13] Sillars, R. W., "The properties of a dielectric containing semiconducting particles of various shapes," *J. Inst. of Elect. Engs.*, (London) **80**, 378–394, 1937.

[14] Sumner, J. S., *Principles of Induced Polarization for Geophysical Exploration*, Elsevier, Amsterdam, New York, 1976.

[15] van Beek, L. K. H., "Dielectric behavior of heterogeneous systems," *Progress in Dielectrics*, **7**, 69–114, 1967 (Heywood Books, London).

[16] Veinberg, A. K., "Permeability, electrical conductivity, dielectric constant, and thermal conductivity of a medium with spherical and ellipsoidal inclusions, " *Sov. Phys. Dokl.*, **11**, 593–595, 1967.

[17] Wait, J. R., "A phenomenological theory of induced electrical polarization," *Can. J. Phys.*, **36**, 1634–1644, 1958.

[18] Wait, J. R., "Electromagnetic response of a medium loaded with coated (spheroidal) conductive particles," *IEEE Trans.*, **GE-20**, 500–504, 1982.

[19] Wait, J. R., "Effective electrical properties of heterogeneous earth models," *Radio Science*, **18**, 19–24, 1983b (errata 18, 796, 1983).

[20] Wong, J. and D. W. Strangway, "Induced polarization in disseminated sulphide ores containing elongated mineralization," *Geophysics*, **46**, 1258–1268, 1981.

[21] Wait, J. R., "Complex conductivity of disseminated spheroidal ore grains," *Gerlands Beitr. Geophysik*, **92**, 49–69, 1983a.

[22] Wait, J. R., "Physical model for the complex conductivity of the earth," *Electronics Letters*, **23**, 979–980, 1987.

1.8 The Electrochemical Perspective

a. Introduction

The actual processes of electrical polarization are found to occur at the microscopic level in the electrolyte filled pores at the grain and pore interfaces. On a macroscopic level, the complex resistivity effect is a consequence of these polarization effects. The frequency dispersive charge separation effects at the grain interfaces can be included explicitly in the model by undertaking a rigorous analysis of the electrochemistry of the metal grain interface such as propounded by Wong [1979] who employed a spherical model of the ore particle. The analysis was extended to spheroidal shaped particles by Wong and Strangway [1981] which allows rock texture into the model. Sometime earlier [Wait 1958, 1959] *, the electrochemistry at the metal-electrolyte interface was described phenomenologically by an interface impedance with any desired frequency dependence. Such an approach was also employed by Komarov [1972]. More recently, the electrochemical charge separation effects around small resistive spheres was treated by Chew and Sen [1982]. They assumed that the diffuse double layer is given by the Boltzmann distribution in terms of the potential following the Guoy-Chapman theory. Chew and Sen obtain analytical results in the limit that the double layer thickness is small compared with the radius of the particle. Their results are highly relevant to the important question of the huge dielectric enhancements observed in clay bearing geological materials.

Common to all of the microscopic theories, the induced electric dipoles in the elemental particles are superimposed to deduce an effective complex conductivity or permittivity of the assemblage. Attempts to generalize such theories to dense packing of the basic particles lead to fundamental difficulties. But some interesting and ingenious suggestions have been made to extend the validity of the mixture formulas beyond the sparse loading idealization. (e.g. see discussions in sections 1.4, 1.5 and 1.6)

Here, we review some of the basic electrochemical concepts which hopefully provide needed insight into the underlying mechanisms. Also, we wish to provide a justification for the phenomenological model which represents the boundary region adjacent to the particle as an in-

* see Section 1.4 references.

terface impedance or surface admittance or some combination thereof.

b. Classical Diffusion Concept

Basically, an electrolytic solution is a mixture of positively charged cations and negatively charged anions. Normally, the solution is regarded as being in equilibrium where the net charge density q is zero. Thus,

$$q = ep - en = 0 \tag{1}$$

where p and n are the densities of the anions and cations, respectively, while e is the unit electrical charge. Now a net movement of charged ions (i.e. current flow) will occur when an electrical field $\overline{E}$ is applied or when the ionic concentrations p and n are perturbed. Thus, the respective cationic and anionic flux densities are

$$\overline{J}_p = -D\nabla p + \mu p\overline{E} \tag{2}$$

and

$$\overline{J}_n = -D\nabla n - \mu n\overline{E} \tag{3}$$

where D is the diffusivity coefficient and μ is the electrical mobility. Clearly, the second terms on the R.H.S. of (2) and (3) describe the migration of the ions under the effect of the electric field. On the other hand, the first terms on the R.H.S. of (2) and (3) describe the diffusion of the ions under a concentration perturbation. For a uniform external field applied to an electrolyte initially at equilibrium, the charge flow is described exclusively by the migration term since there is no initial concentration gradient.

Now, according to the divergence theorem

$$\nabla \cdot \overline{J}_p = \frac{\partial p}{\partial t} \tag{4}$$

and

$$\nabla \cdot \overline{J}_n = \frac{\partial n}{\partial t} \tag{5}$$

Thus, the accumulation of charge with time are described by

$$\frac{\partial p}{\partial t} = D\nabla^2 p - \mu\nabla \cdot (p\overline{E}) \tag{6}$$

and

$$\frac{\partial n}{\partial t} = D\nabla^2 n + \mu\nabla \cdot (n\overline{E}) \tag{7}$$

where $\nabla^2 = \nabla \cdot \nabla$ is the Laplacian operator. In what follows, we also assume that $E = -\nabla V$ when V is a potential. Actually, (2) and (3) are Fick's first law while (6) and (7) are Fick's second law [Bockris and Reddy, 1973]. Here, we have assumed single valence ionic species. Of course, if the divergence terms on the R.H.S. at (6) and (7) are ignored, we have the simple diffusion equation for p and n. This is a good approximation when the ion concentration gradients are sufficiently small.

In actual rock-electrolyte systems, the gradient of the charge densities may be large near interfaces when an electric field is applied. In fact, charge separation may occur spontaneously in the electrolytic pore fluid. When the external field is removed, the charge accumulation may discharge in a manner roughly akin to a charged capacitor. This particular process is known as the Maxwell-Wagner effect [Hasted 1973]. The relaxation is very rapid except in special cases of thin pancake shaped occlusions which exhibit high effective internal capacitances [Sen 1981].

c. *Chew and Sen's Model* *

In the frequency range of interest (0.01 – 1000 Hz), it has been observed that small resistive particles such as clays occurring in sizes of the order of 10^{-8} m in diameter do indeed give a large dielectric enhancement. As indicated by Chew and Sen [1982], the surface conduction around the periphery of the resistive or insulating particle can lead to a major dielectric enhancement. In their electrochemical model, the frequency dependence of the conduction or admittance layer is a consequence of their theory rather than an "ad hoc" supposition.

In the case where the basic particles are metallic ore grains, there are species of ions present in the electrolyte which produce a redox reaction across the given interface during the charging and discharging process. This redox reaction constitutes a leakage current that in effect shunts the equivalent capacitance of the interfacial charge separation. Thus, the rate or time constant of the discharge of the induced

---

* W. C. Chew and P. N. Sen, "Dielectric enhancement due to electrochemical double layer ; thin double layer approximation," *Jour. Chem. Phys.*, **77**, 4683–4692, 1982.

dipole is no longer controlled by this capacitance of the interface as it would be in the Maxwell-Wagner model. Instead, it becomes a complicated function of the diffusion rate of the reactive ionic species in the neighborhood of the grain and of the rate of the associated redox reaction.

d. Butler-Volmer Equation

Here, we will look at the charge transfer mechanism at a metal/ electrolyte boundary. The ore mineral grain will be modeled as a metallic conducting particle while the rock matrix is described as an electrolyte by virtue of the conduction in the fluid field pores. As indicated by Angoran and Madden [1977], only Cu^{++} and S^{--} ions have measurable activity in copper/iron sulfide ore systems. Thus, it seems reasonable to use a model of a single active species electrolyte for the rock pore fluid.

Actually, for a metal particle immersed in such a system, there will be a continuous charge transfer across the interface occurring even at equilibrium conditions. The redistribution of charge on the electrolyte side of the interface sets up a plane of charge a short distance away. A corresponding plane of image charges then appears in the conductor with equal and opposite polarity. The resultant structure is referred to as the "double layer" and it is depicted in Fig. 1.8.1a. The quantity $\Delta\phi_e$ is the equilibrium potential associated with the interface decays from a constant value at the particle surface to zero at a point within the electrolyte as sketched in Fig. 1.8.1b. The position of the closest approach of non-adsorped cations is termed the Outer Helmholtz Plane (OHP) as shown in Fig. 1.8.1c. The charge balancing the image charge in the metal does not all lie in the OHP but some of it is diffused into the bulk electrolyte. The region is the Gouy-Chapman diffuse layer. However, most of the potential drop occurs between the electrode and the Outer Helmholtz Plane. As a consequence, the structure behaves as a capacitor when the intervening dielectric is affected by the oriented water dipoles between the OHP and the metal surface. The water dipoles are shown as arrows in Fig. 1.8.1a. For time varying conditions, the capacitive property will allow passage of displacement current even if there were no charge transfer. This contribution to the total *effective* current flow is described as being "non-faradaic".

Under truly non-equilibrium conditions, there is a net charge flow across the interface and this portion is called "faradaic." It may occur

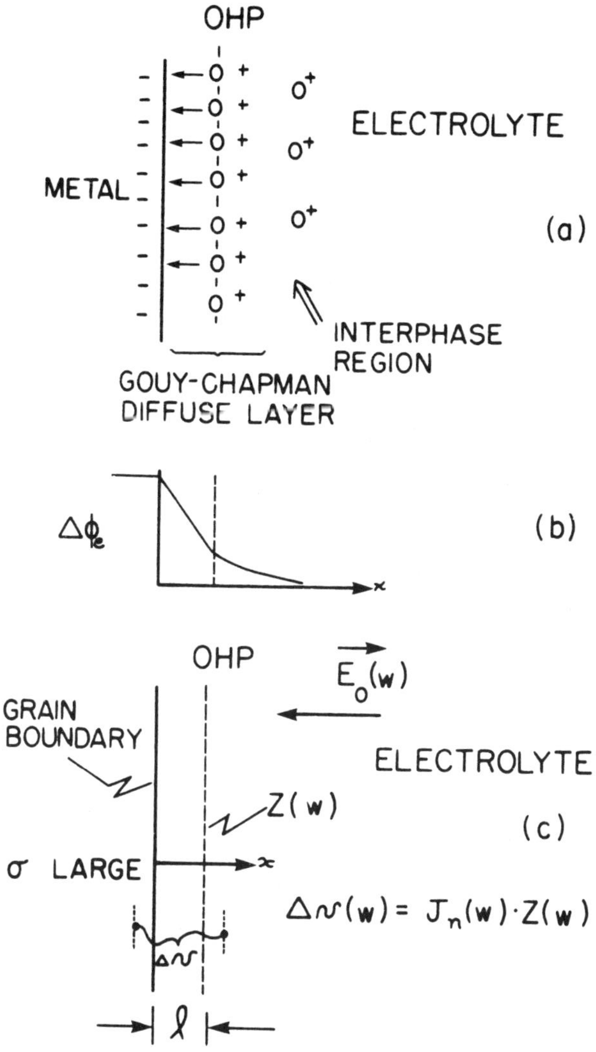

Figure 1.8.1 The electrochemical double layer. In Fig. (c), $J_n(j\omega)$ is the net normal current density at the interface and $\Delta v(j\omega)$ is the voltage drop.

if there is a perturbation in the concentration of the ionic species in the interphase region or alternatively there is an "overvoltage" v applied across the interface. This excess current i_e is described by the Butler-Volmer equation which, followed Bockris and Reddy [1973], is written

$$i_e = i_0^F \exp\left(\frac{ev}{2kT}\right) - i_0^B \exp\left(-\frac{ev}{2kT}\right) \tag{8}$$

where e is the electronic charge, k is Boltzman's constant, T is the absolute temperature and where i_0^F and i_0^B are the forward and backward equilibrium reaction currents. For sufficiently small overpotentials, (8) can be approximated by the linear form

$$i_e = i_0^F(1 + \frac{ev}{2kT}) - i_0^B(1 - \frac{ev}{2kT}) \tag{9}$$

which corresponds to the common assumption that the I.P. response is linear with respect to low current densities in the bulk volume. We can further write

$$i_0^F = eKp \quad \text{and} \quad i_0^B = eKn \tag{10}$$

where K is the forward or backward reaction rate (assumed equal). Then (9) is reduced to

$$i_e = eK\left[(p-n) + (p+n)\frac{ev}{2kT}\right] \tag{11}$$

e. More on the Interface Impedance

We proceed further by noting that the thickness of the double layer (e.g. 10^{-9} m) is certainly small compared with the radius of the typical mineral grain (e.g. 10^{-3} m). Thus, it is safe to assume that the surface of the grain and the interaction region can be described locally in a planar geometry where $x > 0$ represents the electrolyte and where $x < 0$ is the metal. We abstract here from Flanagan and Wait [1985] and Flanagan [1983]*.

With the one dimensional geometry indicated, we may reduce (6) and (7) to

$$\frac{\partial^2 V}{\partial x^2} = -\frac{e}{\epsilon}(p-n) \tag{12}$$

* see references in Section 1.4

$$j\omega p = D\frac{\partial p^2}{\partial x^2} - \mu\frac{\partial p}{\partial x}\frac{\partial V}{\partial x} - \mu p\frac{\partial^2 V}{\partial x^2} \tag{13}$$

$$j\omega n = D\frac{\partial^2 n}{\partial x^2} + \frac{\partial n}{\partial x}\frac{\partial V}{\partial x} + \mu n\frac{\partial^2 V}{\partial x^2} \tag{14}$$

where we have adopted a time factor $\exp(j\omega t)$. Solutions of (12), (13), and (14), which are the basic electrokinetic equations for the problem can be obtained under appropriate boundary conditions. An obvious condition is that the potential V at $x = 0$ on the metal surface is a constant. Then the voltage $v(j\omega)$ across the interface region of thickness ℓ, is related to the net current flow $i_e(j\omega)$ by the linear relation

$$v(j\omega) = Z(j\omega)\, i_e(j\omega) \tag{15}$$

where $Z(j\omega)$ can be aptly defined as the *interface impedance.* On the basis of (11) and (15), we have

$$Z(j\omega) = \frac{1}{ek}\frac{V(\ell)}{(p-n) + (p+n)eV(\ell)/2kT} \tag{16}$$

where $V(\ell)$ is the voltage drop and where p and n are the effective values at $x = \ell$. The point being made here is that the electrochemistry of the double layer can be represented by a frequency dependent impedance which is merely the voltage drop across the layer divided by the normal current density. Linearity of the response is a vital ingredient of this conclusion. But linearity is always something that can be checked in the laboratory or field setting (e.g. by confirming lack of harmonic distortion)

f. Perturbation Approach

An alternative approach is to deal only with first order perturbation of the time varying ion densities. Following Wong [1979], we set

$$p = p_0 + p(x,t) \tag{17}$$

and

$$n = n_0 + n(x,t) \tag{18}$$

where p_0 and n_0 are background concentrations and $p(x,t)$ and $n(x,t)$ are perturbations. When the time variation is according to

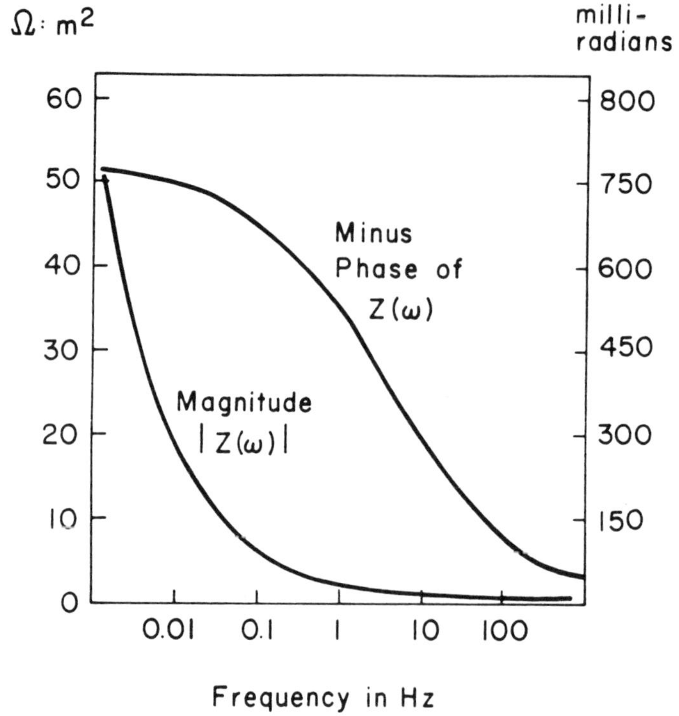

Figure 1.8.2 Fit of Olhoeft's [1982] data for pyrite/electrolyte interface to modified Warburg model.

$\exp(j\omega t)$, we can linearize the resulting equations provided second order quantities are neglected. The interface impedance concept then refers to the proportionality of the net perturbation current flow (at a frequency $\omega/2\pi$) and the corresponding perturbation voltage across the double layer. It is also possible to include the non-faradaic portion of the alternating current flow in the interface impedance current flow because it is basically in parallel with the faradaic portion.

We make no attempt here to calculate the actual interface impedance $Z(j\omega)$ for a given situation. But certainly, such an approach is possible following the lead of Angoran and Madden [1977], Wong [1979], and Wong and Strangway [1982]. Instead, we will argue that $Z(j\omega)$ is an intrinsic parameter of the interface which can be specified at the outset. Fortunately there are good data (e.g. Olhoeft, [1982]) on actual measurements of the frequency dependence of the interface impedance for a pyrite/electrolyte solution. An example of many such data is shown in Fig. 1.8.2 where both the amplitude and phase of $Z(j\omega)$ are shown for a wide frequency range. Olhoeft [1982] found an empirical fit, for the frequency dependence, to be of the form

$$Z(j\omega) = Z_\infty + Z_1/(j\omega)^{\frac{1}{2}} \tag{19}$$

While the particular form of the frequency dependence given by (19) is not universal, it is a convenient reference to employ in further theoretical studies. Also, while it is referred to as a Warburg impedance because of the $1/(j\omega)^{1/2}$ term, we are certainly aware of other electrochemical factors that may contribute to the frequency dependence. For example, Fink [1980] and more recently Klein et al. [1984] have indicated that there are two physical processes responsible for the electrode polarization at mineral/electrode interfaces. For minerals such as chacocite, the response is related to a reversible oxidation-reduction couple with copric ions in solution. But for other minerals such as pyrite and chalcopyrite, diffusion of active species in the electrolyte as propounded by Wong [1979] is not the controlling process. Alternative mechanisms are surface controlled processes such as surface diffusion or adsorption phenomena. In any case, the interface impedance is a valid description from the standpoint of formulating a boundary value problem for a given particle or mineral grain located within the electrolyte.

References

[1] Angoran, Y. and T. R. Madden, "I.P. : A preliminary study of its chemical basis," *Geophysics*, **42**. 788–803, 1977.

[2] Antropov, L., *Theoretical Electrochemistry*, Mir Publishers, Moscow, 1972.

[3] Bockris, J. O'M and A. K. N. Reddy, *Modern Electrochemistry*, Plenum Press, 1970.

[4] Chew, W. C., "Dielectric enhancement and electrophoresis due to electrochemical double layer — a uniform approximation," *J. Chem. Phys.*, **80**, No. 9, 1984.

[5] Fink, J., "Warburg impedance — what it is and what it isn't," *50th. Annual SEG Meeting, Houston,* Nov. 1980.

[6] Hasted, J. B., *Aqueous Dielectrics*, Chapman and Hall, London, 1973.

[7] Johnson, D. L., and P. N. Sen (Editors), *Physics and Chemistry of Porous Media*, American Institute of Physics, New York, 1984.

[8] Kan, R. and P. N. Sen, "Electrolytic conduction in periodic arrays of insulators with charges," *Jour. of Chemical Physics,* **86**, 5748–5756, 1987.

[9] Klein, J. D., T. Biegler, and M. D. Horne, "Mineral interfacial processes in the method of induced polarization," *Geophysics*, **49**, 1105–1114, 1984.

[10] Komarov, V. A., "The importance of the IP method for the exploration of ore deposits," *Econ. Geol. Report, (Geol. Survey of Canada)*, **26**, 1967.

[11] Olhoeft, G. R., "Warburg impedances at single water/mineral interfaces," *52nd. Annual SEG Meeting, Dallas,* Oct. 1982.

[12] Sen, P. N., "Dielectric anomaly in inhomogeneous materials with application to sedimentary rocks," *Appl. Phys. Letters*, **39**, 667–668, 1981.

[13] Shaw, D. J., *Introduction to colloid and surface Chemistry*, Butterworth's, London, 2nd. Ed., 1970.

[14] Tomkiewicz, M. and P. N. Sen (Editors), *Chemistry and Physics of Composite Media*, Electrochemical Society, New Jersey, 1985.

[15] Wait, J. R., "Physical model for the complex resistivity of the earth", *Electronics Letters*, **23**, 979–980, 1987.

[16] Wong, J., "An electrochemical model of the IP phenomena in disseminated sulphide ores," *Geophysics*, **44**, 1245–1265, 1979.

[17] Wong, J. and D. W. Strangway, "IP in disseminated sulphide ores containing elongated mineralization," *Geophysics*, **46**, 1258–1268, 1981.

Appendix — Thin Sheet Boundary Conditions

The transition region between two homogeneous media can often be handled by effective boundary conditions in order to simplify subsequent applications. To illustrate the concept, we present here an analysis of a sector of a cylindrical shell whose thickness is small compared with the radius of curvature a.

The situation is illustrated in Fig. 1.A.1 with reference to cylindrical coordinates (ρ, ϕ, z). We assume here the problem is entirely two dimensional so that $\partial/\partial z = 0$. The shell is composed of two regions such that the conductivity is σ_1 for $a + \ell > \rho > a$, and the conductivity is σ_2 for $a > \rho > a - s$.

We do not need to specify explicitly the regions outside the shell (i.e. $\rho > a+\ell$ and $\rho < a-s$) except to say their properties do not vary with ϕ. Also we will assume that the fields are everywhere solutions of Laplace's equation. This means that the frequency is sufficiently low that all significant distances of the problem are small compared with the effective wavelength in any of the regions [Wait 1985]. Displacement currents can be included in the problem by regarding σ_1 and σ_2 as being complex (e.g. let $\sigma_1 \to \sigma_1 + j\epsilon_1\omega$, $\sigma_2 \to \sigma_2 + j\epsilon_2\omega$; ϵ_1, ϵ_2 are the respective permittivities). Another possibility is to replace σ_1 by $(\rho_1)^{-1}$ and σ_2 by $(\rho_2)^{-1}$ in terms of the complex resistivities of the two shell regions.

For later convenience we specify that the conductivity in the external region (i.e. for $\rho > a + \ell$) is σ and the conductivity for the internal region (i.e. for $\rho < a - s$) is σ_i. We later let ℓ and s become vanishingly small.

Solutions for the potentials Ψ_1 and Ψ_2 in the two regions are

$$\Psi_1 = \left[A\left(\frac{\rho}{a}\right)^m + B\left(\frac{a}{\rho}\right)^m\right]e^{-jm\phi} \tag{1}$$

and

$$\Psi_2 = \left[C\left(\frac{\rho}{a}\right)^m + D\left(\frac{a}{\rho}\right)^m\right]e^{-jm\phi} \tag{2}$$

where m is the order of the harmonic. Here A, B, C and D are constants. The corresponding radial current densities (using $J_\rho =$

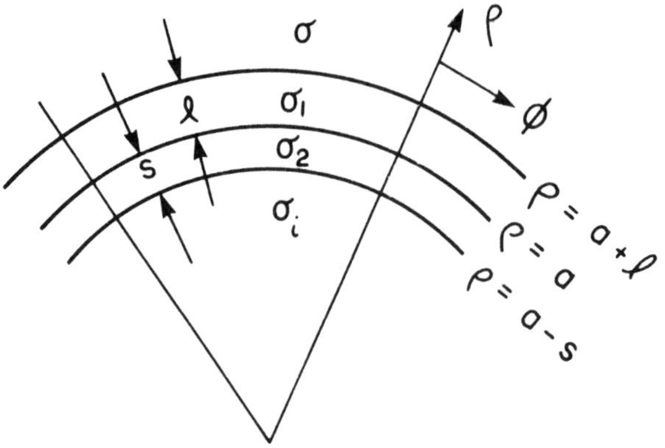

Figure 1.A.1 Segment of the cross section of the cylindrical shell model used to justify the effective boundary conditions for a curved interface.

$-\sigma\partial\Psi/\partial\rho$) are then

$$J_{\rho_1} = -\sigma_1 m \left[A\left(\frac{\rho^{m-1}}{a^m}\right) - B\left(\frac{a^m}{\rho^{m+1}}\right)\right] e^{-jm\phi} \tag{3}$$

and

$$J_{\rho_2} = -\sigma_2 m \left[C\left(\frac{\rho^{m-1}}{a^m}\right) - D\left(\frac{a^m}{\rho^{m+1}}\right)\right] e^{-jm\phi} \tag{4}$$

Continuity of Ψ and J_ρ at $\rho = a$ leads nicely to

$$A\left(1 + \frac{\sigma_1}{\sigma_2}\right) + B\left(1 - \frac{\sigma_1}{\sigma_2}\right) = 2C \tag{5}$$

and

$$A\left(1 - \frac{\sigma_1}{\sigma_2}\right) + B\left(1 + \frac{\sigma_1}{\sigma_2}\right) = 2D \tag{6}$$

which, of course, indicate that C and D are linearly related to A and B.

The "voltage" v across the shell is given by

$$\begin{aligned} v =& \Psi_2(a-s) - \Psi_1(a+\ell) \\ =& \left\{-A\left(\frac{a+\ell}{a}\right)^m - B\left(\frac{a}{a+\ell}\right)^m \right. \\ & \left. + C\left(\frac{a-s}{a}\right)^m + D\left(\frac{a}{a-s}\right)^m\right\} e^{-jm\phi} \end{aligned} \tag{7}$$

Using (5) and (6), we see that

$$v = A\left\{\left[-\left(1+\frac{\ell}{a}\right)^{m} + \frac{1}{2}\left(1+\frac{\sigma_1}{\sigma_2}\right)\left(1-\frac{s}{a}\right)^{m} + \frac{1}{2}\left(1-\frac{\sigma_1}{\sigma_2}\right)\left(1-\frac{s}{a}\right)^{-m}\right]\right\} + B\left\{\left[-\left(1+\frac{\ell}{a}\right)^{-m} + \frac{1}{2}\left(1-\frac{\sigma_1}{\sigma_2}\right)\left(1-\frac{s}{a}\right)^{m} + \frac{1}{2}\left(1+\frac{\sigma_1}{\sigma_2}\right)\left(1-\frac{s}{a}\right)^{-m}\right]\right\} e^{-jm\phi} \tag{8}$$

which so far is exact! We now let ℓ and s become small compared with, say, a. Also we assume $\sigma_2 \gg \sigma_1$ (i.e. inner shell region is a good conductor, relatively speaking). Now (8) reduces to

$$v = -(\ell m/a)(A - B)e^{-jm\phi} \tag{9}$$

The other quantity of interest is the radial current density at the outer surface of the double shell. Clearly, it is given by (3) at $\rho = a+\ell$. That is

$$J_{\rho_1}(a+\ell) = -\sigma_1 m\left[A\frac{(a+\ell)^{m-1}}{a^m} - B\frac{a^m}{(a+\ell)^{m+1}}\right]e^{-jm\phi} = -\frac{\sigma_1 m}{a}\left[A\left(1+\frac{\ell}{a}\right)^{m-1} - B\left(1+\frac{\ell}{a}\right)^{-m+1}\right]e^{-jm\phi} \tag{10}$$

Then, if $\ell \ll a$ (i.e. thin shell),

$$J_{\rho_1}(a+\ell) \simeq -\frac{\sigma_1 m}{a}(A - B)e^{-jm\phi} \tag{11}$$

which is proportional to (9). The latter fact suggests that we define an interface impedance Z_0 according to

$$Z_0 = \frac{v}{J_{\rho_1}(a+\ell)} \tag{12}$$

In terms of the thin shell approximation, we simply have

$$Z_0 = \frac{\ell}{\sigma_1} \tag{13}$$

but, in general, we need to use (8) and (10) in the definition of (12). In that case Z_0, the "interface impedance" depends on the ratio B/A which is a function of the fields internal and/or the external to the shell. As defined, (12) is also a function of m since a general representation of the potentials involve summation over all integer values of m. However, in the thin sheet approximation, Z_0 is independent of m as given by (13).

Another property of the double shell is the difference of the radial current densities at the inner and outer surfaces. Clearly, this quantity is given by

$$\begin{aligned}\Delta J_\rho =& J_{\rho_2}(a-s) - J_{\rho_1}(a+\ell)\\ =&\sigma_1 m\left[A\frac{(a+\ell)^{m-1}}{a^m} - B\frac{a^m}{(a+\ell)^{m+1}}\right]e^{-jm\phi}\\ &-\sigma_2 m\left[C\frac{(a-s)^{m-1}}{a^m} - D\frac{a^m}{(a-s)^{m+1}}\right]e^{-jm\phi}\end{aligned} \tag{14}$$

Using (5) and (6) again, we have

$$\begin{aligned}\Delta J_\rho = \Bigg\{ & A\left[\frac{\sigma_1 m}{a}\left(1+\frac{\ell}{a}\right)^{m-1} - \frac{\sigma_2 m}{2a}\left(1+\frac{\sigma_1}{\sigma_2}\right)\left(1-\frac{s}{a}\right)^{m-1}\right.\\ &\left.+\frac{\sigma_2 m}{2a}\left(1-\frac{\sigma_1}{\sigma_2}\right)\left(1-\frac{s}{a}\right)^{-(m+1)}\right]\\ &+B\left[-\frac{\sigma_1 m}{a}\left(1+\frac{\ell}{a}\right)^{-(m+1)} - \frac{\sigma_2 m}{2a}\left(1-\frac{\sigma_1}{\sigma_2}\right)\left(1-\frac{s}{a}\right)^{m-1}\right.\\ &\left.+\frac{\sigma_2 m}{2a}\left(1+\frac{\sigma_1}{\sigma_2}\right)\left(1-\frac{s}{a}\right)^{-(m+1)}\right]\Bigg\} e^{-jm\phi}\end{aligned} \tag{15}$$

In passing to the thin shell limit, we need be a bit more subtle here. First of all, we consider only that $\sigma_2 \gg \sigma_1$ (i.e. inner shell is a relatively good conductor as before) but there is no restriction on geometry. Then (15) reads

$$\begin{aligned}\Delta J_\rho \simeq Ae^{-jm\phi}\Bigg\{ & -\frac{\sigma_2 m}{2a}\left[\left(1-\frac{s}{a}\right)^{m-1} - \left(1-\frac{s}{a}\right)^{-(m+1)}\right]\\ &+\frac{\sigma_1 m}{a}\left[\left(1+\frac{\ell}{a}\right)^{m-1} - \frac{1}{2}\left(1-\frac{s}{a}\right)^{m-1} - \frac{1}{2}\left(1-\frac{s}{a}\right)^{-(m+1)}\right]\Bigg\}\end{aligned}$$

$$+ Be^{-jm\phi}\left\{-\frac{\sigma_2 m}{2a}\left[\left(1-\frac{s}{a}\right)^{m-1}-\left(1-\frac{s}{a}\right)^{-(m+1)}\right]\right.$$
$$\left.-\frac{\sigma_1 m}{a}\left[\left(1+\frac{\ell}{a}\right)^{-(m+1)}-\frac{1}{2}\left(1-\frac{s}{a}\right)^{m-1}-\frac{1}{2}\left(1-\frac{s}{a}\right)^{-(m+1)}\right]\right\} \tag{16}$$

Now, we regard s/a and ℓ/a as small such that terms containing $(s/a)^2$ and $(\ell/a)^2$ and higher powers can be neglected. Then (16) looks like

$$\Delta J_\rho \simeq Ae^{-jm\phi}\left\{-\frac{\sigma_2 m}{2a}\left[1-\frac{s}{a}(m-1)-1-\frac{s}{a}(m+1)\right]\right.$$
$$\left.+\frac{\sigma_1 m}{a}\left[1+\frac{\ell}{a}(m-1)-\frac{1}{2}\left(1-\frac{s}{a}(m-1)+1+\frac{s}{a}(m+1)\right)\right]\right\}$$
$$+ Be^{-jm\phi}\left\{-\frac{\sigma_2 m}{2a}\left[1-\frac{s}{a}(m-1)-1-\frac{s}{a}(m+1)\right]\right.$$
$$\left.-\frac{\sigma_1 m}{a}\left[1-\frac{\ell}{a}(m+1)-\frac{1}{2}\left(1-\frac{s}{a}(m-1)+1+\frac{s}{a}(m+1)\right)\right]\right\} \tag{17}$$

which, after cancellation of terms, gives

$$\Delta J_\rho \simeq (A+B)\frac{s}{a^2}m^2\sigma_2 e^{-jm\phi} \tag{18}$$

noting that $\partial/\partial\phi = -jm$, we see that the effective boundary condition for the discontinuity of the normal current density is

$$J_{\rho_2}(a-s) - J_{\rho_1}(a+\ell) \simeq -\frac{s\sigma_2}{a^2}\frac{\partial^2\Psi_1(a)}{\partial\Phi^2} \tag{19}$$

where

$$\Psi_1(a) = (A+B)e^{-jm\phi} \tag{20}$$

is the potential at $\rho = a$. For operational convenience, we note that

$$\Psi_2(a-s) = \left[C\left(1-\frac{s}{a}\right)^m + D\left(1-\frac{s}{a}\right)^{-m}\right]e^{-jm\phi} \simeq \Psi_1(a) \tag{21}$$

which is simply a statement that the potential drop across the inner shell is negligible. Thus, the boundary condition (19) is replaced by

$$J_\rho(a-s) - J_\rho(a+\ell) \simeq -\frac{s\sigma_2}{a^2}\frac{\partial^2\Psi}{\partial\phi^2}(a-s) \tag{22}$$

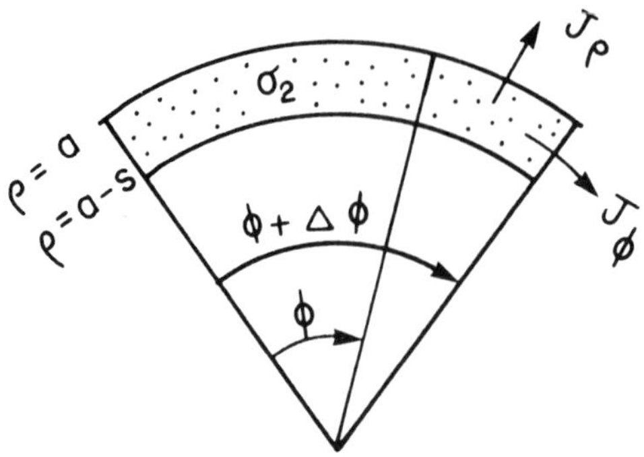

Figure 1.A.2 Portion of inner shell and elemental volume.

which, with the use of Laplace's equation, is equivalent to

$$J_\rho(a-s) - J_\rho(a+\ell) \simeq \frac{Y_0}{a^2}\left[\rho\frac{\partial}{\partial\rho}\rho\frac{\partial\Psi}{\partial\rho}\right]_{\rho=a-s} \tag{23}$$

where $Y_0 = s\sigma_2$ is the conductance of the inner shell.

It is useful to consider the physical aspects of the boundary condition given by (23). Here, we just focus on the inner highly conducting shell of thickness s which is shown in Fig. 1.A.2 (not drawn to scale). We now apply the divergence theorem $\Delta\cdot\overline{J} = 0$ to the volume bounded by $a < \rho < a - s$ and by ϕ and $\phi + \Delta\phi$. We can write approximately

$$\Delta\phi(a-s)J_\rho(a-s) - \Delta\phi a J_\rho(a) + \int_{a-s}^{a}\Big[J_\phi(\phi) - J_\phi(\phi+\Delta\phi)\Big]d\rho = 0 \tag{24}$$

or, using differentials,

$$a\Big[J_\rho(a-s) - J_\rho(a)\Big]d\phi + s\left(-\frac{\partial J_\phi}{\partial\phi}\right)d\phi = 0 \tag{25}$$

Noting $J_\phi = -(\sigma_2/a)(\partial\Psi/\partial\phi)$ within the shell, we obtain

$$J_\rho(a-s) - J_\rho(a) = -\frac{s\sigma_2}{a^2}\frac{\partial^2\Psi(a)}{\partial\phi^2} \tag{26}$$

or

$$J_\rho(a-s) - J_\rho(a) = \frac{s\sigma_2}{a^2}\left[\rho\frac{\partial}{\partial\rho}\rho\frac{\partial\Psi}{\partial\rho}\right]_{\rho=a} \tag{27}$$

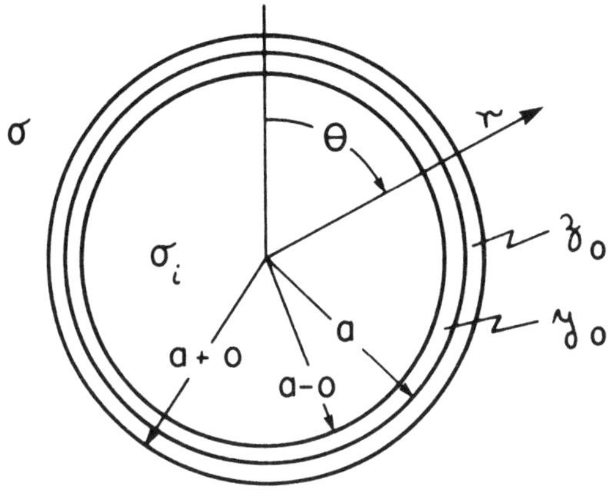

Figure 1.A.3 The double shell spherical model for the solid particle of conductivity σ_i immersed in a medium of conductivity σ.

which, not surprisingly, are consistent with (21) and (23). Again, we note that $Y_0 = s\sigma_2$.

It is now clear that in the limit where ℓ and $s \to 0$, the boundary conditions for the double shell model can be succinctly written

$$\left(\sigma\frac{\partial\Psi}{\partial\rho}\right)_{\rho=a+0} = \left[\sigma_i\frac{\partial\Psi}{\partial\rho} + \frac{Y_0}{a^2}\left(\rho\frac{\partial}{\partial\rho}\rho\frac{\partial\Psi}{\partial\rho}\right)\right]_{\rho=a-0} \tag{28}$$

and

$$\left(\Psi - Z_0\sigma\frac{\partial\Psi}{\partial\rho}\right)_{\rho=a+0} = (\Psi)_{\rho=a-0} \tag{29}$$

where σ_i and σ are the conductivities inside and outside the shell respectively.

The thin sheet boundary conditions are useful to solve problems involving particles where the source of the field is outside the particle. We have attempted to show the development and justification for a cylindrical model. The corresponding case in spherical geometry is illustrated in Fig. 1.A.3. The spherical particle has a radius a. To account for bound ions or other related surface conduction mechanisms, we imbue the particle with a surface admittance Y_0. Then, the outer more resistive region (i.e. diffuse ionized layer) just adjacent to the particle is represented by an interface impedance Z_0. The effective boundary conditions, which connect the interior potential for $r < a$ and the exterior potential for $r > a$, are then written as follows

$$\left(\sigma\frac{\partial\Psi}{\partial r}\right)_{r=a+0} = \left[\sigma_i\frac{\partial\Psi}{\partial r} + \frac{Y_0}{a^2}\frac{\partial}{\partial r}\left(r^2\frac{\partial\Psi}{\partial r}\right)\right]_{r=a-0} \tag{30}$$

and

$$\left(\Psi - Z_0\sigma\frac{\partial\Psi}{\partial r}\right)_{r=a+0} = (\Psi)_{r=a-0} \tag{31}$$

These two equations are completely analogous to (28) and (29). Their application was already indicated (e.g. Chap. 2 in [Wait 1985]).

References

[1] Wait, J. R., *Electromagnetic Wave Theory*, Harper & Row/ Wiley, 1985.

Additional I.P. References

[1] Aguirre, G. and J. R. Wait, "Resistivity and I. P. response of a thin sheet," *Pure and Appl. Geophysics*, **123**, 882–892, 1985.

[2] Appario, A. and V. S. Sarma, "A modified pseudo-depth section as a tool in resistivity and I. P. prospecting," *Geophysical Research Bulletin (India)*, **19**, 187–208, 1981.

[3] Balasanyar, S. Yu., "The mechanism of I. P. of geological medium Izvestia," *Earth Physics*, **16**, 543–545, 1980, (an editorial on the "fallacy" of the common approach to the mechanism of I. P. by transplantation of the theories of physical chemistry into geophysics).

[4] Barnett, C. T., "Theoretical modelling of I. P. effects due to arbitrarily shaped bodies," *Ph.D. thesis,* Colorado School of Mines, Golden, Colorado, 1972.

[5] Belluigi, A., "The need of revising the apparatus employed in the measurement of the electrochemical effect of the ground," *Beitr. z. angew. Geophys.*, **5**, 169–177, 1936.

[6] Bertin, J. and J. Loeb, *Experimental and theoretical aspects of I. P.*, Gebrŏder Borntraeger, Berlin, 1976.

[7] Bleil, D. F., "I. P.: A method of geophysical surveying," *Geophysics,* **18**, 636–661, 1953.

[8] Buchheim, W., "Prinzipielles zur I. P.-Method mit Frequenz Variation," *Geophysik u. Geologie*, **No. 5**, 14–16, 1963.

[9] Buchheim, W. and G. Irmer, "Theory of induced galvanic polarization in solids with electrolyte-filled pores," *Gerlands Beitr. Geophys.,* **88**, 53–72, 1979.

[10] Chew, W. C. and P. N. Sen, "Dielectric enhancement due to electrochemical double layer: Thin double layer approximation," *Journal Chemical Physics,* **77**, 4683–4692, 1982.

[11] Collett, L. S., "Laboratory investigations at overvoltage," (i.e. I. P.) in *Overvoltage Research and Geophysical Applications*, (ed. by J. R. Wait), Pergamon, 1959.

[12] Daknov, V. N., M. G. Latishova, and V. A. Ryapolov, "Investigations of wells by the I. P. method," *The Log Analyst*, 46–82, Nov./Dec. 1967.

[13] Daniels, J. J., "3-D resistivity and I. P. modelling using buried electrodes," *Geophysics*, **42**, 1006–1019, 1977.

[14] DeWitt, G. W., "Parametric studies of I. P. spectra," *unpublished M.S. thesis*, University of Utah, 1978.

[15] Dias, C. A., "A non-grounded method for measuring induced electric polarization and conductivity," *Ph.D. thesis*, University of California, Berkeley, 1968.

[16] Dieter, K., N. R. Patterson, and F. S. Grant, "I. P. and resistivity type curves for 3-D bodies," *Geophysics,* **34**, 615–632, 1969.

[17] Edwards, L. S., "A modified pseudosection for resistivity and I. P.," *Geophysics,* **42**, 1020–1035, 1977.

[18] Eskola, L., E. Floranta, and R. Puranen, "A method for calculating I. P. anomalies with surface polarization," *Geophysical Prospecting,* **32**, 79–87, 1984.

[19] Fink, J., "Warburg Impedance — what it is and what it isn't" *S. E. G. 50th Annual Meeting,* Houston, November, 1980.

[20] Freedman, R. and J. P. Vogiatzis, "Theory of I. P. logging in a borehole," *Geophysics,* **57**, 1830–1849, 1986.

[21] Fricke, H. and S. Morse, "An experimental study of the electrical conductivity of disperse systems, I.," *Phys. Rev.,* **26**, 361–367, 1925.

[22] Fricke, H., "A mathematical treatment of the electrical conductivity and capacity of disperse systems, II," *Phys. Rev.* , **26**, 678–681, 1925.

[23] Fricke, H., "The electric conductivity and capacity of disperse systems," *Physics,* **1**, 106–115, 1931. (Deals with suspensions of biological cells whose interior is a well conducting salt solution but, at its surface, there is an insulating membrane. He discusses a postulated equivalent circuit and shows measured effective capacitance of the suspension at audio frequencies.)

[24] Fricke, H., "The Maxwell-Wagner dispersion in a suspension of ellipsoids," *J. Physical Chemistry,* **57**, 934–937, 1953.

[25] Grisseman, C., "Examination of the frequency-dependent conductivity of ore-containing rock or artificial models," *Sci. Report,* **No. 2**, Electronics Lab, University of Innsbruck, July 15, 1971. [U. S. Government Contract F-44620-71-C-007, NTIA Doc. AD 73131225.] (Employed "Wait model" to predict I. P. response of coated mineral grains with a Warburg interface impedance and showed reasonable agreement with laboratory data.)

[26] Grow, L. M., "I. P. for geophysical exploration," *Leading Edge,* (SEG publication), **1**, 55–56, 69–70, 1982.

[27] Gruszka, T. P. and J. R. Wait, "EM coupling "removal" from I. P.," *URSI Open Symposium on Wave Propagation,* Durham, New Hampshire, July 1986.

[28] Guptasarma, P., "Effect of surface polarization or resistivity modelling," *Geophysics,* **48**, 98–106, 1983. [Eqn. (1) seems invalid because it refers to a pure dielectric.]

[29] Hallof, P. G. and W. H. Pelton, "The removal of inductive coupling effects from spectral I. P. data," *WEG 50th Int'l. Meeting,* November 1980.

[30] Hohmann, G. H., P. P. Kintzinger, G. D. van Voorhis, and S. H. Ward, "Evaluation of the measurement of induced electric polarization with an inductive system," *Geophysics*, **35**, 901–915, 1970.

[31] Hohmann, G. W., "3-D I. P. and EM modelling," *Geophysics*, **40**,

309–324, 1975.

[32] Iliceto, V., A. Santarato, and S. Veronese, "An approach to the identification of fine sediments by I. P. laboratory measurements," *Geophysical Prospecting,* **30**, 331–347, 1982.

[33] Kan, R. and P. N. Sen, "Electrolytic conduction in periodic arrays of insulators with charges," *J. Chemical Physics*, **86**, 5748–5756, 1987.

[34] Karous, M., "I. P. anomaly above a sphere," *GeoExploration*, **21**, 49–63, 1983.

[35] Klein, J. K., T. Biegler, and M. D. Horne, "Mineral interfacial processes in the method of I. P.," *Geophysics*, **49**, 1105–1114, 1984. (Using the rotation electrode method, shows that Warburg impedance model is not valid for pyrite/electrolyte interface.)

[36] Lee, T., "The Cole-Cole model in time domain I. P.," *Geophysics,* **46**, 860–868, 1981.

[37] Levislaaya, T. S. M., "Dielectric relaxation in rock," *Izvestiya Earth Physics,* **20**, 777–780, 1984. (bridge measurements of ϵ and tan δ from 50 to 29,000 Hz. for NaCl saturated samples.)

[38] Lewis, R. G. and T. B. Lee, "The detection of I. P. with a transient EM system," *IEEE Trans.,* **GE-22**, 69–80, 1984.

[39] Major, J. and J. Silic, "Restrictions on the use of Cole-Cole dispersion models in complex resistivity interpretation," *Geophysics,* **46**, 916–931, 1981.

[40] Maxwell Garnett, J. C., "Colours in metal glasses and in metallic films," *Phil. Trans. Roy. Soc. A*, **203**, 385–420, 1904. (extends J. Clerk Maxwell's mixture formulae to optical scattering from particulates)

[41] McChadran, R. C. and D. R. McKenzie, "The conductivity of spheres I, the simple cubic lattice," *Proc. Roy. Soc. London,* **A-359**, 45–63, 1978. (gives corrections to Rayleigh's 1892's formulations)

[42] Losito, G. and G. Finizi-Cortini, "Laboratory instrumentation to study changes of electrical conductivity of rocks with changes of frequency, temperature and pressure," *Geophysical Prospecting,* **29**, 923–931, 1981.

[43] Madden, T. R. and D. J. Marshall, "I. P., a study of its causes," *Geophysics,* **24**, 790–816, 1959.

[44] Mahon, M. K., J. D. Redman, and D. W. Strangway, "Complex resistivity of synthetic sulphide bearing rock," *Geophysical Prospecting*, **34**, 743–761, 1986.

[45] Mandel, M., "The dielectric constant and Maxwell-Wagner dispersion of a suspension of oriented prolate spheroids," *Physics*, **27**, 827–840, 1961.

[46] Muller, M., "Ergebnisse geoelactrischer Polarisationsmessungen," *Zeits. f. Geophysik,* **16**, 274–284, 1940.

[47] Nabighian, M. S. and C. L. Elliot, "Negative I. P. effects from layered media," *Geophysics*, **41**, 1236–1255, 1976.

[48] Ogilivy, A. and E. N. Kuzmina, "Hydrogeologic and engineering — geologic possibilities of employing method of I. P.," *Geophysics,* **37**, 839–861, 1972.

[49] Olhoeft, G. R., "Electrical properties of natural clay permafrost," *Can. J. Earth Science,* **Vol. 14**, 16–24, 1977.

[50] Olhoeft, G. R., "Electrical properties of rocks, Chapter 9," in *Physical Properties of Rocks,* (ed. by Y. S. Youlovkiana and C. Y. Ho), McGraw-Hill, 1981.

[51] Parra, G. O., "Effects of pipelines on spectral I. P. surveys," *Geophysics*, **49**, 1979–1992, 1984.

[52] Patella, D. and D. Schiavone, "On the determination of the apparent frequency-effect in the frequency domain I. P. method," *GeoExploration*, **14**, 81–91, 1976.

[53] Patella, D., "Tutorial interpretation of MT measurements over an electrically dispersive 1-D earth," *Geophysical Prospecting*, **35**, 1–11, 1987.

[54] Pelton, W. H., L. Rijo, and C. H. Swift, Jr., "Inversion of 2-D resistivity and I. P. data," *Geophysics,* **43**, 788–803, 1978.

[55] Ramachandyan, M. and N. Sanyal, "EM coupling in IP measurements using some common electrode arrays over a uniform half-space," *GeoExploration,* **18**, 97–109, 1980.

[56] Rayleigh, Lord, (J. W. Strutt), "On the influence of obstacles arranged in rectangular order for the properties of a medium," *Phil.*

Magazine, **34**, 481–502, 1892.

[57] Schwan, H. P., G. Schwarz, J. Maczuk, and H. Pauly, "On the low frequency dielectric dispersion of colloidal particles in electrolyte solution," *J. Physical Chemistry,* **66**, 2626–2635, 1962.

[58] Schwarz, G., "A theory of the low frequency of dielectric dispersion," *Journal of Physical Chemistry,* **66**, 2636–2642, 1962.

[59] Scott, W. J. and G. F. West, "I. P. of synthetic, high-resistivity rocks containing disseminated sulphides," *Geophysics,* **34**, 87–100, 1969.

[60] Seigel, H. O. and A. W. Howland-Rose, "The magnetic I. P. method," Chapter 3, in *Developments in Geophysical Exploration Methods,* (ed. by A. A. Fitch), Appl. Sci. Publishers, Essex, 1982.

[61] Seigel, H. O., "The magnetic I. P. method," *Geophysics,* **39**, 321–339, 1974.

[62] Sidorov, V. A. and A. M. Yakin, "IP in rocks with inductive excitation," *Izvestia Earth Physics,* **15**, 810–814, 1979.

[63] Sihvola, A. and J. A. Kong, "Effective permittivity of dielectric mixtures," *Proc., URSI Commission F Symposium,* University of New Hampshire, Durham, July 1986.

[64] Snyder, Y. Y., K. H. Merkel, and J. F. Williams, "Complex formation resistivity — the forgotten half of the resistivity log: Soc. Prof. Well Log." *Analysts Meeting (18th Annual Symposium),* Houston, 1977.

[65] Stoyer, C. H., "Consequences of I. P. in MT interpretation," *Pure and Appl. Geophys.,* **114**, 435–449, 1976.

[66] Sumner, J. S., "The I. P. exploration method," in *Geophysics and Geochemistry in the search for Metallic Dyes,* (ed. by P. J. Hood). *Geological Survey of Canada Report,* 123–133, 1979.

[67] van Voorhsi, G. D., P. H. Nelson, and T. L. Drake, "Complex resistivity spectra of porphyry copper mineralization," *Geophysics,* **38**, 49–60, 1973.

[68] Vinegar, H. J. and M. H. Waxman, "I. P. of shaly sands," *Geophysics,* **49**, 1267–1287, 1984.

[69] Wait, J. R., H. O. Seigel, L. S. Collett, W. E. Bell, and A. A. Brant, "Method and apparatus for geophysical exploration," U.

S. Patent 2,766,421, 1956. (Application, No. 272,422, February, 1952.)

[70] Wait, J. R., "Relaxation phenomena and I. P.," *GeoExploration*, **Vol. 22**, 107–127, 1984.

[71] Wait, J. R. and P. Debroux, "I. P. in EM inductive schemes," *Geophysical Prospecting*, **32**, 1147–1154, 1984.

[72] Wait, J. R., "Extension to the phenomenological theory of I. P.," *IEEE Trans.*, **GE-24**, 409–414, 1986.

[73] Wait, J. R., "Physical model for the complex resistivity of the earth," *Electronic Letters*, **23**, 979–980, 1987.

[74] Wong, P. Z., "Fractal surfaces in porous media," in *Physics and Chemistry of Porous Media*, 300–318, (ed. by J. Benavar) A.I.P., New York, 1987.

[75] Wong, J., "An electrochemical model of the I. P. phenomenon in disseminated sulphide ores," *Geophysics*, **44**, 1245–1265, 1979.

[76] Wynn, J. C. and K. L. Zonge, "EM coupling, its intrinsic value, its removal and the cultural coupling problem," *Geophysics*, **40**, 831–850, 1975. [Unfortunate errors in equations (8), (9) and (12) which invalidate anisotropic model calculations.]

[77] Xiong, Z., Y. Luo, S. Wang, and G. Wu, "I. P. and EM modelling of a 3-D body buried in a 2-layered anisotrophic earth," *Geophysics*, **51**, 2235–2246, 1986.

[78] Zonge, K. L. and J. C. Wynn, "Recent advances and applications in complex resistivity measurements," *Geophysics*, **40**, 851–864, 1975.

Epilogue

This (survey) deals with the electromagnetic theory and the phenomenology of the complex resistivity of geophysical media. The phenomena are important in understanding the bases of subsurface probing for mineral and petroleum resources using electrical methods. An attempt is made to lead the reader from the basic potential theory to quasi-static concepts and culminating in comprehensive discussions of the prevailing macroscopic theories from the current literature.

The material presented has been used as lecture notes in first year graduate courses in geoelectromagnetism at the University of Arizona. The author has benefitted from many useful suggestions from students and interested colleagues in the geophysical community in the U.S., Canada and abroad.

While the topic selection may seem rather specialized, the results are relevant to other fields such as sub-surface telecommunications, earth conduction and corrosion, non-destructive testing, and hyperthermia. However most of the illustrative examples are shown in a geophysical context to permit an orderly flow of information.

I am grateful to the eminent Professor Jin Au Kong of M.I.T. for the invitation to publish this survey in his series on PIER. I must also thank Dr. Tom Gruszka and Peter Flanagan, former graduate students, for their input which is fully acknowledged in the text.

Most of the writing of this survey was done while I was a visiting professor at the University of British Columbia in Vancouver during the latter half of 1987. I am particularly appreciative here for the hospitality extended by Professors Ed Jull in Electrical Engineering and Doug Oldenburg in Geophysics.

Finally, I wish to thank the diligent Son Nghiem for numerous and useful suggestions.

James R. Wait
Regents Professor
University of Arizona

Mail address:
2210 East Waverly
Tucson, AZ 85719

2

SYNTHESIS OF OPTIMUM MICROWAVE ANTENNA APPLICATORS FOR USE IN TREATING DEEP LOCALIZED TUMORS

Carey M. Rappaport

2.1 Electromagnetic Power Deposition in Lossy Media and Hyperthermia

a. Introduction

Electromagnetic interaction with matter, though less widely studied than radiowave communication and guidance, has many important technological applications. Waves penetrate through media in well-understood ways, with sensitive phase and amplitude information changing depending on the size, shape, and electrical characteristics of obstacles encountered. Electromagnetic waves find use in nondestructive evaluation, remote sensing, tomographic imaging, and medical evaluation and treatment. Higher power electromagnetic fields are also useful for sealing in the plastics industry, for bulk drying and sterilizing, and for controlled medical heating. Radio-frequency diathermy devices are used for deep heating of injured or afflicted muscles. Perhaps the most challenging but beneficial use of medical heating is hyperthermia cancer treatment [1]. The optimized delivery of power below the surface of idealized lossy structures is examined in this chapter. Specifically addressed is the problem of determining the limitations and best strategies for non-invasively heating tumors without harming the surrounding healthy tissue.

Cancer is the second greatest cause of death in the developed world. About one out of every four persons will eventually contract the disease. Cancer takes many forms, from solid tumors in organs, bone, and muscle; to systemic neoplasias of blood and lymph; to superficial lesions such as melanoma of the skin. Survival rates vary greatly, depending on type of cancer and time of detection.

Currently, there are three primary means of treating cancer: surgery, radiation, and chemotherapy. Of the approximately 700,000 diagnosed life-threatening cases of the disease in 1977, 190,000 did not respond to any of these treatments and an additional 155,000 patients had a high risk of cancer recurrence [2].

Cancer surgery is the major means of treating localized, solid tumors. For surgical treatment of primary cancer, a macroscopic, recognizable tumor is resected along with a margin of surrounding tissue [3]. However, it has been estimated that about 70% of solid tumors will have micrometastases, or extensions into normal tissue [2], which often lead to recurrences of the disease.

Radiation therapy makes use of two main types of ionizing radiation: electromagnetic, including X-rays and γ-rays; and particle radiation including beta, alpha, neutron, proton, and negative π-meson [4,5]. Therapy can take the form of either bachytherapy, where a radioactive device is implanted in or near a tumor, or teletherapy, where a beam is externally aimed toward the tumor.

Radiation has its greatest effects on cells which are most rapidly reproducing. Normal tissues which reproduce to function, such as the skin, bone marrow, gastrointestinal mucosa, and reproductive organs, are most sensitive to radiation. Radiation damage to these tissues must be carefully considered prior to treatment.

Chemotherapy treats the systemic spread of the cancer using toxic drugs which harm both neoplastic and normal cells, with the hope that the former are more sensitive. Most of the chemotherapeutic drugs are primarily effective in killing dividing cancer cells as opposed to resting cells. Unfortunately, as tumors grow, a greater percentage of cancerous cells convert from actively dividing to a resting state, and thus avoid the cytotoxicity of the drugs. Methods to accelerate the onset of the dividing cycle improve the performance of chemotherapy.

The use of heat to treat cancer is a relatively new modality, still in its experimental stages despite knowledge of its benefits as early as Roman times [6]. The first published account of hyperthermia was in 1866: W. Busch reported the selective effect of heat on tumors over normal tissue with the cure of facial sarcomas following a high fever caused by a streptococcal skin infection [7].

Starting in 1971, several studies established therapeutic temperature data [8–10]. The minimum temperature for therapeutic benefit is most often cited to be 41.8° C [7]. Although this relatively low temperature kills tumor cells slowly, it may have a preferential effect on malignant cells, leaving normal cells unaffected [11].

In living humans, tumors tend to overheat more easily than normal surrounding tissue "because their neovascularity is physiologically unresponsive and cannot augment blood flow to adapt to thermal stress" [12]. Whereas the blood flow in a tumor will not change noticeably when heated to 43° C, it will increase 160–360% in the surrounding tissue [13,14]. And since the cooling effect due to blood perfusion is usually 4 times that of conduction to neighboring tissue (and may be as high as 100 times in certain structures, such as kidney), a large difference in perfusion rate would prevent a tumor from cooling as much as healthy tissue.

Hyperthermia is effective in conjunction with other conventional modalities of cancer treatment, particularly ionizing radiation [16–18]. There is an additive, or synergistic effect between heat and X-rays [18]. It takes two-thirds the radiation dose to achieve the same fraction of cell survival when hyperthermia is also used [6]. Heat tends to overcome radioresistance of some tumors, and may prevent the repair of sub-lethal damage or cause mis-repair of DNA [13,19]. Hyperthermia also appears to sensitize tumors to chemotherapeutic agents. Heating tissue by 4° C to the level of 41° C increases the tumor cell killing effect of the drug BCNU by 10,000 times [2].

b. Non-Invasive Generation of Hyperthermia

In most cases, it is preferable to generate heat within tissue without physically penetrating the body surface. Complicating effects such as anesthetizing, sterilizing, cutting, implanting a device, and then suturing the ill patient are serious drawbacks to surgical implantation. Also, a non-invasive applicator can be designed to surround the body part containing the tumor, taking advantage of constructive interference and focussing to concentrate more heat at the tumor than in the surrounding normal tissue.

The two broad classes of hyperthermia applicators are electromagnetic and ultrasound. Electromagnetic applicators radiate waves which propagate at the speed of light, 3×10^8 m/sec in vacuum, or slower in matter, and are characterized by both a propagation direction and a vector polarization. Ultrasound makes use of acoustic waves which cannot propagate in a vacuum and has a much slower velocity (about 1500 m/sec in soft biological tissue), and since it is a compressional wave, does not have a polarization attribute.

The entire spatial behavior of either type of wave is governed by the attenuation rate, α, and the propagation constant, β. A convenient representation in terms of the complex wave number is

$$k = \beta - j\alpha \qquad (1)$$

where $j = \sqrt{-1}$. Any spatial features, such as peaks or troughs, will extend for a distance on the order of a wavelength. A major distinction between electromagnetic and ultrasound radiation is the much greater resolution and focussing ability of the latter. Since the propagation velocity of ultrasound is 200,000 times slower than that of electromagnetic waves, the propagation constant β is much bigger for a given attenuation constant α. Thus spatial resolution and focussing is much greater for ultrasound wave then for electromagnetic waves.

Ultrasound, while having a much higher power peak resolution, faces considerable impedance mismatch at soft tissue boundaries with air or bone [20]. Air is relatively impenetrable, while reflections from bones produce large standing wave effects. Thus, areas such as the lung, gastrointestinal tract, pelvis, rib cage, and head are difficult to treat with ultrasound.

The frequency range used for electromagnetic heating is much broader than for ultrasound. Six frequencies are reserved for industrial and medical uses: 13.56, 27.12, 40.68, 433, 915, and the standard used in microwave ovens 2450 MHz. Although it is clear that with increasing frequency, penetration decreases, attenuation and resolution increase, and the ability to focus increases, the relative changes are quite nonlinear.

Different tissue responds in different ways to electromagnetic stimulation. Wave behavior is governed by the conductivity σ and the dielectric constant ϵ' of the medium. These parameters are often joined in a single expression for the complex dielectric permittivity

$$\epsilon = \epsilon_0(\epsilon' - j\epsilon'')$$

where

$$\epsilon'' = \frac{\sigma}{\omega\epsilon_0}$$
$$\epsilon_0 = 8.85 \times 10^{-12}\text{F/m} \qquad (2)$$

The values of ϵ' and σ are determined experimentally as a function of radian frequency ω. Tissue electrical parameters can be divided into two groups: high water content, including muscle, blood, and organs; and low water content, including fat and bone [21]. These two groups are generally referred to as the muscle or the fat tissue cases.

A plot of the most recently established values of ϵ' and σ [12], [22–28] is shown in Fig. 2.1.1. Dissipated power obeys the equation

$$P = \frac{\sigma}{2}|\overline{E}|^2 \qquad (3)$$

where $\overline{E}$ is the electric field vector. Thus, tissue with higher conductivity, for the same electric field, dissipates proportionally greater power.

The electrical and thermal properties of biological tissue *in vivo* in the frequency range 27 – 2450 MHz have been successfully modeled

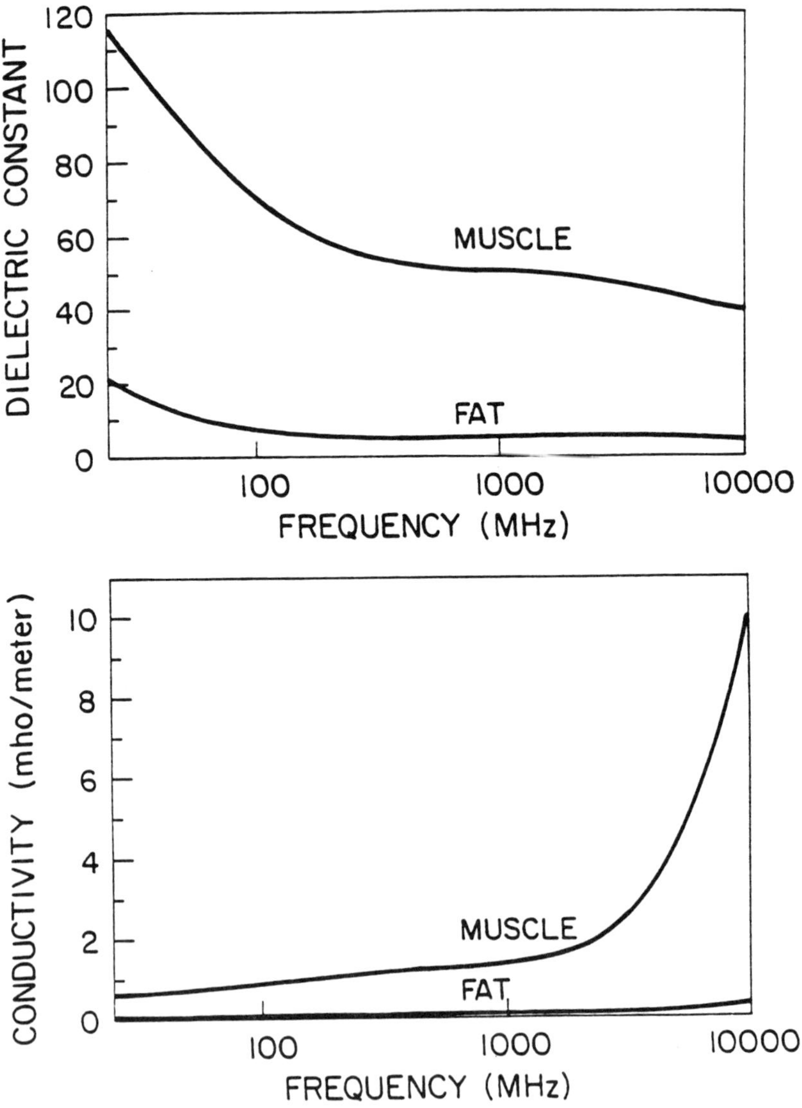

Figure 2.1.1 Electrical characteristics of high and low water content biological tissue as function of frequency (Data from Storm [12]).

with combinations of readily available substances. Guy [29] proposes a phantom recipe of 84.8% laminac polyester, 0.45% catalyst, 0.25% acetylene black, and 14.5% aluminum powder to simulate fat; and for a muscle phantom: 75.5% water, 0.9% salt, 15.2% polyethylene powder, and 8.4% TX–150 (a jelling agent). With slight proportion adjustments to compensate for different frequency variations, the resulting phantoms are very accurate dielectric models of living tissue.

From these values of the complex dielectric permittivity, the spatial parameters of the wave number of (1) can be determined. By definition, the dispersion relation for electromagnetic waves is

$$k = \omega\sqrt{\mu\epsilon} \tag{4}$$

where $\omega = 2\pi f$ is the radian frequency and μ is the magnetic permeability (taken in a biological tissue as the constant for free space, $4\pi \times 10^{-7}$ H/m). From (1) and (2), using the speed of light $c = 1/\sqrt{\mu_0\epsilon_0}$, and solving for the real and imaginary parts of the square of (4) yield

$$\beta = \frac{\omega}{c\sqrt{2}}\left[\sqrt{\epsilon'^2 + \left(\frac{\sigma}{\omega\epsilon_0}\right)^2} + \epsilon'\right]^{1/2}$$

$$\alpha = \frac{\omega}{c\sqrt{2}}\left[\sqrt{\epsilon'^2 + \left(\frac{\sigma}{\omega\epsilon_0}\right)^2} - \epsilon'\right]^{1/2} \tag{5}$$

Since $\lambda = 2\pi/\beta$, it is now possible to determine the muscle tissue wavelength for the frequencies of interest. Table 2.1.1 [30] lists the tissue wavelengths as well as the ratio α/β, which determines the amount of decay in each cycle. Also listed in the last two columns are maximum dimensions for heating idealized muscle volumes, which will be referred to in sections 2.3 and 2.4.

An important point is the matching of impedances from the applicator to tissue. Unless the waveguide is loaded with dielectric with the same complex permittivity as the tissue being heated, there will be an electrical impedance mismatch. The mismatch causes reflection which prevents the complete transfer of power into tissue and contributes strongly to stray radiation. The mismatch may be reduced by inserting a matching layer composed of lossless dielectric of intermediate constant with specific thickness between the applicator and the tissue. If the tissue were lossless, the ideal layer would be a one-quarter wavelength thick section of material with dielectric constant equal to the geometric mean of that of the tissue and applicator material. With actual lossy tissue, the thickness and dielectric constants are slightly different.

Table 2.1.1 Attenuation to propagation ratio, tissue wavelength, and maximum penetration slab thickness and diameter, as a function of frequency, for muscle tissue.

f (MHz)	α/β	WAVELENGTH IN TISSUE (cm)	SLAB THICKNESS	MAX D
27	.76	68	20.6	36
41	.73	51	16.2	29
100	.65	27	11.2	19
433	.40	8.8	5.6	11.5
750	.27	5.3	3.8	11.8
915	.23	4.5	3.4	12.6
1500	.20	2.8	2.2	9.6
2450	.17	1.8	1.5	7.7

This matching layer is often incorporated into a bolus. A bolus filled with a circulating liquid or high thermal conductivity powder would conform to irregular body contours and provide skin surface cooling.

The advantages and disadvantages of the various non-invasive methods of generating hyperthermia are summarized in Table 2.1.2 [21,13].

Table 2.1.2 Comparison of methods of using waves to generate hyperthermia.

Ultrasound

Advantages

Simple to use, fat not preferentially heated, does not interfere with temperature measurements, tumors absorb ultrasound better than normal tissue; no significant reflection from muscle/fat boundaries; precise localization possible; focussing depth as great as 15 cm

Disadvantages

Differential heating of bone; Coupling to uneven surface difficult; air/tissue interface causes reflections; inability to heat lung, abdomen; volume to be heated limited by size and number of transducers or scanning time

Electromagnetic, Capacitive

Advantages

Can design applicator for specific patient needs; can heat large volumes; skin easily cooled

Disadvantages

Skin and fat preferentially absorb power and heat; heating pattern hard to predict; metal temperature probe perturbs field, gives false readings

Electromagnetic, Inductive

Advantages

Simple coils; preferential heating of muscle; can heat large volumes; direct contact unnecessary.

Disadvantages

Weak tissue coupling, non-uniform field; heat not easily localized; metal temperature probes are corrupted

Electromagnetic Microwave

Advantages

Muscle tissue preferentially heated; can heat in presence of air, i.e., lung, stomach, bowel, rectum, pelvis; can heat near bones, brain; penetrates well in fat, high resolution focus; localization can be readily directed

Disadvantages

Limited penetration depth; shielded room may be needed to contain hazardous leakage radiation, focussing not useful for low frequencies; metal temperature probes are corrupted; potential danger to pacemakers; applicator may be awkward

c. Microwave Source Synthesis for Producing Constructive Interference at Depth

The goal of any hyperthermia source synthesis is simple to state: to generate a power pattern with more power in the vicinity of a focal target than anywhere else, and thus heat the tumor without overheating healthy tissue.

Attenuation rate varies directly (though nonlinearly) with frequency while field resolution decreases with increasing frequency. In human muscle tissue volumes, the "focus" is relatively wide and of low intensity. Generally, the sharpest focus — or highest resolution — corresponds to the highest possible frequency that deposits at least as much power at the tumor as at any point in the surrounding volume or on its surface. surface. Exceeding this frequency may produce higher resolution, but the actual safe maximum penetration depth into the tissue will be less.

More complicated than frequency selection is the determination of optimal source distribution. For constructive interference at a focus, electric field at the tissue surface must be properly aligned and phased so that waves propagating along all paths in the entire tissue volume arrive in the same fashion. While this simple condition is almost always met with planar applicators, cylindrical sources are often designed to produce a circumferential electric field, E_ϕ [31,32]. This choice of polarization is relatively easy to generate using a single turn, flat loop of conductor. Also, the circumferential polarization has the advantage of being transverse to tissue boundaries. However, if the aim is to produce power along the cylinder axis, the E_ϕ polarization is absolutely wrong. By intuitive reasoning, there can be no rotating field at the axis, $\rho = 0$, so E_ϕ, and hence dissipated power, vanishes at the axis.

Transverse polarization is preferable to normal (radial, in the case of a cylinder) polarization [33–35]. Capacitive electrodes, for example, are inefficient because the normal electric field heats subcutaneous fat. The normal electric field changes across a medium boundary in proportion to the ratio of the dielectric constants of the two media. Fat, with nine times less dielectric permittivity than muscle at 915 MHz, has 9 times the normal electric field. Transverse electric field, on the other hand, is continuous across media boundaries. According to (3), muscle tissue, with about 15 times the conductivity of fat, dissipates 15 times more power for the transverse polarization. But because of the squaring of electric field in (3), the normal polarization produces 81/15 as much power in the fat as in the muscle. Since tumors are usually found embedded in muscle tissue which in turn is surrounded by fat, it is much better to use tangential polarization which concentrates power on the muscle side of the muscle/fat boundary. The attenuation rate is much lower in fat than in muscle because of its lower conductivity. For typically thin subcutaneous fat layers, the power pattern is relatively constant. Furthermore, fat surrounds muscle, so keeping power at at reduced level in the fat reduces the overall problem to minimizing power maxima in a volume of muscle tissue.

Even though the thermal conductivity of fat tissue is relatively low, it still acts to cool a heated muscle tissue surface. Thus, for a first level analysis, as long as the electric field polarization is kept almost transverse to the muscle/fat boundary (and hence transverse to the skin), the fat layer can be ignored. Design considerations can thus be based on prescribing the source distribution on the muscle surface, assuming a constant temperature muscle surface boundary condition.

Merely adjusting phase, polarization, and amplitude for maximum focussing does not necessarily produce an acceptable power density distribution. Consider Fig. 2.1.2 with a planar source on the left, with phase adjusted to focus on the point labeled F, three tissue wavelengths into the lossy medium. That is, the phase at any source point is set to the negative of the physical distance from F to that point, times the tissue propagation constant β. This focussing is often referred to as a conjugate phase source distribution. Although F is the only point of coherence in the tissue, having significant gain over a uniform phase distribution as shown in Fig. 2.1.2b, the power there is still much lower than at points closer to the source, Fig. 2.1.2c, which experience less exponential wave decay. These plots were calculated for muscle tissue at 915 MHz, with electrical parameters taken from Storm [12], using 32-point Gauss-Legendre integration of the diffraction integral.

Several papers have dealt with this conjugate phase concept [36–39]. The frequencies used in their papers were all 433 MHz or above. The low electromagnetic frequencies would not be suitable for this focussing method because the source-to-focus path length variations

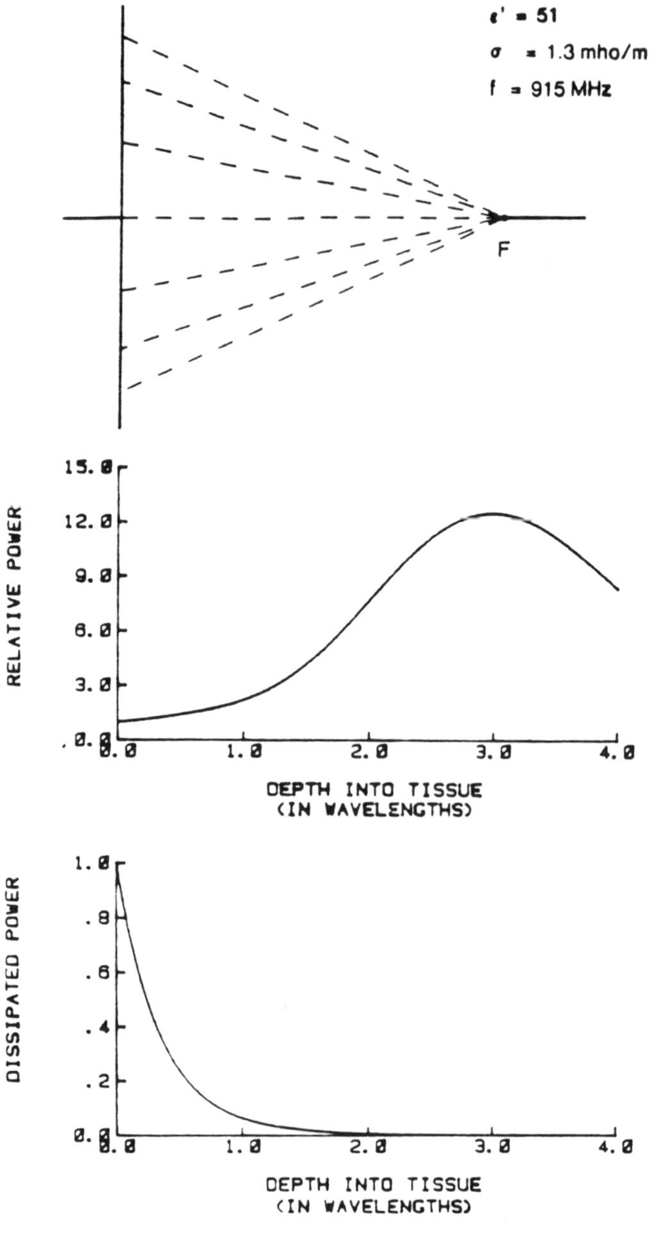

Figure 2.1.2 Focussing at depth: Fallacy of maximum intensity at focus in a lossy medium, (a) geometry, (b) relative power density, (c) power gain over uniform plane wave.

are less than a single tissue wavelength. Despite improvements of up to 9 dB at 2.7 wavelengths into muscle (6 cm) at 1500 MHz [39], the power is still 20dB below the power at the surface.

The conjugate method used by Gee et al., [37] is a useful proof-of-principle that irregular and inhomogeneous tissue samples can be illuminated with a focussed field almost as well as uniform tissue. The method involves measuring the field at the source points due to a test source placed at the focal point. The resulting phase represents the effective electrical path length through the irregular tissue volume. Unfortunately, it is difficult to place a test source in a tumor without facing the risks of invasive treatment outlined previously. Also, the method neglects the possibilities of intervening "hot-spots," or power maxima that may be generated at tissue discontinuities. The general idea, though, is a good one and should be studied further.

In view of the inability to use any frequency of planar electromagnetic radiation to produce a power maximum below the tissue surface, it is clear why this single surface applicator modality is currently used only to treat surface tumors. The solution to this dilemma is to surround the tissue with radio frequency and microwave sources. This can either be done with opposing planar sources [40], perpendicular planar sources [41], or with a cylindrical or even spherical source [42].

d. Planar Arrays

For two infinite, uniform, coherent planar sources, each facing an infinite slab of muscle tissue, the dissipated power yields a significant local maximum at the slab mid-plane. The electrical field distribution is a superposition of two counterpropagating plane waves, each represented by

$$E_{\pm} = \mathrm{Re}\left\{\hat{y}E_0 e^{\mp\alpha z \mp \beta z} e^{j\omega t}\right\} \tag{6}$$

where waves propagating in the positive z-direction follow the upper signs with a complex source value of E_0 at $z = 0$, and in the negative z-direction with the lower signs. Solving for power according to equation (3) gives:

$$P = \frac{\sigma {E_0}^2}{2}(\sinh^2 \alpha z + \cos^2 \beta z) \tag{7}$$

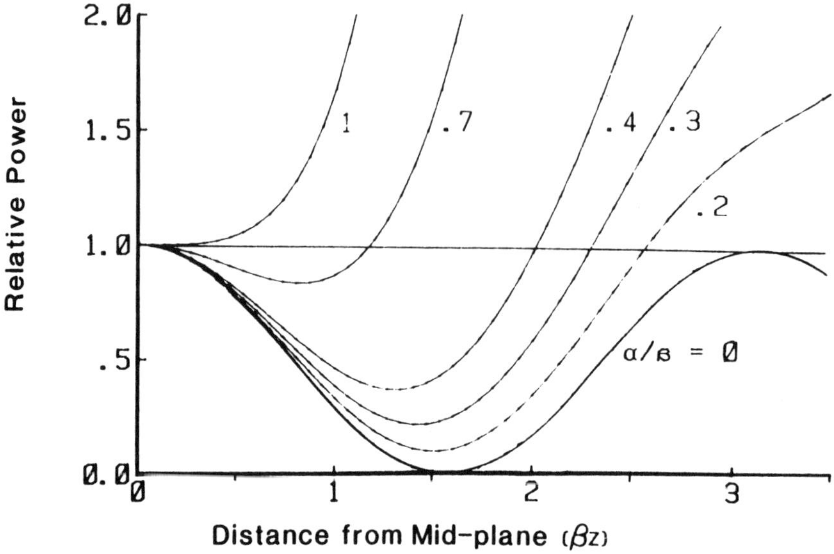

Figure 2.1.3 EM power density distribution normalized to the mid-plane density of an infinite planar slab of thickness $2z$, with uniform coherent source excitation, for typical values of α/β.

The power pattern is depicted in Fig. 2.1.3, where the curve for each value of the ratio α/β is plotted as a function of electrical length, βz, from the mid-plane ($\beta z = 0$) out, and normalized to the power at the mid-plane. The requirement of equal power at the mid-plane and the surface determines the greatest uniform source separation distance. This distance ranges from more than $\beta z = 3.14$ for the lossless case to $\beta z = 1.2$ for $\alpha/\beta = 0.7$. This latter case, $\alpha/\beta = 0.7$, corresponds to a layer of thickness $2.4\lambda/2\pi$. These are universal curves in that frequency and physical dimensions have not yet been specified. The α/β ratios listed in Table 2.1.1 can now be used to determine the appropriate curve in Fig. 2.1.3 to represent the power pattern in actual muscle tissue. The wavelength must also be specified to determine the proper physical scale of slab thickness.

The constructive interference of two opposing uniform sources gives power patterns which are much better than the single source arrangements represented by Fig. 2.1.2. These sources do not make

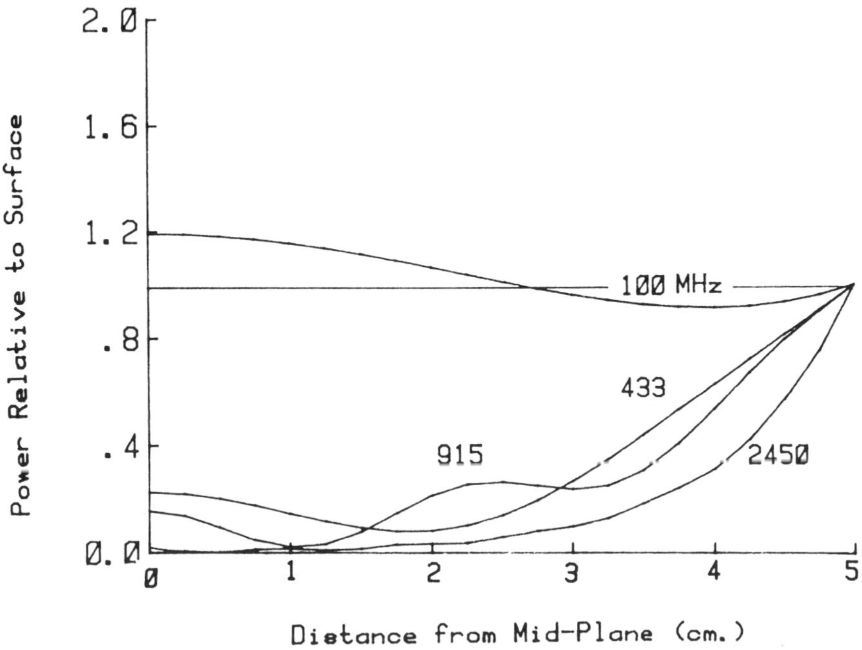

Figure 2.1.4 Power patterns for one half of a 10 cm thick planar slab of muscle tissue, normalized to the power at the surface.

use of phase adjustments to focus on a particular target; rather, they simply produce standing wave maxima in the lossy slab. These curves also indicate the very limited flexibility of constructive interference.

Figure 2.1.4 shows several of these same curves normalized to a surface 5 cm from the mid-plane with a physical distance scale. A 915 MHz arrangement could only be used with slabs less than 3.4 cm thick. For a slab 10 cm thick even the 433 MHz frequency is too high to safely heat the center.

The best method for designing opposing planar arrays is to start with the assumption that power on the tissue surface should be uniform and equal to the maximum tolerable level. Because of the strong exponential decay in muscle tissue, the source surfaces tend to have the greatest deposited power. When this level is the same as that at the focus, the tissue slab thickness is maximum.

e. Circular Cylindrical Arrays

Circular geometry admits wave equation solutions in terms of Bessel functions. The field due to a uniform, axially polarized electric field source on the surface of an infinite cylinder, as depicted in Fig. 2.1.5, is represented by

$$E(\rho) = \hat{z}\,\mathrm{Re}\left\{E_0 I_0[(\alpha + j\beta)\rho]e^{j\omega t}\right\} \tag{8}$$

where I_0 is the zeroth-order modified Bessel function, which grows with increasing ρ. This equation corresponds to (6) from the Cartesian geometry, with polarization changed to align with the z-axis. Figure 2.1.6 is the corresponding set of power patterns, similar to Fig. 2.1.3, but with up to four times the maximum allowable penetration thickness. The curves are spread apart more for low α/β values than with the planar case. This is due to the enhanced phase effects of higher frequency waves. Just as in the case with constructively interfering planar source, the larger wavelengths do not interfere to the same degree over the tissue volume as do smaller wavelengths. At 915 MHz, a uniform infinite cylinder of muscle tissue 6.3 cm in radius can have as much power deposited on its axis as on its surface.

The curves of Fig. 2.1.6 are useful for visualizing overall pattern resolution and maximum cylinder radius determination. For a more practical, physical picture, an unnormalized radial scale must be used. Figure 2.1.7 shows the deposited power patterns within a muscle tissue cylinder of typical adult torso radius of 14 cm for various frequencies. The power is normalized to that at the cylinder surface. Note that the highest frequency depicted, 433 MHz, has insignificantly little power at the axis, $\rho = 0$. For a cylinder as large as 14 cm, the higher frequencies simply cannot be used. Instead, resolution must be sacrificed in favor of penetration, and the 41 MHz frequency choice is optimal. It is not necessary to use a frequency as low as 27 MHz. For this choice, the power level is higher at the axis than at 14 cm. While the lower frequencies cannot provide for a high resolution narrow spot at the tumor, they can still be used for a uniform baseline heating, with another modality, such as ultrasound, used to heat the tumor a few remaining degrees into the therapeutic range.

There is much more literature on pattern synthesis in the cylindrical geometry than with the planar geometry [43–46]. Actual phase and

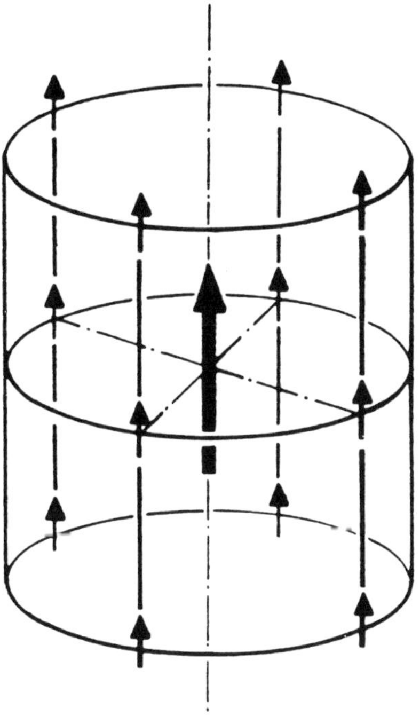

Figure 2.1.5 Electromagnetic polarization alignment for cylindrical source focussing.

amplitude optimization for focussing at various positions throughout the cylinder is also examined [47–49]. Arcangeli et al. [48] and Morita et al. [49] both use a method of optimizing a weighted sum error function. In each case, the tissue volume is assumed to be two-dimensional. The calculated and desired intensity values are determined for several sample points throughout the tissue volume and this error is minimized by least squares. Archangeli then repeats the process by relaxing the limits of the desired power pattern iteratively until the error is minimized. Although this method yields good results, great care must be used in choosing an appropriate desired pattern and sampling grid. An overly optimistic pattern may not be realizable, while a poorly specified grid may either miss fine pattern details or be computationally excessive. Another difficulty is that iterative solutions may not converge, or may converge to a non-physical solution. The Morita paper uses a simplified model of the human torso cross section for their synthesis. Using only uniform muscle, a circular bone region and several

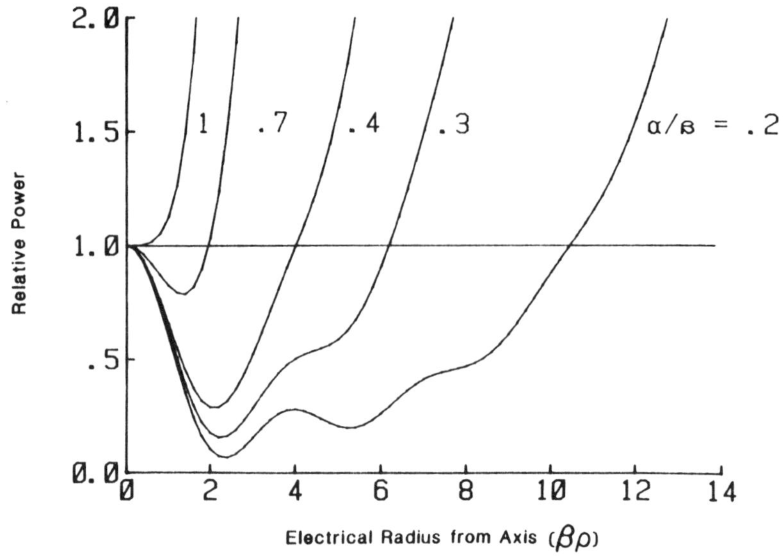

Figure 2.1.6 EM power density distribution normalized to axis density for cylinder of radius ρ with uniform source excitation for typical values of α/β.

circular air-filled bowel regions, with 41 MHz radiation, their derived power patterns are fairly smooth. The authors conclude that because of the transverse polarization of the electric field, the inhomogeneities do not significantly effect the pattern. This important conclusion implies that it may be more useful to optimize the sources for a uniform tissue volume first, and then perturb the results to accommodate the fine structural inhomogeneities.

Arcangeli [48] takes the opposite view by breaking the torso cross section up into discrete regions, each with a representative complex dielectric constant. Using 915 MHz demands a much finer sampling grid, since the sample points must be less than one-half wavelength apart, and a body structure may have a width of 15 wavelengths. The computational expense is much higher for higher frequencies. Although the method is rather sophisticated, the results show focussing very near the surface, with a derived excitation of only 8 of the 50 discrete sources. In other words, the full constructive interference effect of the full cylindrical aperture is ignored in favor of a 60° sector near the

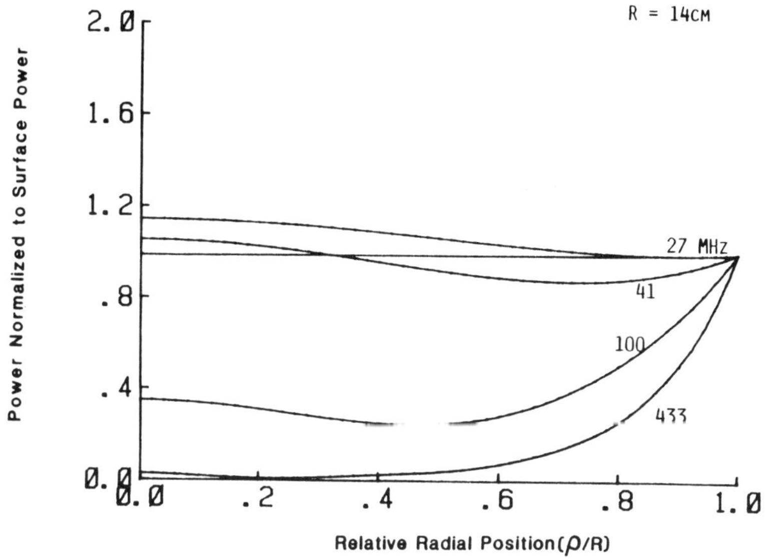

Figure 2.1.7 Power patterns for a 14 cm radius cylinder of muscle tissue normalized to the power at the surface.

region of greatest curvature.

The concept of three dimensional focussing with a cylindrical geometry is fairly new. In a paper by Jouvie et al. [50], the source is focussed in both ϕ- and z-directions. The polarization choice is E_ϕ, lowering the possible efficiency compared to E_z and only a small sector of the neck is irradiated, but their results show a considerable improvement over the uniform source case.

Much work has been done on the Annular Phased Array (APA) system [47], [51–57]. Perhaps the most thoughtfully designed and clinically tested applicator system, the APA uses the entire cylindrical surface surrounding a tissue volume, maintains an axial (z-directed) polarization, and has the capability of rudimentary phase and amplitude adjustment. The APA consists of one or two sets of eight dielectrically loaded, flared, parallel plate TEM waveguides, each facing the cylindrical tissue volume. The edge fringing effects of the TEM wave launchers and the eight plane-wave approximations to a true circular source yield a power pattern slightly worse than the ideal represented by Fig. 2.1.6.

The APA's rudimentary phase control and low frequency operation make it less useful for steering the point of maximum interference throughout the tissue volume. However, it is significantly more flexible than the helical coil applicator [54] which cannot be adjusted after construction.

One more issue not previously considered with the APA system is that of focussing in the axial direction. For maximum power on axis, circumferential distribution is set at uniform phase and amplitude. Adjustments in the axial variation take advantage of the only remaining degree of freedom. This variation is conceptually similar to the planar focussing in the previous section. Amplitude at the source is held constant and the phase is adjusted to compensate for both the electrical path length to the focal point and the complex propagation constant of the medium. The details of this optimization are mathematically presented in section 2.4.

2.2 Optimal Source Distribution for Spherical Tissue Geometry

In section 2.1, it was shown that using various schemes to allow waves to interfere constructively produce much better results than merely radiating from a single, unfocussed applicator. Specifically, surrounding a volume of tissue with sources produced localized maxima in the center. The question arises as to the maximum size of any volume for which as much power can be deposited in the center as elsewhere. This chapter answers this question and presents the optimum source distribution function. It will be demonstrated that the optimum geometry for power generation at depth is a sphere, and that with proper focussing, the maximum penetration depth represents a fundamental heating limit.

In contrast to several previously reported studies, the current examination considers the synthesis of the particular best distribution. A large amount of work has been done analyzing the power pattern in a sphere illuminated by a plane wave [58–60]. For this scattering problem, the deposition pattern is unsymmetrical, since power is incident from only one side. Placing sources on all sides — in effect surrounding the sphere with sources — makes use of the entire aperture and therefore provides the greatest constructive interference at the center. The symmetric distribution synthesized has a much greater focal-point-to-surface distance than the plane wave pattern. The optimal synthesis problem, though conceptually more complex, is less computationally difficult than the plane wave scattering analysis, which is usually solved using numerical approximations. Since the former specifies a source polarization, phase and amplitude distribution on the sphere surface, it is not easily analyzed using ray paths or wave fronts. However, since the solution is ultimately determined in terms of spherical harmonics, it is specified in closed form.

There are relatively few biological structures which can be modelled by the spherical geometry. The head (with adequate eye protection) and the breast, making use of the higher conductivity muscle "ground plane" backing layer, are possible candidates. Despite the limited applications, knowledge of the absolute maximum heating depth provides a rule of thumb which specific microwave applicator designs might strive to approach, but never exceed.

a. Lossy Sphere Field Solutions

Since attenuation in dissipative media is roughly exponential with depth, the volume geometry most suitable for maximum constructive interference is one which minimizes the distance from a focal point to the source on its surface. By this criterion, a sphere is optimum, since any other volume can always be inscribed by a sphere. Thus, given the minimum focal point to surface distance R and the maximum tolerable surface power, a sphere of radius R will yield the greatest power maximum at that focal point, the origin.

The polarization of the source on the surface of the sphere is another important characteristic. As shown in Fig. 2.2.1, it must be parallel to some axis, say the z-axis. By symmetry, these z-directed source components are independent of circulating angle (ϕ), and have polar angle (θ) dependence which is symmetric about the equator $(\theta = \pi/2)$. Any additional radially symmetric (r) polarization component cancels itself at the center, whereas any circulating component vanishes there. An additional polar angle component can be decomposed into an antisymmetric radial component and a z-directed component for overall power pattern imrovement, and may be a desirable

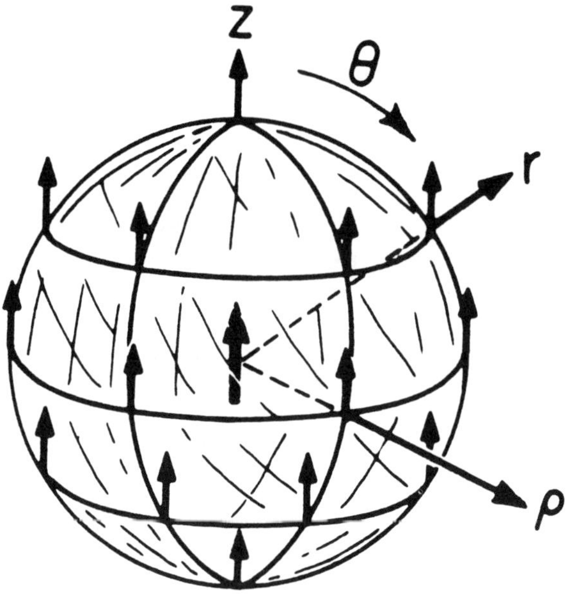

Figure 2.2.1 Currents polarized in the z-direction on the surface of a sphere, and the resulting maximum constructive interference of electric field at the center.

component, but cannot be easily used for preliminary source synthesis. If possible, the first step for a source design would be to maintain polarization parallel to the z-axis.

Specifying the optimal phases and amplitudes of these source components is the next step in this derivation. Contrary to established intuition, the best focussed source distribution is not uniform amplitude and phase, as it would be for an acoustic compression wave. To examine the deficiencies of this uniform distribution, we first solve the surface current diffraction integral and then improve this solution using a modal analysis to yield the optimum distribution.

The uniform z polarized surface current at a radius R is represented as

$$\overline{J}(\overline{r}') = \delta(r - R)\hat{z}$$

where $\delta(r)$ is the three-dimensional spatial impulse function. This current is inserted in the dyadic Green's function diffraction formula [61]

$$\overline{E} = -j\omega\left(\overline{\overline{I}} + \frac{1}{k^2}\nabla\nabla\right)\cdot\overline{A}(r) \tag{1}$$

$$\overline{A}(r) = \mu\int_{V'} dv' \frac{e^{-jk|\overline{r}-\overline{r}'|}}{4\pi|\overline{r}-\overline{r}'|}\overline{J}(\overline{r}') \tag{2}$$

where the left hand side of (1) is the vector electric field at the observer point $\overline{r}$, $\overline{A}(\overline{r})$ is the vector potential, ω is radian frequency, $\overline{\overline{I}}$ is the identity dyadic, $k = \beta - j\alpha$ is the complex wave number, and $|\overline{r}-\overline{r}'|$ is the observer-to-integration-point distance. The law of cosines is used to represent $|\overline{r}-\overline{r}'|$, where $\overline{r}$ is assumed to lie on the z-axis. This assumption is justified since the source is uniform and symmetric and the only requirement on polarization is that it is parallel to any arbitrary axis. That is , the vector potential $\overline{A}$ is independent of observer coordinates θ and ϕ. As would be the case with a scalar wave, $\overline{A}$ represents a family of concentric, spherical equipotentials-potentials. The source polarization enters only as a linear multiplier to this scalar function. The integral in (2) becomes a surface integral

$$\overline{A}(r) = \mu\int_0^{\pi} d\theta' \int_0^{2\pi} d\phi' R^2 \sin\theta' \frac{e^{-jk\sqrt{R^2+r^2-2rR\cos\theta'}}}{4\pi\sqrt{R^2+r^2-2rR\cos\theta'}}\hat{z} \tag{3}$$

which is readily integrated, giving

$$\overline{A} = \hat{z}\frac{\mu}{2}\left(\frac{R}{-jkr}\right)\left(e^{-jk|R+r|} - e^{-jk|R-r|}\right) \tag{4}$$

Inside the sphere, $r < R$, the vector potential can be represented by a spherical Bessel function of radial position, $j_0(kr)$,

$$\begin{aligned}\overline{A}(r) &= \hat{z}\mu\,\Re\left\{e^{-jkR}\frac{\sin kr}{kr}\right\} \\ &= \hat{z}\mu\,\Re\left\{e^{-jkR}j_0(kr)\right\}\end{aligned} \tag{5}$$

It is at this point that the vector nature of the electromagnetic field presents itself. Substituting this expression back into (1) and taking derivatives in cylindrical coordinates, (ρ, ϕ, z), give

$$\overline{E}(z,\rho) = -j\omega\mu\,\Re\left\{e^{-jkR}\left\{\hat{z}\left[\left(\frac{\rho}{r}\right)^2\left(j_0(kr) - \frac{j_1(kr)}{kr}\right) + \left(\frac{z}{r}\right)^2\left(2\frac{j_1(kr)}{kr}\right)\right] + \hat{\rho}\left(\frac{z\rho}{r^2}\right)\left[-j_0(kr) + 3\frac{j_1(kr)}{kr}\right]\right\}\right\} \quad (6)$$

with

$$r = \sqrt{z^2 + \rho^2}.$$

Several features of (6) are worth mentioning. First, there is the addition of the ρ-component, and the z-component has three terms instead of just one (as might be expected for a scalar field). Second, for small values of radius, since $j_0(kr) \sim 1$ and $j_1(kr) \sim kr/3$ [62], the electric field simply reduces to

$$\begin{aligned}\overline{E}(z,\rho) &\approx -\hat{z}j\omega\mu\,\Re\left\{\frac{2}{3}e^{-jkR}\right\} \\ &\approx \hat{z}\left(-j\frac{2}{3}\right)\eta k\,\Re\left\{e^{-jkR}\right\}, \quad kr \ll 1 \qquad (7)\end{aligned}$$

Next, for large values of radius $|j_n(kr)| \sim |\sin(kr + \frac{n\pi}{2})/kr|$ [63], so (6) becomes

$$\overline{E}(z,\rho) \sim -j\eta k\,\Re\left\{e^{-jkR}j_0(kr)\frac{\rho}{r}\left(\hat{z}\frac{\rho}{r} - \hat{\rho}\frac{z}{r}\right)\right\}, \qquad kr \gg 1 \quad (8)$$

Note that in (8), there is an appreciable E_ρ component away from the equator. Thus, even though the surface currents are polarized parallel to the z-axis, it is recognized that the resulting E-field for large radii is primarily polarized in the $\hat{\theta}$-direction. Equation (8) is no longer valid near the poles where $\rho \approx 0$. Reconsidering (6), again for small ρ gives

$$\overline{E}(z,\rho) \sim -j\eta k\,\Re\left\{e^{-jkR}\left(\hat{z}\,2\frac{j_1(kr)}{kr} - \hat{\rho}\,k\rho\frac{j_0(kr)}{kr}\right)\right\}, \quad \rho \ll z \quad (9)$$

Equation (9) indicates that there exists normal z-directed electric field components at the poles, but that these are $1/kr$ times smaller than the tangential field near the equator.

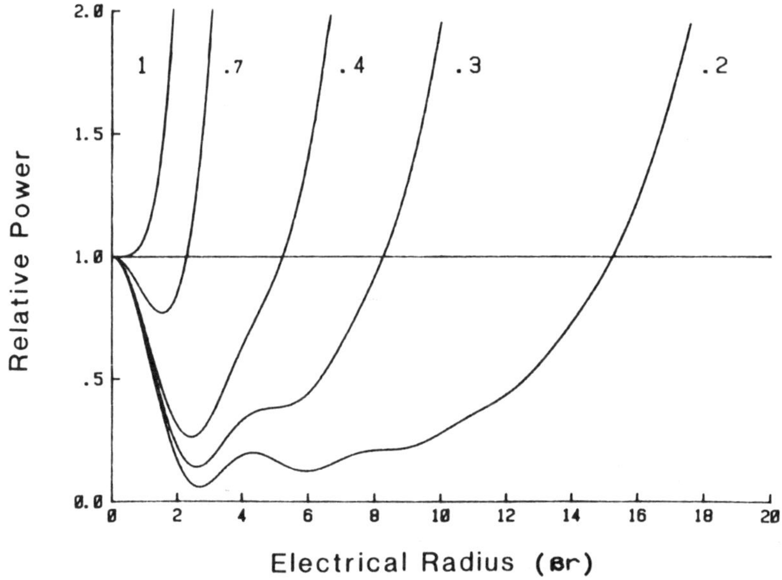

Figure 2.2.2 EM power density distribution normalized to axis density for a sphere of radius r for typical values of α/β.

The dissipated power in a medium of conductivity σ is $\sigma|E|^2/2$. From (8), this is

$$P(z,r) = \frac{\sigma\eta^2}{2}|kr|^2 e^{-2\alpha R}|j_0(kr)|^2 \left(\frac{\rho}{r}\right)^2 \tag{10}$$

which is maximized at $z = 0$, $\rho = r$, at the equator. The exact power deposition function, based on (6) normalized to the power at the origin, is computed and plotted as a function of radius at $z = 0$ in Fig. 2.2.2. These curves are of the same type as the previous universal dimensionless curves. Figure 2.2.3 plots the same set of data on a centimeter scale for several standard frequency values. It is the intersection of each curve with unity which determines the maximum allowable sphere radius. As expected, the lowest frequency curve has a greater penetration depth due to its larger physical wavelength; whereas the higher frequency patterns have much greater resolution with a rather narrow peak at the origin for 2450 MHz. An interesting anomaly exists for 915 MHz, where the penetration depth is actually greater than for 433 MHz. This

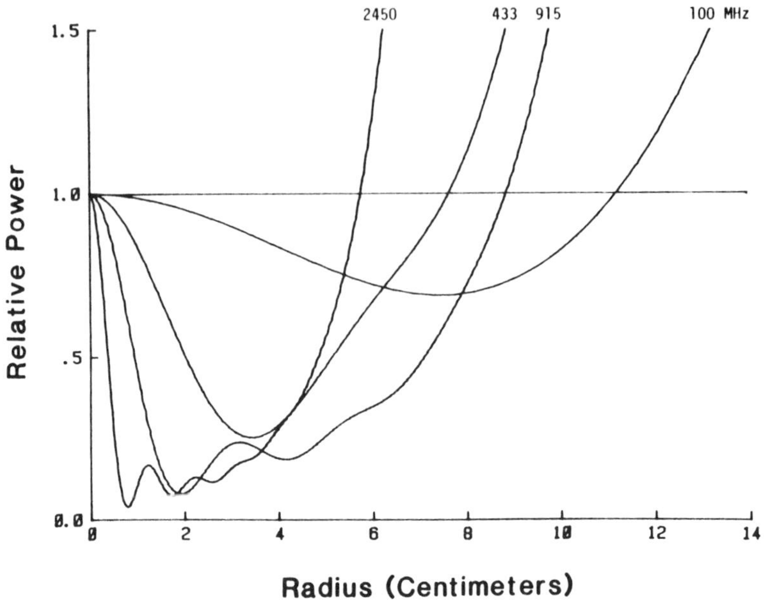

Figure 2.2.3 Power in a sphere of muscle tissue as a function of radius at $\theta = \pi/2$ for four standard hyperthermia frequencies: Uniform current distribution.

is due to the more prominent constructive and destructive phase effects for the higher frequency.

For a given (large) radius, the power varies as $\sin^2\theta$. Arising from the $\sin\theta$ dependence of the infinitesimal element of spherical area, this indicates a much higher power deposition at the equator than at regions near the poles. This can be understood intuitively by considering the limiting cases of planar current distributions on the surface of a sphere of infinite radius. The equatorial positions degenerate to tangential planes of current, while the polar positions degenerate to planes of normal currents, which do not radiate.

The $\sin^2\theta$ non-uniform surface power distribution is corrected using modal analysis in the next section to yield the optimum vector wave solution to the power deposition problem.

To test the error incurred with the assumption of large radius made for (8) for the worst case, 433 MHZ, where the maximum radius is approximately $.9\,\lambda$, $|kr| = 6.3$, the neglected power, as given by computing the squared magnitude of the electric field in (9), is of the order of 4% at the pole. For 915 MHz, this error is only about 1%.

b. Modal Analysis

Solutions to the wave equation for spherical geometry are harmonics of index n, represented by [64]

$$\overline{E}(r) = \sum_{n=1}^{\infty} \left\{ \hat{r} A_n \frac{n(n+1)}{kr} j_n(kr) P_n(\cos\theta) + \hat{\theta} A_n \left[j_{n-1}(kr) - \frac{n}{kr} j_n(kr) \right] \right.$$
$$\left. \cdot [n \cos\theta P_n(\cos\theta) - n P_{n-1}(\cos\theta)] \left(\frac{1}{\sin\theta} \right) \right\} \quad (11)$$

where P_n is the Legendre polynomial of order n, and A_n is the constant of the nth harmonic. By setting n equal to 1, $A_1 = -j\omega\mu Re^{-jkR}$, and recalling that $P_0(\cos\theta) = 1$ and $P_1(\cos\theta) = \cos\theta$, it is evident that (6) is the first harmonic spherical wave solution as given by (11).

It is the fundamental point of this analysis that for higher order harmonics, $n = 2, 3, 4 \cdots$, the E-field at $r = 0$ vanishes. Since $j_n(kr)$ varies as $(kr)^n$ for small r, these spherical Bessel functions approach zero at the origin. Only the first harmonic, with $j_0(kr)$ and $j_1(kr)/kr$ factors, remains non-zero there, taking the form of (7). This property will now be used to reduce undesirably high power levels at the surface of the sphere without affecting the power at the origin. Specifically, a distribution can be synthesized from these higher order modes with appropriately chosen phase and amplitude to partially cancel the first order mode field at the surface, near the equator, and thereby increase the maximum sphere size.

Once again, for large radius values, the θ-component of E-field is the dominant contribution to the dissipated power function. The θ-dependence of the nth mode of E_θ can be expressed as

$$E_{\theta n}(\theta) \sim \hat{\theta} \frac{\partial P_n(\cos\theta)}{\partial\theta} \quad (12)$$

Since for even order n, $P_n(\cos\theta)$ is an even function of $\cos\theta$, its derivative will be an odd function of $\cos\theta$, with a total value of zero at $\theta = \pi/2$. Since the goal is to find the harmonics with non-zero values at the equator to be subtracted from the $\sin\theta$ dependence of the first harmonic. Thus only odd order harmonics need be considered for this optimization.

The derivatives of Legendre polynomials comprising the first three odd harmonics are, from (12)

$$\begin{aligned} E_{\theta 1} &\sim -\sin\theta \\ E_{\theta 3} &\sim -\frac{15}{2}\cos^2\theta\sin\theta + \frac{3}{2}\sin\theta \\ E_{\theta 5} &\sim -\frac{315}{8}\cos^4\theta\sin\theta + \frac{105}{4}\cos^2\theta\sin\theta - \frac{15}{8}\sin\theta \end{aligned} \tag{13}$$

with values -1, $3/2$, and $-15/8$, respectively, at $\theta = \pi/2$.

The three equations in (13) can be rewritten in terms of sines of odd multiples of θ

$$\begin{aligned} E_{\theta 1} &\sim -\sin\theta \\ E_{\theta 3} &\sim -\frac{15}{8}\sin 3\theta - \frac{3}{8}\sin\theta \\ E_{\theta 5} &\sim -\frac{315}{128}\sin 5\theta - \frac{105}{128}\sin 3\theta - \frac{30}{128}\sin\theta \end{aligned} \tag{14}$$

The θ-dependence of the sum of the three harmonics from (10) may now be represented by

$$f(\theta) = \sin\theta + B_1\sin 3\theta + B_2\sin 5\theta \tag{15}$$

where B_1 and B_2 are selected to minimize the maximum value of power, $[f(\theta)]^2$, over the domain $0 \le \theta \le \pi$. Solving this transcendental formula interactively yields $B_1 = .2365$, and $B_2 = .0640$. The resulting surface power variation has three equal peaks of .685 at $\theta = .294\pi$, $.5\pi$, and $.706\pi$.

Note that a standard infinite Fourier series composed of higher odd harmonics of the fundamental $\sin\theta$ can produce a square pulse of height $\pi/4$, giving a power (amplitude squared) of .617. However, for any finite sum of Fourier terms, there is always a Gibbs Phenomena at the step edge increasing the amplitude by a factor of about 1.09. This would increase the maximum power to .733, which is greater than the previously derived maximum.

What remains is the derivation of the actual mode amplitudes and phases which multiply each harmonic so that their sum equal $f(\theta)$. Define $\mathcal{J}_n = A_n[j_{n-1}(kR) - (n/kR)j_n(kR)]$ as the multiplicative constant of E_θ from (11) for a given sphere radius R. Then, combining the three equations (14) into the form of (15) gives the system:

$$\mathcal{J}_1(-1) + \mathcal{J}_3\left(-\frac{3}{8}\right) + \mathcal{J}_5\left(-\frac{30}{128}\right) = 1$$
$$\mathcal{J}_3\left(-\frac{15}{8}\right) + \mathcal{J}_5\left(-\frac{105}{128}\right) = B_1$$
$$\mathcal{J}_5\left(-\frac{315}{128}\right) = B_2. \qquad (16)$$

The solution is $\mathcal{J}_1 = -.9509$, $\mathcal{J}_3 = -.1148$, and $\mathcal{J}_5 = -.0260$.

Although the third and fifth order modes do not contribute to power at the origin of the sphere, the first order mode is slightly reduced in order to compensate for the small $\sin\theta$ contributions in these higher modes. Thus, the amplitude in the center is reduced to $-.9509$ of the uniform current value. Renormalizing the coefficients to $\mathcal{J}_1' = 1$, $\mathcal{J}_3' = .1207$, and $\mathcal{J}_5' = .0288$ compensates for this reduction. The surface power is plotted in Fig. 2.2.4, along with the uniform current power. The peak power level is reduced to .757 of the single mode power. The magnitude of the ripple is about $\pm 1.5\%$. Addition of higher order terms would reduce the ripple and lower the maximum surface power slightly. The maximum improvement would be less than 2%, and since from Fig. 2.2.2, the increase in maximum allowable radius varies approximately logarithmically with power, the minute increase in radius does not justify the additional computational complexity required for seventh and higher order harmonies.

Finally, the coefficients A_n from (11) are determined from the $\mathcal{J}_n'$ constants and the magnitudes and phases of the spherical Bessel functions at the unity crossing points of the power pattern for each frequency. Since the pattern changes for each choice of the A_n set, this is an iterative process. For 915 MHz, $A_1 = 1.5e^{j(.6368)}$, $A_2 = 1.364\,e^{j(-2.8608)}$, and $A_3 = 1.136\,e^{j(-.3618)}$, determined at $\beta R = -13.2$, or $R = 9.45$ cm.

The power patterns at $\theta = \pi/2$, for the same frequencies as before are plotted in Fig. 2.2.5. Note the increases in maximum allowable

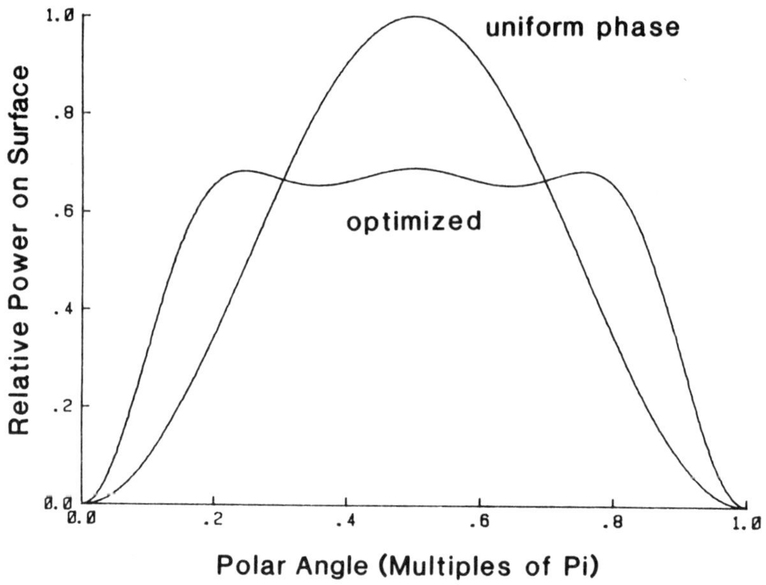

Figure 2.2.4 Surface power as a function of θ for single mode (uniform current) and three modes (approximate uniform power) distributions.

radius, as indicated by the intersection points for the uniform current distribution, taken from Fig. 2.2.2. For 100, 433, 915, and 2450 MHz, the increases amount to 1.72, 0.84, 0.57, and 0.32 cm., respectively.

From (11), it can be seen that for the large radius dominant E_θ term, the radial and polar angle dependencies are separable. Thus, the $\hat{\theta}$-directed power behavior always has the same shape as Fig. 2.2.5 scaled by the θ-dependence of Fig. 2.2.4. Figure 2.2.6 plots the same power curves normalized to the power on the surface at a radius of 9.45 cm. These curves represent the total ($\hat{r}$- and $\hat{\theta}$-directed) power taken at $\theta = \pi/2$. For all other angles except $\theta \approx 0, \pi$, the curves would be similar, with the levels at $r = 0$ and $r = 9.45$ cm being the same as at $\theta = \pi/2$. Note the superiority of the 915 MHz curve both in penetration and resolution.

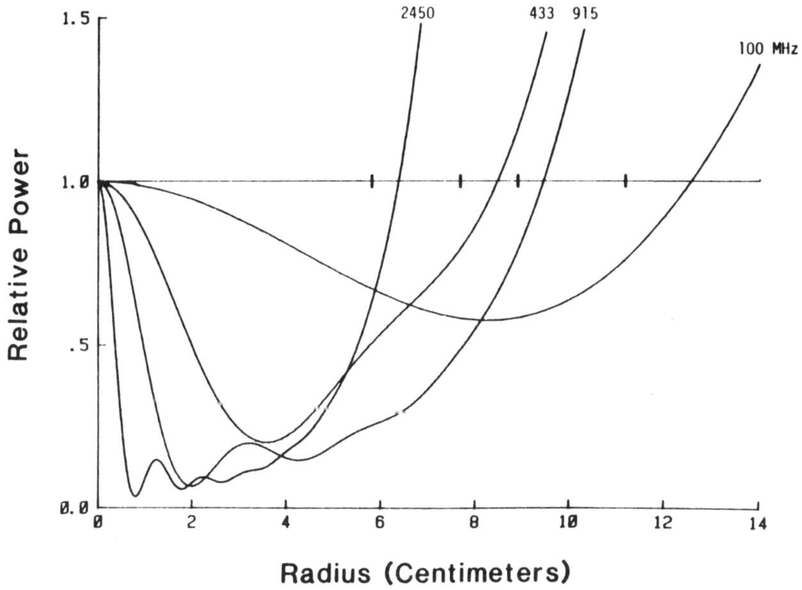

Figure 2.2.5 Power in a sphere of muscle tissue for approximate uniform surface power distribution.

c. *Summary*

The preceeding discussion has determined the dimensions of the largest convex volume of muscle tissue which can be efficiently heated non-invasively. Constraints have been imposed such that as much power is deposited in the center as elsewhere in the volume or on its surface. These limits are the theoretical best cases (within 1.5%): it is not possible to improve on them by altering the surface phase or amplitude distribution. For other tissue geometries, the maximum penetration depth will, of course, be lower. Although the determination of heating patterns depends on the thermal characteristics of the tissue which are governed by the bio-heat equation, it is reasonable to assume that this solution is also the best for heating the center without overheating any intervening tissue. Heating pattern improvements are possible by making use of surface cooling, but this effect is limited to a few millimeters *in vitro*, or to several centimeters in living animals [65].

The novel aspect of this problem is the vector nature of the source.

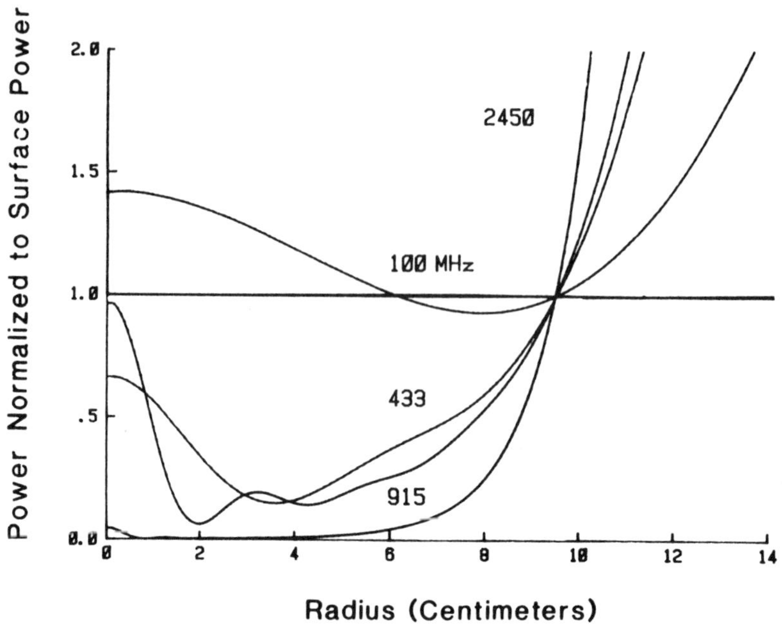

Figure 2.2.6 Power pattern for a 9.45 cm radius sphere of muscle tissue, normalized to the power at the surface.

For scalar waves, such as acoustic waves, the uniform surface excitation would be optimal. For electromagnetic waves, however, polarization is an essential consideration. The method of adding additional modes which do not contribute to power in the center simultaneously provides for adjustments to amplitude, phase, and polarization of the source.

The power optimization makes use of just the θ-component of electric field. As mentioned for the large radii spheres being considered, the radial field contribution is more than an order of magnitude smaller than the tangential $\hat{\theta}$-directed component. Upon squaring the field magnitudes, the radial components contribute only 1% to the surface power, only at the poles (where the tangential contribution is zero). Thus on the sphere surface, the *total* power behavior can be well represented by $|E_\theta|^2$.

This optimization makes use of only one polarization of electric field. Other surface field distributions which are combinations of rotations of the above formulation can be added to the $\hat{z}$-directed maximum at the center. These other combinations will be orthogonal to the original distribution, and hence simply uniformly scale the power curves of Figs. 2.2.5 and 2.2.6. Some amount of rotated field (to form, say, an $\hat{x}$-directed central maximum) may be used to increase the power near the poles up to the same level at the equator. This added power is proportional to the marginal surface area covered by this second source orientation — roughly a $2(0.1\pi)^2/4\pi$ or 0.5% increase.

It is clear, however, that any modifications symmetric to the $\hat{z}$-axis will only worsen the situation. These modifications only provide constructive, $\hat{z}$-directed interference in the manner of the surface distributions discussed herein, yet they will add to surface power, making it too large at some locations. The uniform E_θ surface distribution is the best.

Although penetration depth increases with decreasing frequency below 433 MHz, the resolution of the focal spot at the center decreases. However, due to the non-linear dependence of complex dielectric constant on frequency, increasing the frequency does yield an increase in penetration depth for a limited range, as shown by the plot of 915 MHz power curves. For 433 MHz, $\alpha/\beta = 0.40$, whereas for 915 MHz, $\alpha/\beta = 0.23$. There is a small advantage to using a more uniform power surface distribution than the uniform current distribution. The improvements are more pronounced for the lower frequencies since wavelengths are longer and the slopes of the power curves are shallower. It is clear that for muscle tissue geometries of less than 10 cm radius, 915 MHz is the best standard frequency for a producing a well focussed, high-resolution power peak. Of course, if a larger sphere is to be safely heated electromagnetically, the lower frequencies must be used.

2.3 Optimal Power Deposition in a Planar Slab

The method used for deriving the best power pattern in a planar slab is more elaborate than for the sphere. Despite its simpler geometry, there is no single non-vanishing Cartesian harmonic corresponding to the lowest order spherical harmonic. Every harmonic-mode solution to the rectangular wave equation contributes to the power deposition at the focal target. Instead of using a modal algorithm, a direct integration of the surface field must be used to synthesize the best pattern.

a. Uniform Amplitude Current Method

The most popular approach used to generate radiated electric field is integrating the surface currents across an antenna aperture equation (1) in section 2 [66–70]. Although this equation is easily approximated as a Fourier Transform for large source-to-observer distances, it is intractable in the reactive nearfield. Approximating the integral with a Fourier Transform, as done by Johnson and Guy [22], introduces large errors, particularly near the surface. Only with careful numerical integration can the actual field and power pattern be computed.

The optimization process involves determining the phase and amplitudes of an infinite ensemble of surface currents which maximizes the field amplitude at the focus. This must be done subject to the constraint of keeping the maximum power elsewhere below a specified threshold. The greatest difficulty is that it is not known where in the slab the maxima will occur. From the power pattern of the uniform planar case (Figure 2.1.3), the first assumption would be that maxima would fall close to the planes $\beta z = \pm\pi$, corresponding to one half a wavelength from the focus. Waves from each of these source planes add constructively at the focus. Sources placed along these planes are then assigned modified conjugate phase, corresponding to the negative of the electric path length from the source to the focus at $z = 0$ plus a correction factor for lossy media. The correction is necessary to account for the non-exponential factor in the integrand, which is a function of the wave number k. In lossy media, k is complex, so the true phase of the integrand includes a term equal to the arctangent of the imaginary part divided by the real part of this factor, in addition to the usual electrical distance from source to focus. For a current source with phase opposite to this resulting phase, the integrand is purely real for the observation point at the focus.

It is harder to choose the best source current amplitude distribution than phase distribution. Contrary to intuition, uniform source current amplitude is a poor choice. With non-uniform phase, a uniform amplitude source current does not produce uniform surface power. The contributions from these currents to the electric field on the surface add constructively or destructively depending on the observation position. Figure 2.3.1 shows the wide surface power variation generated by a phase-focussed uniform amplitude current source. The power on the

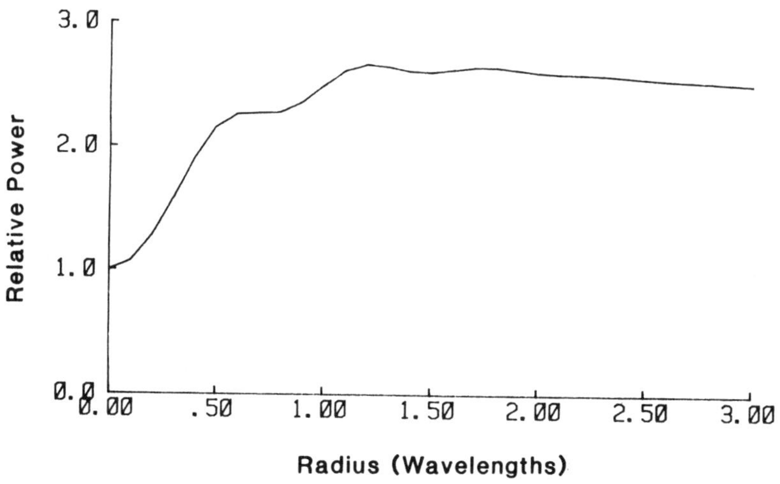

Figure 2.3.1 Radial power distribution at the source surface for focussed surface current.

source planes is plotted as a function of radius from the axis of phase symmetry. Since heating is directly related to dissipated power, it is essential to minimize the maximum power peaks on the source surfaces to prevent overheating. Also, the lower power levels can be increased to add power at the focus, until they approach the maximum surface power level. Thus, a uniform surface *power* rather than current distribution is desirable. Clearly the almost three-to-one radial variation in power in Fig. 2.3.1 makes optimal synthesis based solely on the power along the symmetry axis $(\rho = 0)$ infeasible.

b. Uniform Surface Power Method

Having shown that the uniform amplitude surface current produces non-uniform surface power, it becomes clear that it is very hard to generate an optimum choice.

If instead of specifying the electrical surface current, $\overline{J}$, the surface field, $\overline{E}$, is chosen first, the results are preferable. As with the case of the sphere in section 2.2, it is assumed that the power at the surface is the limiting constraint, and should therefore be made uniform. The only mechanism for power dissipation in biological tissue is ohmic loss with dependence $\sigma|\overline{E}|^2/2$. It is the electric field that must be optimized. Magnetic currents at the source will either be included in the electric field by the equivalence principle, or will be extraneous. Optimizing just the electric field optimizes the volume power dissipation distribution. Unfortunately, the modal solution for the planar geometry cannot be used for this synthesis.

Alternatively, it is possible to use the Kirchhoff integral formula [61] to generate the electric field everywhere in a half-space bounded by a plane with specified electric field, and having decaying wave solutions which vanish at infinity. By assuming the field on this source plane has uniform amplitude, the synthesis problem reduces to finding the best phase distribution. Determining the phase function is similar to the previously mentioned current case: choose it so as to make the integrand purely real for the observation field-point at the focus. It is considerably more difficult, however, to do this since a non-linear differential equation must be solved.

The integral formula can be written as

$$\overline{E}(\overline{r}) = \int_{A'} da' \left[\overline{E}(\overline{r}') \frac{\partial g(\overline{r}, \overline{r}')}{\partial n'} - g(\overline{r}, \overline{r}') \frac{\partial \overline{E}(\overline{r}')}{\partial n'} \right] \tag{1}$$

$$g(\overline{r}, \overline{r}') = \frac{e^{-jk|\overline{r}-\overline{r}'|}}{4\pi|\overline{r} - \overline{r}'|}$$

where $g(\overline{r}, \overline{r}')$ is the spherical Green's function, $\overline{r}$ is the observer field-point (ρ, z), $\overline{r}'$ is the source field point vector (ρ', z'), $\partial/\partial n'$ is the derivative normal to the planar source surface A' with the normal pointing towards the direction of propagation and $\overline{E}(\overline{r}')$ is the specified electric field on that surface.

For a planar geometry, using cylindrical coordinates with the source plane perpendicular to the z-axis crossing through $z = \mp d/2$, (1) reduces to

$$\overline{E}(\overline{r}) = \int_0^{2\pi} d\phi' \int_0^{\infty} \rho d\rho' \left[\overline{E}(\overline{r}') \left(\pm \frac{\partial g(\overline{r}, \overline{r}')}{\partial z'} \right) - g(\overline{r}, \overline{r}') \left(\pm \frac{\partial \overline{E}(\overline{r}')}{\partial z'} \right) \right]_{z'=\mp d/2} \tag{2}$$

with

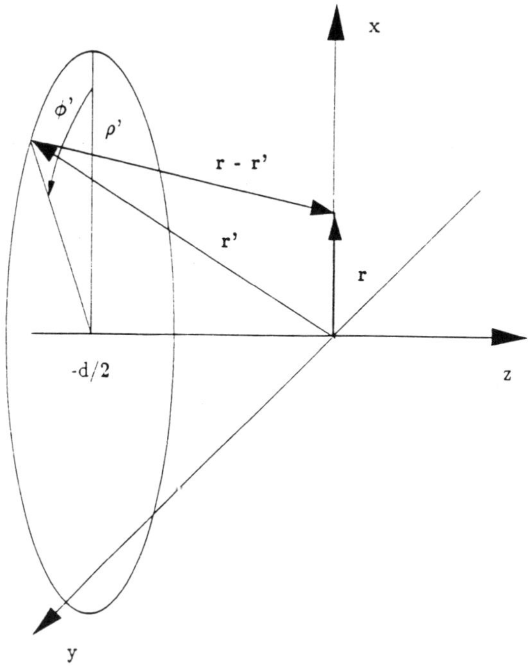

Figure 2.3.2 Coordinate geometry for planar surface integration.

$$|\overline{r} - \overline{r}'| = \sqrt{\rho^2 + \rho'^2 - 2\rho\rho' \cos\phi' + (z - z')^2}\,\bigg|_{z'=\mp d/2}$$

Equation (2) is used for both opposing planar sources, with $d/2 + z$ being the distance along the z-axis from observer point z to the source at $z = -d/2$, and $d/2 - z$ for distance from z to the source at $z = d/2$. The normal derivative is chosen to always have the normal point to the origin. The first three terms in the expression for $|\overline{r} - \overline{r}'|$ result from the Law of Cosines, using $\phi = 0$ as the observation point, as shown in Fig. 2.3.2.

To specify the electric field on the source plane, both the tangential (radial) and longitudinal (axial) dependences must be used. Although only the radial dependence of phase exists on the source plane, the normal derivative at the surface requires phase terms proportional to z. Assuming a y-polarized wave with amplitude held constant at unity, the electric field can be represented with a phase function power series in increasing powers of z

$$\overline{E}(\overline{r}) = \hat{y}\, e^{j\Phi(\rho) - jk_z(\frac{d}{2} \pm z) - jk_{z2}(\frac{d}{2} \pm z)^2 - jk_{z3}(\frac{d}{2} \pm z)^3 - \cdots} \tag{3}$$

where $\Phi(\rho)$ is the surface phase radial dependence at $z = \mp d/2$, and the choice of sign preceding each z is the positive for the source at $z = -d/2$ with wave propagation to the right, and negative for source at $z = d/2$ propagating to the left. The entries into the integral of (2) are

$$\begin{aligned} \overline{E}(\overline{r}')\Big|_{z'=\mp d/2} &= \hat{y}\, e^{j\Phi(\rho')} \\ \pm\frac{\partial \overline{E}(\overline{r}')}{\partial z'}\Big|_{z'=\mp d/2} &= \hat{y}\,(-jk_z)e^{j\Phi(\rho')} \end{aligned} \tag{4}$$

where k_z is yet to be determined.

Using one further identification to simplify notation, $R = |\overline{r} - \overline{r}'|$, the normal derivative of the Green's function is

$$\begin{aligned} \pm\frac{\partial g(\overline{r},\overline{r}')}{\partial z'}\Big|_{z'=\mp d/2} &= (-jk - 1/R)\frac{e^{-jkR}}{4\pi R}\frac{\partial R}{\partial z'}\Big|_{z'=\mp d/2} \\ &= (jk + 1/R)\frac{e^{-jkR}}{4\pi R}\left(\frac{d/2 \pm z}{R}\right) \end{aligned} \tag{5}$$

c. *Wavefront Determination*

Appealing to the theory of the eikonal for large k, let the wave phase be represented by $L(\rho, z)$. The eikonal equation is then [71]

$$\|\nabla L(\rho, z)\| = k \tag{6}$$

where the double magnitude indicates the complex vector length, or the square root of the inner product with itself. The normal to a constant phase wavefront is

$$\hat{n} = \frac{\nabla L}{\|\nabla L\|} \tag{7}$$

Using the expression for phases in (3) gives

$$\nabla L \bigg|_{z'=\mp d/2} = \hat{\rho}\frac{\partial \Phi}{\partial \rho'} - \hat{z}k_z \tag{8}$$

Note that $\partial\Phi/\partial\rho'$ can be defined as $-k_\rho$.

Substituting (8) into (6) yields two equations, the real and the imaginary parts being

$$\begin{aligned} \Phi'^2 + \beta_z^2 - \alpha_z^2 &= \beta^2 - \alpha^2 \\ -2\alpha_z\beta_z &= -2\alpha\beta \end{aligned} \tag{9}$$

where $\Phi' = \partial\Phi/\partial\rho'$ and $k_z = \beta_z - j\alpha_z$.

Solving the two equations of (9) for β_z and α_z gives

$$\begin{aligned} \beta_z^2 &= \frac{\beta^2 - \alpha^2 - \Phi'^2}{2} + \sqrt{\left(\frac{\beta^2 - \alpha^2 - \Phi'^2}{2}\right)^2 + \alpha^2\beta^2} \\ \alpha_z^2 &= -\frac{\beta^2 - \alpha^2 - \Phi'^2}{2} + \sqrt{\left(\frac{\beta^2 - \alpha^2 - \Phi'^2}{2}\right)^2 + \alpha^2\beta^2} \end{aligned} \tag{10}$$

The positive values of β_z and α_z are chosen to give waves propagating toward the focus.

The values of β_z and α_z join $-k_\rho$ to represent the shape of an inhomogeneous incoming wave as it crosses the $z = \mp d/2$ planes. This wave has many features of a spherical wave, and yet it is different in an important way. A spherical wave, far from its source, approaches an e^{-jkr}/r dependence and yields a hyperbolic phase function on the $z = \mp d/2$ surface $\sqrt{\rho'^2 + (d/2)^2}$. This will be shown shortly to be very close to the phase for the currently-addressed focussed solution. However, a spherical wave has a tapering amplitude as well. As seen with the uniform focussed current analysis, improper amplitude choice can lead to unexpectedly bad results.

There are several important features of (10) worth examining. First, consider the behavior of α_z and β_z as Φ' increases from zero (as it will soon be shown to do). Initially at zero, it follows that $\alpha_z = \alpha$, and $\beta_z = \beta$. Alternatively, for large Φ', α_z approaches Φ', and β_z

approaches $\alpha\beta/\Phi'$, which vanishes for infinite Φ'. Furthermore, taking the derivative with respect to Φ'^2 of each equation in (10) results in the pair

$$\frac{\partial\beta_z^2}{\partial\Phi'^2} = -\frac{1}{2} - \frac{\beta^2 - \alpha^2 - \Phi'^2}{2\sqrt{\left(\beta^2 - \alpha^2 - \Phi'^2\right)^2 + 4\alpha^2\beta^2}}$$

$$\frac{\partial\alpha_z^2}{\partial\Phi'^2} = \frac{\partial\beta_z^2}{\partial\Phi'^2} + 1 \tag{11}$$

Setting (11) to zero to find an extremum for α_z or β_z has no real solutions for Φ'. Thus β_z starts at β and decreases monotonically to zero; while α_z starts at α and monotonically increases with increasing Φ'.

Physically, the above mathematics can be interpreted as follows. The attenuation rate in the direction of wave penetration is minimum for a uniform phase distribution ($\Phi' = 0$) with the value of the medium, α. The propagation constant is either the value of the medium, β, for uniform phase, or less for varying phase. Therefore, the penetration wavelength increases for higher spatial frequency variation, $\Phi' > 0$. These two effects, greater attenuation rate and greater wavelength, both conspire to prevent high resolution penetration. In fact, the α_z/β_z ratio — which for non-uniform waves replaces α/β as the relevant pattern parameter — grows much faster with Φ' than with frequency. It can be concluded that the higher the spatial frequency variation of the source phase, the less the wave penetrates the medium. For the optimum source distribution, it would be expected that a majority of the source would have slowly varying phase, with a limitation on the maximum spatial variation. This will next be shown to be the case.

d. Phase Function Solutions

Combining (2), (3), (5), and the notation of (10) gives a revised integral

$$\overline{E}(\overline{r}) = \int_0^{2\pi} d\phi' \int_0^{\infty} \rho' d\rho' \left[j\left(\frac{d/2 \pm z}{R}\beta + \beta_z\right) + \frac{d/2 \pm z}{R}\left(\alpha + \frac{1}{R}\right) + \alpha_z \right] \frac{e^{-j[kR - \Phi(\rho')]}}{4\pi R} \tag{12}$$

for the electric field due to sources on the plane at $z = \pm d/2$. The remaining task is to determine $\Phi'(\rho')$ so that the integral is maximized at the focal point, $(\rho, z) = (0, 0)$. This condition occurs when the integrand is entirely real and positive for all values of integration arguments ϕ' and ρ'. The integrand can be written in magnitude/phase form as

$$\frac{\rho'}{4\pi R}\sqrt{\left[\frac{d/2 \pm z}{R}\beta + \beta_z\right]^2 + \left[\frac{d/2 \pm z}{R}(\alpha + 1/R) + \alpha_z\right]^2}$$
$$\cdot \exp\left\{ -\alpha R + j\left[-\beta R + \Phi(\rho') + \tan^{-1}\frac{\beta(d/2 \pm z) + \beta_z R}{(d/2 \pm z)/R + \alpha(d/2 \pm z) + \alpha_z R}\right]\right\} \tag{13}$$

Setting the imaginary part of the exponent to zero at the focus, $z = 0$, yields the nonlinear differential equation for Φ:

$$\Phi(\rho') = \beta R_0 - \tan^{-1}\frac{\beta d/2 + \left[\frac{\beta^2 - \alpha^2 - \Phi'^2}{2} + \sqrt{\left(\frac{\beta^2-\alpha^2-\Phi'^2}{2}\right)^2 + \alpha^2\beta^2}\right]^{1/2} R_0}{d/2R_0 + \alpha d/2 + \left[-\frac{\beta^2 - \alpha^2 - \Phi'^2}{2} + \sqrt{\left(\frac{\beta^2-\alpha^2-\Phi'^2}{2}\right)^2 + \alpha^2\beta^2}\right]^{1/2} R_0} \tag{14}$$

where

$$R_0 = \sqrt{\rho'^2 + (d/2)^2}$$

Equation (14) must be solved numerically. Fortunately, the variations due to the arctangent function are small, so good approximations for small values of ρ' are possible using only the first few terms of a power series. A solution including a term proportional to $1/\rho'$ is needed in the series expansion for large values of ρ'. The blending region between the two series approximations ends up being near a point of inflection of the arctangent function.

Using trial and error adjustments on coefficients a_i, the value of the arctangent is compared to $\Phi(\rho') - \beta R_0$ with the phase approximated by the even power series

$$\Phi(\rho') - \beta R_0 = a_0 + a_1\rho'^2 + a_2\rho'^4 + a_3\rho'^6 + a_4\rho'^8 \tag{15}$$

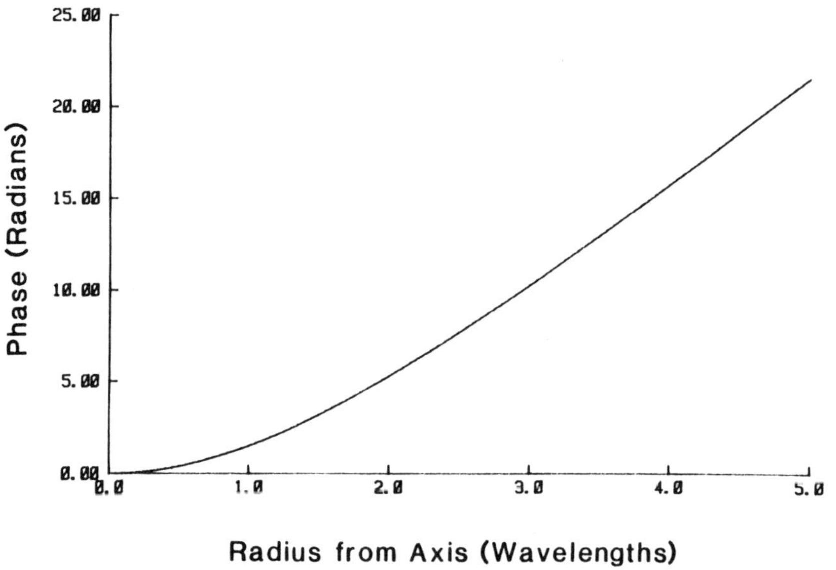

Figure 2.3.3 Phase function for focussing at a point 1.225λ from the source plane, at 915 MHz.

The coefficient a_0 is equal to $-\tan^{-1}[\beta d/(1+\alpha d)]$. Next, starting with a_1, each coefficient is adjusted until the best possible fit occurs. The larger ρ' values are minimized with adjustments to progressively higher order coefficients. After the crossover point, between 1.4 and 3 wavelengths — depending on frequency and $d/2$ — another series of the form

$$\Phi(\rho') - \beta R_0 = b_0 + b_1/\rho' + b_2\rho' + b_3\rho'^2 \tag{16}$$

is used. Table 2.3.1 lists the best phase coefficients and crossover points for several values of $d/2$ at 915 MHz. This frequency is chosen to take the greatest advantage of phase effects without compromising penetration. As shown in section b with the sphere analysis, 915 MHz is preferable to both 433 and 1200 MHz. Figure 2.3.3 plots the phase function $\Phi(\rho')$ for $d/2 = 1.225\lambda$ (wavelength), for $0 \leq \rho \leq 5\lambda$ radius. Note that this function is primarily hyperbolic with the functional dependence seen previously. Also, $\Phi(\rho')$ does indeed start from 0 and increases monotonically, as expected. If this hyperbolic term is subtracted out, Fig. 2.3.4 results. This curve corresponds to the additional effects of the complex amplitude. It is clear that this addition of almost one half a wavelength to the phase must not be neglected. One further plot, Fig. 2.3.5 subtracts the series approximations of $\Phi(\rho')$ (15) and (16) from the resulting $\Phi(\rho')$ given in (14) using the derivatives of these series. This curve shows that the series gives a very good approximate solution to the differential equation (14), with error of only several thousandths of a wavelength.

Table 2.3.1 Phase coefficients and crossover points for complex amplitude integrand.

$d/2$ (λ)	a_0	a_1 $\div .01$	a_2 $\div .001$	a_3 $\div .001$	a_4 $\div .0001$	b_0	b_1 $\div .01$	b_2 $\div .01$	b_3 $\div .001$	X-pt
1.0	5.0135	1.08	3.5	−2.5	−2.3	4.9765	−0.8	2.81	−3.1	1.2
1.2	5.0014	.88	−.07	−.045	−.21	4.584	−3.4	1.50	−1.2	1.9
1.225	5.0002	.84	.005	−.012	−.35	5.1200	−3.4	1.51	−1.2	2.0
1.25	4.9989	.80	.07	−.045	−.20	5.1129	−3.4	1.52	−1.2	2.1
1.3	4.9967	.765	−.02	−.095	.06	5.1776	−4.7	1.15	−0.8	2.5
1.5	4.9892	.57	.25	−.130	−.10	5.4062	−10.0	−0.001	0.3	2.9
2.0	4.9768	.35	−.02	−.006	0.0	4.7193	2.32	2.47	−1.7	2.5

e. Optimum Planar Power Pattern

With a focussed source, the electric field and dissipated power pattern depend on the thickness of the planar slab. This contrasts with the uniform source examples of section 2.1, where the maximum thickness is selected based on where a particular curve intersects unity. It is difficult to determine the thickest possible slab directly. Instead, this limit must be found iteratively.

Consider first a two-wavelength thick slab, $d = 2\lambda$. Figure 2.3.6 shows the power pattern along $\rho = 0$, the axis of symmetry of the phase function, through the focal point at $(\rho, z) = (0, 0)$. The focussed power at this focal point is 10.4 times greater than for the unfocussed opposing plane case of Fig. 2.1.3, which is redrawn for comparison

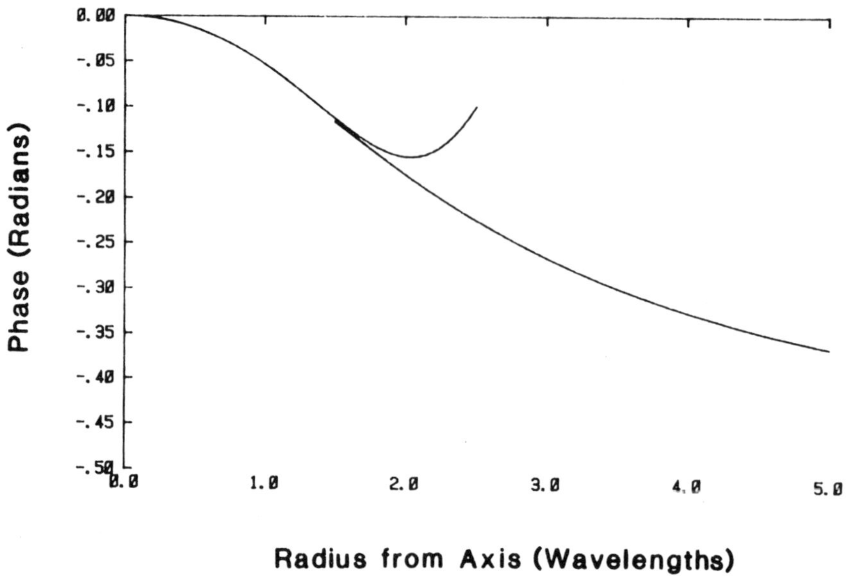

Figure 2.3.4 Correction function for complex integrand amplitude for focussing at 1.225λ depth.

in Fig. 2.3.6 normalized to the power at the source planes. In fact, the power is greater at the focus than anywhere else. The secondary maximum at about one half wavelength from the focus has been moved out to .67λ and is only 89% of the focal point power. Whereas this secondary maximum height established the largest slab thickness for the unfocussed example, it is no longer a problem with this particular phase-focussed case. The constructive effects of phase focussing are much less prevalent at points close to the source than at the mid-plane. Already, the maximum thickness has been doubled by using focussing.

Since no part of the power pattern exceeds the focal point level, it is possible to extend the slab thickness even further. An upper limit to the thickness can be extrapolated from the two-wavelength example. If it is assumed that the growth of the focussed power pattern near a source edge behaves like the unfocussed pattern, then it suffices to extend the focussed pattern until it reaches the level of the focal point power. This occurs at a thickness of 2.59λ. Unfortunately, for this thickness, the secondary maximum becomes excessively high — almost 1.3 times the

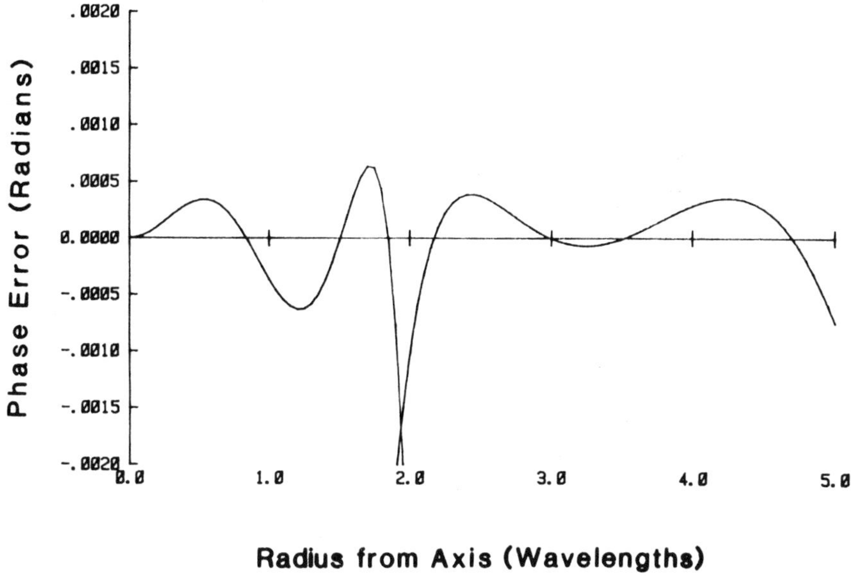

Figure 2.3.5 Approximation error for 1.225λ series.

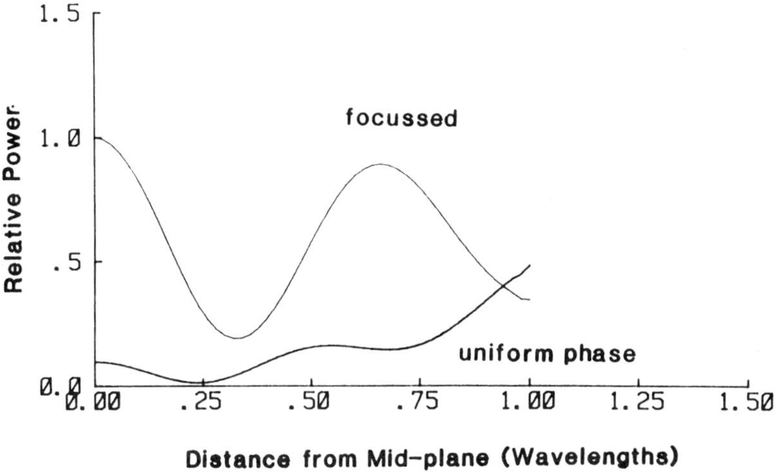

Figure 2.3.6 Power pattern for a focussed 2λ thick planar slab.

level of the focal point power. The rise in this secondary maximum is due to the increase in the constructive interference at this location, now farther from the source. That is, the thicker the slab, the less the hyperbolic phase minus the electrical path length to $(\rho, z) = (0, .67\lambda)$ varies with radial source position. It is concluded that the limiting thickness falls between $2\,\lambda$ and $2.59\,\lambda$.

There are two competing constraining effects limiting the planar slab thickness: the high power levels at the source surface and at the secondary maximum. This is a different situation from the sphere synthesis, and, as will be shown in the next section, from the focussed cylinder synthesis. An additional optimizing method has been derived to reduce the secondary maximum power level to a greater extent than the focal point power without raising the surface power. The method involves decreasing the amplitude of the source in particular regions. The phase is kept the same as from the above synthesis so that the focus still receives coherent contributions from the entire source. The secondary maximum lies along the axis of phase symmetry, $\rho = 0$, and receives the greatest portion of its power from sources close to this axis. At this secondary maximum, contributions from these closest sources are greater than the non-coherent destructive interference from more remote sources. For observer points farther from the source than this secondary maximum, including, in particular the focus, the path length variations (and hence the exponential attenuation multiplier) are smaller for a given change in ρ. Mathematically, if the source-to-secondary maximum distance is represented by s, it is clear that

$$\left| \sqrt{\rho_1^2 + (d/2)^2} - \sqrt{\rho_2^2 + (d/2)^2} \right| < \left| \sqrt{\rho_1^2 + s^2} - \sqrt{\rho_2^2 + s^2} \right|$$

$$d/2 > s, \quad \text{for any two radial positions } \rho_1 \neq \rho_2$$

Thus, if the amplitude is reduced in a narrow region about the $\rho = 0$ axis, the destructive effects of the remote sources become more important. The secondary maximum level will be reduced more than the focal point level. Furthermore, since the source amplitude is only being reduced, there will be no increase in surface power anywhere.

The choice of the amplitude reduction function is somewhat arbitrary. One function that works well is based on a Gaussian, and is

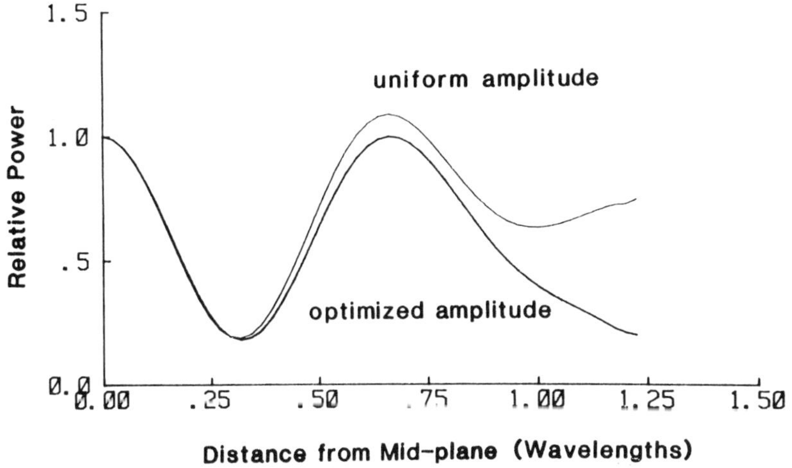

Figure 2.3.7 Focussed unit amplitude and optimal power patterns for 915 MHz focussed planar source.

stated

$$A(\rho) = \frac{1}{1 + ae^{-b\rho^2}} \tag{17}$$

By iterative optimization, the best attainable power is for a planar slab 2.45 λ thick, with the amplitude dependence of (17) with $a = 1.0$ and $b = 7$. It is interesting to note that this amplitude variation is in the opposite manner to that of a spherical wave, discussed previously. Optimization involves testing a set of a and b parameters to obtain the largest focal point power while simultaneously keeping the secondary maximum below this power level. In general, as a increases the power is reduced at both positions, while decreasing b causes a preferential reduction of the secondary maximum power level. If despite all combinations of a and b the greatest value of focal point power is still below either the highest source surface power level or the secondary maximum power level, the thickness must be reduced. The effects of the second opposing source on a source surface must be taken into account when computing the power there. The power is proportional to the square of the sum of $A(\rho)$ and the intensity from the other source. The unit amplitude pattern and adjusted amplitude pattern are plot-

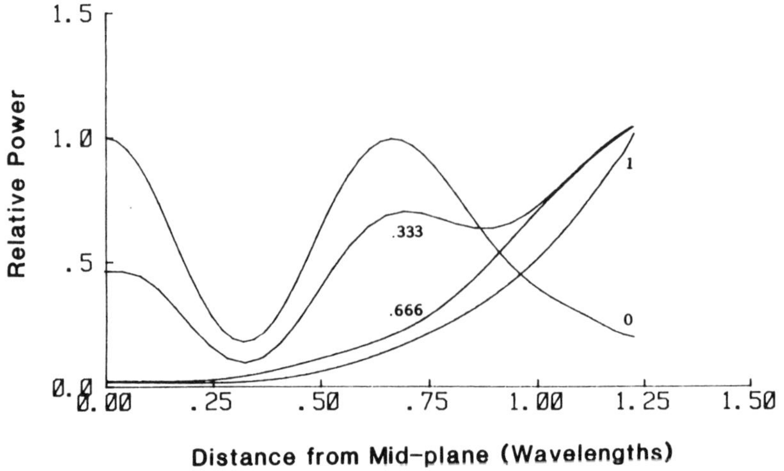

Figure 2.3.8 Power patterns for 915 MHz focussed planar source on axis and at .333, .666, and 1λ radius.

ted in Fig. 2.3.7. For the adjusted amplitude pattern, the focal point power and secondary maximum power levels are equal. These levels, in turn, are equal to the surface power for large ρ, as shown in Fig. 2.3.8, which repeats the amplitude adjusted pattern of Fig. 2.3.7 and shows the power pattern along parallel lines $(\rho, \theta) = (i0.333\lambda, 0), i = 1$, 2, 3. This figure also demonstrates that there are no other secondary maxima: the largest power levels are along the symmetry axis, $\rho = 0$.

Figure 2.3.8 represents the best possible power pattern for EM radiation at 915 MHz in an infinite slab of muscle tissue, in that a maximum amount of power is delivered to the deepest point without risking overheating any other regions. It is conceivable that a different choice of amplitude distribution might increase the ratio of focal point power to secondary maximum power without significantly suppressing the former below the surface power. However, this improvement would not exceed one or two percent, and since the measured values of the electrical parameters have a much wider variation, it is not useful to pursue further improvement.

The maximum thickness of 2.45λ corresponds to 11.1 cm at 915

MHz. This value compares well with the maximum diameter for any high resolution cylinder pattern, which is 12.6 cm from Table 2.1.1. Also, the focussed maximum thickness is more than three times the 3.6 cm maximum for the uniform source case of section 2.1.

Section 2.4 explores focussing the source for the cylindrical geometry where both modal analysis and an integral focussing are used.

2.4 Optimal Power Deposition in a Circular Cylinder

The cylinder is the last of the three simple geometries to be analyzed. There are many more applications for a cylindrical source since body parts, such as the limbs, neck, torso, and even the head can be modelled as cylinders. As expected, the penetration limits fall between those of the sphere and the planar slab.

It is interesting that the optimum cylindrical source synthesis case makes use of both of the design methods of the other two cases. As with the spherical analysis, the modal solutions to the wave equation restrict the distribution to a reduced set of functions. But instead of there being only one mode which contributes power at the focus, there is a family of cylindrical solutions which constructively add at the axis. These solutions are characterized by each being independent of circumferential variation. Although there are an infinite number of possible functions within this family, the synthesis problem is reduced from a two-dimensional one, as with the planar source, to a one-dimensional focussing design. The same phase/amplitude optimization method used with the plane can be applied to this reduced problem.

a. Optimal Source Derivation

The harmonic wave solutions for the cylindrical geometry are of the form [61]

$$E_{nm} = \hat{z} A_{nm} J_m(k_{\rho_n}\rho) e^{-jm\phi} e^{-jk_{z_n}z} \tag{1}$$

where J_m is the mth order cylindrical Bessel function, and with dispersion relation

$$k_{\rho_n}^2 + k_{z_n}^2 = \beta^2 - \alpha^2$$

The analysis problem involves summing a series of appropriately weighted functions of the form of (1) to give a maximum at $(\rho, z) =$

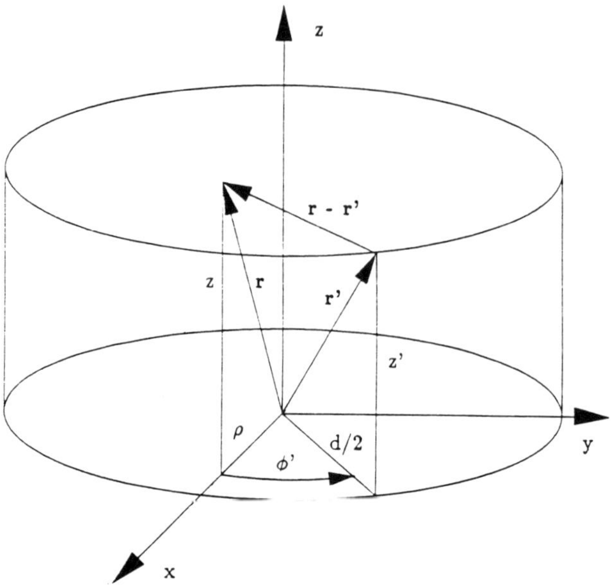

Figure 2.4.1 Coordinate geometry for cylindrical surface integration.

$(0,0)$. Recalling that $J_m(0)$ vanishes for all positive values of m, and $J_0(0) = 1$, it is clear that only harmonic solutions independent of the circumferential coordinate, ϕ, contribute to power on the axis. This is the restricted family of functions to be considered. For determination of the source distribution on a surface of constant radius, only the z-dependence remains to be determined.

To derive the optimum power pattern, the axial dependence of the best distribution is found by a focussed integral of the form of (1) in section 2.3. Analysis proceeds parallel to that of section 2.3. Figure 2.4.1 shows the geometry used for the integration over a cylindrical surface. An important difference between this geometry and the planar geometry is the choice of polarization. As discussed in section 2.1, the only acceptable source polarization is axial, or z-directed.

For a cylinder with radius $d/2$ and normal derivative taken with respect to $-\rho'$ so that waves propagate inwards to the focal target, the operative formula for a cylinder is

$$\overline{E}(\overline{r}) = \int_0^{2\pi} \rho' d\phi' \int_{-\infty}^{\infty} dz' \left[\overline{E}(\overline{r}') \left(-\frac{\partial g(\overline{r}, \overline{r}')}{\partial \rho'} \right) \right.$$
$$\left. - g(\overline{r}, \overline{r}') \left(-\frac{\partial \overline{E}(\overline{r}')}{\partial \rho'} \right) \right]_{\rho' = d/2} \tag{2}$$

with

$$|\overline{r} - \overline{r}'| = R$$
$$= \sqrt{\rho^2 + \rho'^2 - 2\rho\rho' \cos\phi' + (z - z')^2} \Bigg|_{\rho' = d/2}$$

Note that the cylinder power pattern optimization is in terms of radius ρ, from the z-axis rather than distance z, from the mid-plane, as in the planar case. The source must, in this case be determined as a function of z'. Also, since there is only one cylindrical source surface rather than two opposing sources, only one integral needs to be computed.

Next, assume a source phase distribution similar to that in (3) of section 2.3 with surface phase $\Phi(z)$ to be determined, and an added power series in terms of ρ. This series is used to determine the normal field dependence $\partial \overline{E} / \partial \rho$. The series coefficients are once again found by using the eikonal equation (6) in section 2.3

$$\overline{E}(\overline{r}) = \hat{z} e^{j\Phi(z) - jk_\rho(\rho - \frac{d}{2}) - jk_{\rho 2}(\rho - \frac{d}{2})^2 - jk_{\rho 3}(\rho - \frac{d}{2})^3 - \cdots} \tag{3}$$

Equation (3) is evaluated on the integration surface, resulting in

$$\overline{E}(\overline{r}') \Bigg|_{\rho' = d/2} = \hat{z}\, e^{j\Phi(z')}$$
$$-\frac{\partial \overline{E}(\overline{r}')}{\partial \rho'} \Bigg|_{\rho' = d/2} = \hat{z}\, (jk_\rho) e^{j\Phi(z')} \tag{4}$$

with k_ρ to be determined in a similar manner as k_z in section 2.3.

The cylindrical wavefront normal is given by

$$\hat{n} = \frac{\hat{z} \frac{\partial \phi}{\partial z'} - \hat{\rho} k_\rho}{\sqrt{\left(\frac{\partial \Phi}{\partial z'} \right)^2 + k_\rho^2}} \tag{5}$$

and the eikonal equation (6) in section 2.3 results in

$$\begin{aligned} \Phi'^2 + \beta_\rho^2 - \alpha_\rho^2 &= \beta^2 - \alpha^2 \\ -2\alpha_\rho\beta_\rho &= -2\alpha\beta \end{aligned} \tag{6}$$

Note the similarity with (9) in section 2.3. The solutions for α_ρ and β_ρ are identical to those for α_z and β_z in (10) of section 2.3 with each z and ρ parameter interchanged.

Continuing the derivation, the normal derivative of the Green's function is

$$\begin{aligned} -\frac{\partial g(\overline{r}, \overline{r}')}{\partial \rho'}\bigg|_{\rho'=d/2} &= (-jk - 1/R)\frac{e^{-jkR}}{4\pi R}\left(-\frac{\partial R}{\partial \rho'}\right)\bigg|_{\rho'=d/2} \\ &= (jk + 1/R)\frac{e^{-jkR}}{4\pi R}\left(\frac{d/2 - \rho\cos\phi'}{R}\right) \end{aligned} \tag{7}$$

Inserting (4) and (7) into (2) yields

$$\begin{aligned} \overline{E}(\overline{r}) = \frac{d/2}{4\pi}\int_0^{2\pi} d\phi' \int_{-\infty}^{\infty} dz' \Bigg[j\left(\frac{d/2 - \rho\cos\phi'}{R}\beta - \beta_\rho\right) \\ + \frac{d/2 - \rho\cos\phi'}{R}\left(\alpha + \frac{1}{R}\right) - \alpha_\rho \Bigg] \frac{e^{-j[kR - \Phi(z')]}}{R} \end{aligned} \tag{8}$$

At the focal point, $(\rho, z) = (0, 0)$, $R_0 = \sqrt{z'^2 + (d/2)^2}$, and the maximum value of the integral of (8) occurs when the integrand is purely real and positive for all values of ϕ' and z'. The integrand is essentially the same form as in (12) of section 2.3, with z' replacing ρ', and dz' replacing $\rho' d\rho'$. The phase function, which makes the integrand positive at the focus, is thus identical in form to the phase functions derived in Table 2.3.1.

For $d/2 = 2\lambda$, at 915 MHz, the phase function correction due to complex amplitude and the series approximation error are plotted in Figs. 2.4.2 and 2.4.3. The phase function plot itself is so close to that presented in Fig. 2.3.3 for the function for $d/2 = 1.225\lambda$, that it is not repeated again here.

Determination of the maximum radius cylinder is considerably more straightforward than the similar problem for the plane. As with

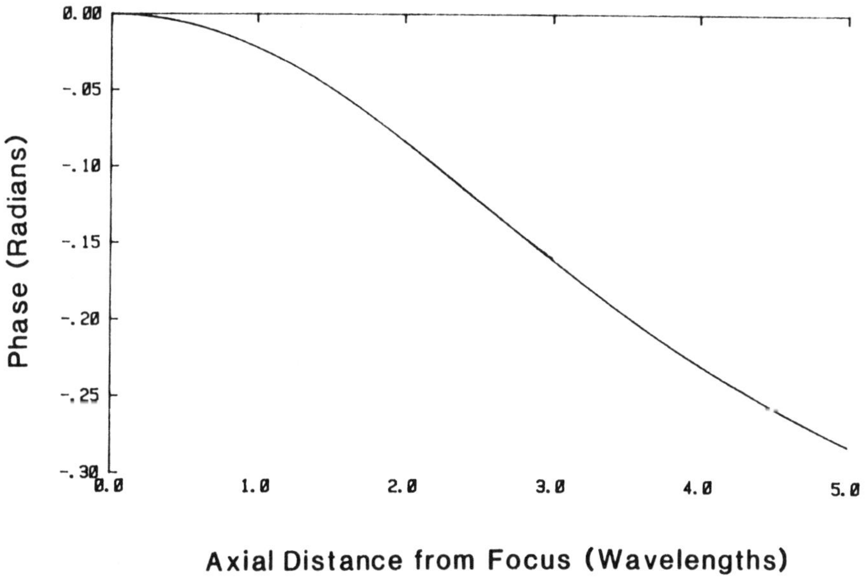

Figure 2.4.2 Correction function for complex integrand amplitude for focussing at 2.0λ depth.

the sphere, the secondary maxima are much lower than the focal point power maximum. This is the case even in the uniform pattern, Fig. 2.1.6. The limiting radius is once again arrived at iteratively, first by finding the focussed power at the focal point when focussed sources are at an arbitrary radius, say the uniform cylinder radius limit, $d/2 = 1.4\lambda$. This power value is 3.22. Extrapolating the uniform cylinder pattern out to the radius where the power level is 3.22 gives the first radius limit approximation. This radius value is used as the next $d/2$ parameter to recompute the focussed phase function. Then the unit amplitude, phase focussed source distribution is imposed at this radius, and the power at the focus is checked. Extrapolating the radius to the 3.22 power level turns out to be too large, since for larger radii cylinders, the focussing effects in lossy media are less pronounced. The $d/2$ radius is reduced in a third iteration step so that the power at a third $d/2$ radius is the same as at the axis. The iteration repeats until the desired accuracy is attained.

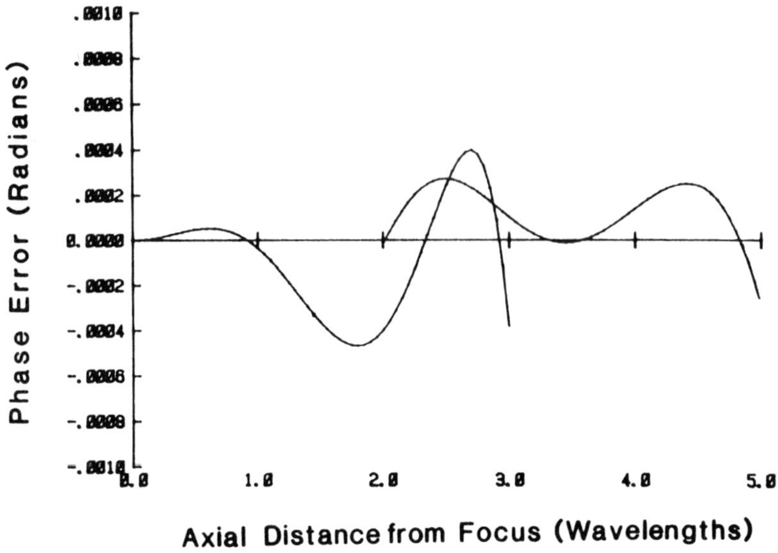

Figure 2.4.3 **Approximation error for** 2.0λ **series.**

The resulting focussed power pattern is depicted in Fig. 2.4.4, with the uniform pattern juxtaposed. These curves are computed by performing the integration in (8), which is most accurate near the focus, but begins to lose accuracy near the source surface. There are no direct ways to confirm the calculation of the focused pattern near the surface. The uniform pattern, however, can be represented in closed form (as in section 2.1, with a Bessel function) to compare to the integrated solution. As seen from Fig. 2.4.4, the agreement is excellent until about $.3\lambda$ from the source.

The resolution near the axis is much greater and the power is 3.09 times higher than it is for a uniformly-exposed cylinder of the same radius; the largest allowable diameter is 36% larger, at 1.92 λ, or 17.3 cm at 915 MHz. This is large enough to be used in human limbs and neck, as well as in torsos of children. Compared to the sphere limits, it is seen that cylinder focussing approaches the fundamental limit of 18.9 cm. The maximum cylinder radius is only 3% less than the uniform current sphere radius, and 8.5% less than the best possible of all source configurations.

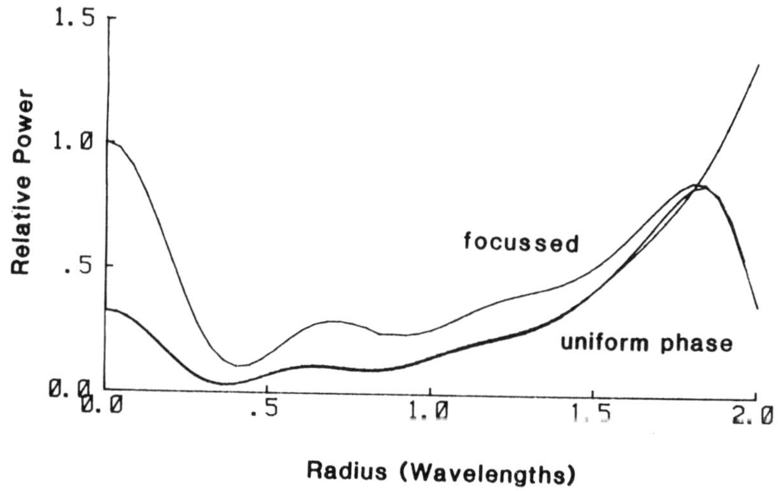

Figure 2.4.4 Optimized, focussed power pattern for a cylinder of muscle tissue at 915 MHz.

b. Summary and Results

The results of sections 2.2, 2.3, and 2.4 show that there is great potential for increasing the penetration limits of electromagnetic power, even in highly constrained geometries, over the currently accepted maxima. The results of the sphere analysis provide a fundamental limitation on the maximum depth for generating a local power peak. No other volume geometry will support as great a peak at the center. Although focussing produces much less concentration in lossy tissue than in free space, it still is prominent enough to significantly increase (up to a factor of 3 with a planar slab) the distance to the deepest point that can be safely heated.

The significant increase in maximum cylinder radius from 1.4 to 1.92λ at 915 MHz shows the effectiveness of axial focussing. This extension cannot be directly scaled to lower frequency, deeper penetrating waves. Although focussing always improves the pattern, higher α/β ratio keeps the electrical distance percentage increase smaller for lower frequencies. Figure 2.4.5 compares the focussed 100 MHz and 915 MHz patterns on a centimeter scale. The dashed lines indicate the extrapolations near the source where the integration approximations break

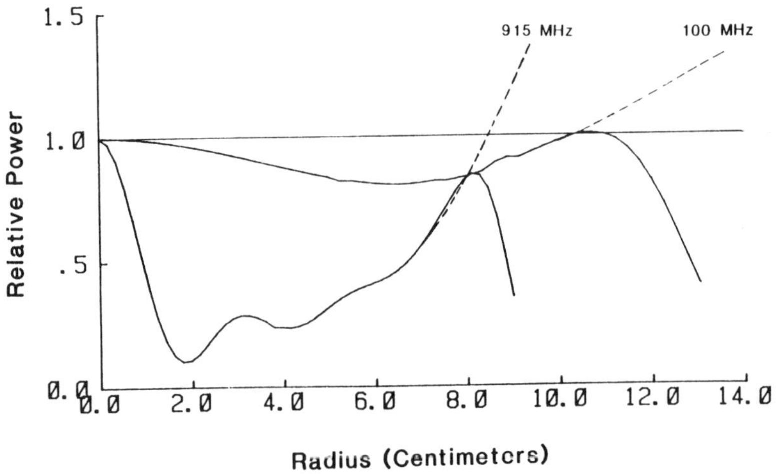

Figure 2.4.5 Comparison of radial power patterns for 915 MHz and 100 MHz focussed sources.

down. Focussing increases the 100 MHz maximum diameter from 19.0 cm (taken from Table 2.1.1) to about 21.0 cm, only a 10.5% increase. The lower frequency penetrates more deeply and would be preferable except for two difficulties. First, the power pattern resolution is not as sharp, with higher minima than with the 915 MHz pattern. Second and more importantly, few tumors are located in a cylindrical body part long enough to have the full aperture necessary for the full focussing effect. At 100 MHz, a 3-wavelength cylinder is 81 cm long.

The effects of truncating the infinite cylinder to represent a more practical antenna are examined in Fig. 2.4.6. Power dissipation curves as a function of radius are shown for 915 MHz, axially polarized, uniform surface power, 1.92λ radius sources with axial lengths 2λ, 4λ, 6λ, 10λ, and infinitely long. There is no perceptible difference between the 10λ and the infinitely long cylinder patterns. The truncation effects are significant for the narrow 2λ source, but the pattern rapidly converges to the ideal case for sources longer than the cylinder diameter.

At 915 MHz, the 17.3 cm maximum muscle diameter limitation allows high resolution heating for cancer cases in the head, neck and

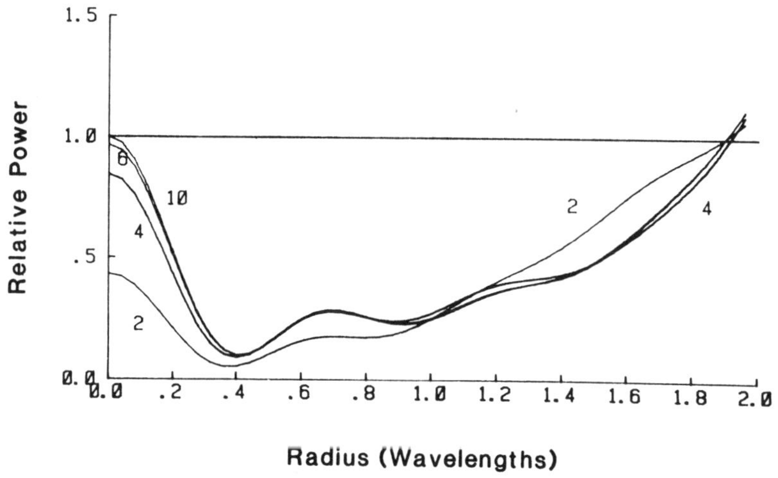

Figure 2.4.6 Power patterns for 915 MHz finite and infinite axial length focussed cylindrical sources.

limbs. Based on 1977 statistics [7], 11,600 brain; 9000 thyroid; 10,400 larynx; and 6,400 bone and soft tissue cancer cases per year may be suitable focussed cylindrical microwave source candidates. In each of these types of malignancy, the tumor is embedded in body structures which can readily be surrounded by electromagnetic sources, allowing a maximum geometric aperture advantage. Ultrasound, though much easier to focus in general, is impractical in these cases due to interference from bones. The last group of 6400 cases includes sarcomas that are prevalent in children, usually lead to limb amputation [3], would be a natural choice for EM applicators.

Electromagnetic hyperthermia offers a promising modality for treating cancer in humans. Important aspects including polarization, depth of heating, power profile resolution, secondary power maxima, and source distribution must be given careful consideration. Attempts have been made to synthesize power patterns and specify the maximum planar, cylindrical and spherical muscle dimensions for practical heating. Using a modal scheme for simplified geometries, followed by phase correction improvement — rather than experimental conjugate phase specification as suggested in the literature — leads to a pattern less likely to have undesirable hot-spots.

The recurrent theme with focussing in each geometry is the necessity of maintaining uniform surface power at the source *a priori*. The focussed source fields were derived starting by specifying uniform amplitude to ensure no isolated power maxima near the source. This is a novel concept, not previously discussed in the literature. The phase functions are solutions to the non-linear differential equation which makes the Kirchhoff integral formula purely real and positive. The optimum planar source has phase variation $\Phi(\rho)$ according to equations (15) and (16) in section 2.3. The optimal cylindrical source was shown to be polarized in the axial, z-direction, have no phase variation with ϕ and have a longitudinal phase variation function $\Phi(z)$ identical to the radial phase variation function $\Phi(\rho)$ as the planar case. Focussing tends to concentrate the interference effects on the axis (or plane) of phase symmetry for the planar (cylindrical) case which includes the focal point and intercepts the source.

The absolute power penetration limitations for muscle tissue have been determined and presented above. The fundamental limit for all geometries is a sphere of maximum radius as depicted in Fig. 2.2.5. For planar geometries, the best pattern has a slight amplitude reduction near the axis of symmetry which suppresses the secondary peak. The power pattern has equal power level at the focus, the distant source planes, and the secondary peaks. The maximum penetration depth is three times that of the uniform case. For cylindrical geometries, the source has uniform amplitude and a similar phase distribution to the planar case. The best focussed pattern has a maximum radius 35% larger than the uniform pattern radius.

The theoretical optimizations in this study are intended to serve as a set of idealized goals. Biological tissue is representative of dissipating media which is neither very conductive nor very insulating. Although few simplifying assumptions can be made for these intermediate cases, it is clear that focussing provides significant improvement over uniform illumination. This work is useful in many electromagnetic heating cases where the maximum surface power must be kept below a set limit, while deep power dissipation is maximized. Although many body structures may be modeled by the simple geometries considered above, the complete optimum antenna specification must also include tissue inhomogeneities, which in turn demands moment method analysis. The focussing methods described herein are only a guide to be used for feasibility and initial applicator design. Hyperthermia cancer treatment is an exciting but challenging application of antenna synthesis, with much research still to be done.

Acknowledgments

I acknowledge the many helpful technical (and social) discussions with my advisor and mentor, Professor Frederic Morgenthaler. The advising help of the other members of my committee, Profs. Lele, Kong, and Kyhl is also appreciated.

This work was supported by NIH grant number CA 31303. This funding is gratefully acknowledged.

References

[1] Rappaport, C. M., *PhD Thesis*, Massachusetts Institute of Techonology, 1987.

[2] Devita, V. T., S. Hellman, and S. A. Rosenberg, *Cancer, Principles and Practice of Oncology*, Lippincott, Philadelphia, 1982.

[3] Pace, O. T. and B. Cady, "Overal principles of cancer management, II. Surgery," in *Cancer: A Manual for Practitioners*, American Cancer Society, Boston, 1978.

[4] Holland, J. F. and E. Frei, *Cancer Medicine*, Lea and Rebiger, Philadelphia, 1982.

[5] Moosa, A. R., M. C. Robson, and S. C. Sahimoff, (ed.), *Comprehensive Textbook of Oncology*, Williams and Wilkins, Baltimore, 1986.

[6] Hahn, G. M., "Hyperthermia for the engineer: A short biological primer," *IEEE Biomedical Engineering*, **BME-31**, No. 1, 3–8, January 1984.

[7] Short, J. G. and P. F. Turner, "Physical hyperthermia and cancer therapy," *Proceedings of the IEEE,*, **68**, No. 1, 133–142, January 1984,

[8] Strohbehn, J. W. and E. B. Douple, "Hyperthermia and cancer therapy: A review of biomedical engineering contributions and challenges," *IEEE Trans. on Biomedical Engineering*, **BME-31**, No. 12, 779–787, December 1984.

[9] Westra, A. and U. C. Dewey, "Heat shock during the cell cycle of Chinese hamster cells *vitro*," *International J. of Radiation Biology*, **19**, 467–477, 1971.

[10] Palzer, R. J. and C. Heidelberger, "Studies on the hyperthermic killing on HeLa cells," *Cancer Research,* **33**, 415–421, 1973.

[11] Connor, W. G., E. W. Gerner, R. C. Miller, and M. L. M. Boone, "Prospects for hyperthermia in human cancer therapy, Part II," *Radiation Biology*, 197–503, May 1977.

[12] Storm, F. K., ed., *Hyperthermia in Cancer Therapy*, Prentice-Hall, Boston, MA, 1983.

[13] Hahn, G. M., *Hyperthermia and Cancer*, Plenum Press, New York, 1982.

[14] Song, C. W., "Effects of local hyperthermia on blood flow and microenvironment: a review," *Cancer Research*, **44**, 4721s–4730s, October 1984.

[15] Thrall, D. E., "Clinical requirements for localized hyperthermia in the patient," *Radiation and Environmental Biophysics*, **17**, 224–237, 1980.

[16] Meyer, J. L., "The clinical efficacy of localized hyperthermia," *Cancer Research*, **44**, 4745s–4751s, October 1984.

[17] Lai, P. K., C. A. Cain, and H. S. Ducoff, "Interaction between 2450 MHz microwaves and ionizing radiation in tribolium confusum," *IEEE Trans. Microwave Theory and Techniques*, **MTT-26**, No. 8, 530–534, August 1987.

[18] Crile, G., "The effects of heat and radiation on cancer implanted in the feet of mice," *Cancer Research*, **23**, 372–380, 1963.

[19] Dewey, W. C., "Interaction of heat with radiation and chemotherapy," *Cancer Research*, **44**, 4714s–4720s, October 1984.

[20] Lele, P. P. and J. K. Parker, "Temperature distributions in tissues during local hyperthermia by stationary or steered beams of unfocussed or focussed ultrasound," *British J. of Cancer*, **45-supplement V**, 108–121, 1982.

[21] Cheung, A. Y. and A. Neyzari, "Deep local hyperthermia for cancer therapy: External electromagnetic and ultrasound techniques," *Cancer Research*, **44**, 4736s–4744s, October 1984.

[22] Johnson, C. C. and A. W. Guy, "Nonionizing electromagnetic wave effects in biological materials and systems," *Proceedings of the IEEE*, **60**, No. 6, 692–718, June 1972.

[23] Guy, A. W., J. F. Lehmann, J. A. McDougall, and C. C. Sorenson, "Studies of therapeutic heating by electromagnetic energy," *Thermal Problems in Biotechnology*, 26–45, 1973.

[24] Guy, A. W., J. F. Lehmann, and J. B. Stonebridge, "Therapeutic applications of electromagnetic power," *Proceedings of the IEEE*, **62**, No. 1, 55–75, January 1974.

[25] Schwan, H. P. and K. R. Foster, "RF-field interactions with biological systems: Electrical properties and biophysical mechanisms," *Proceedings of the IEEE*, **68**, No. 1, 104–113, January 1980.

[26] Schwan, H. P., "Interaction of micrewave and radio frequency radiation with biological system," *IEEE Trans. on Micrewave Theory and Techniques,* **MTT-19**, No. 2, 146–152, February 1971.

[27] Stuchly, M. A. and S. S. Stuchly, "Dielectric properties of biological substances – tabulated," *J. of Microwave Power*, **15**, No. 1, 19–24, January 1980.

[28] Schwan, H. P., "Radiation biology, medical applications, and radiation hazards," in *Microwave Power Engineering*, E. C. Okress, ed., Academic Press, New York, **2**, 215–234, 1968.

[29] Guy, A. W., "Analyses of electromagnetic fields induced in biological tissyes by thermographic studies on equivalent phantom models," *IEEE Trans. on Microwave Theory and Techniques,* **MTT-19**, No. 2, 205–214, February 1971.

[30] Rappaport, C. M., "Localized hyperthermia with electromagnetic arrays and the leaky wave troughguide applicator," *IEEE Trans. on Microwave Theory and Techniqhes*, **MTT-34**, No. 5, May 1986.

[31] Storm, F. K., R. S. Elliot, W. H. Harrison, and D. L. Morton, "Clinical RF hyperthermia by magnetic loop induction: a new approach to human cancer therapy," *IEEE Trans. and Microwave Theory and Techniques,* **MTT-30**, No. 8, 1149–1157, August 1982.

[32] Ho, H. S., A. W. Guy, R. A. Sigelmann, and J. F. Lehmann, "Microwave heating of simulated human limbs by aperture sources," *IEEE Trans. on Microwave Thoery and Techniques,* **MTT-19**, No. 2, 224–231, February 1971.

[33] Morita, N. and J. B. Andersen, "Near-field absorbtion in a circular cylinder from electric and magnetic line sources," *Bioelectromagnetics*, **3**, No. 2, 253–274, 1982.

[34] Raskmark, P. and J. B. Andersen, "Focused electromagnetic heating of muscle tissue," *IEEE Trans. on Microwave Theory and Techniques*, **MTT–32**, No. 8, 887–888, August 1984.

[35] Andersen, J. B., "Focusing in lossy media," *Radio Science*, **19**, No. 5, 1195–1198, September 1984.

[36] Andersen, J. B., "Theoretical limitations on radiation into muscle tissue," *J. of Hyperthermia*, **1**, No. 1, April 1985.

[37] Gee, W., S. W. Lee, N. K. Bong, C. A. Cain, R. Mittra, and R. L. Magin, "Focused array hyperthermia applicator: theory and experiment," *IEEE Trans. on Biomedical Engineering,* **BME-31**, No. 1, 38–46, January 1984.

[38] Hand, J. W., J. L. Cheetham, and A. J. Hind, "Absorbed power distribution from coherent microwave arrays for localized hyperthermia," *IEEE Trans. on Microwave Theory and Techniques,* **MTT-34**, No. 5, 484–489, May 1986.

[39] Loane, J., H. Ling, B. F. Wang, and S. W. Lee, "Experimental investigation of a retro-focusing microwave hyperthermia applicator: conjugate-field matching scheme," *IEEE Trans. on Microwave Theory and Techniques,* **MTT-34**, No. 5, 490–494, May 1986.

[40] Hannemann, R. J. and J. E. Robinson, "Thermal analysis and design considerations for a dual-beam microwave applicator for hyperthermia research," *J. of Biomechanical Engineering,* **101,** 151–156, May 1979.

[41] Melek, M., B. Eng, and A. P. Anderson, "Theoretical studies of localised tumor heating using focused microwave arrays," *IEEE Proceedings,* **127F,** No. 4, 319–321, August 1980.

[42] Knoechel, R., "Capabilities of multiapplicator systems for focussed hyperthermia," *IEEE Trans. on Microwave Theory and Techniques,* **MTT-31,** No. 1, 70–73, January 1983.

[43] Massoudi, H., C. H. Durney, and C. C. Johnson, "A geometrical optic and an exact solution for internal fields in and energy absorption by a cylindrical model of a man irradiated by an electromagnetic plane wave," *Radio Science,* **14,** No. 65, 35–42, December 1976.

[44] Van Putten, M. H. P. M. and P. M. Van Den Berg, "A three-dimensional model for the 'coaxial TEM' deep body hyperthermia through layered tissue," *IEEE Trans. on Microwave Theory and Techniques,* **MTT-34,** No. 5, 526–531, May 1986.

[45] Sato, G., C. Shibata, S. Sekimukai, H. Wakabayoshi, K. Mitsuka, and K. Giga, "Phase controlled circular array heating equipment for deap-seated tumors: preliminary experiments," *IEEE Trans. on Microwave Theory and Techniques,* **MTT-34,** No. 5, 520–525, May 1986.

[46] Cudd, P. A., A. P. Anderson, M. S. Hawley, and J. Conway, "Phase array design considerations for deep hyperthermia through layered tissue," *IEEE Trans. on Microwave Theory and Techniques,* **MTT-34,** No. 5, 531–536, May 1986.

[47] Wait, J. R., "Analysis of the radiation leakage for a four aperture phased-array applicator in hyperthermia therapy," *IEEE Trans. on Microwave Theory and Techniques,* **MTT-34,** No. 5, 539–551, May 1986.

[48] Arcangeli, G., P. P. Lombardini, G. A. Lovisolo, G. Marsiglia, and M. Piattelli, "Focusing of 915 MHz electromagnetic power on

deep human tissues: a mathematical model study," *IEEE Trans. on Biomedical Engineering,* **BME-31**, No. 1, 47–52, January 1984.

[49] Morita, N., T. Hamasaki, and N. Kumagai, "An optical excitation method in multi-applicator systems forming a hot zone inside the human body," *IEEE Trans. on Microwave Theory and Techniques,* **MTT-34**, No. 5, 523–538, May 1986.

[50] Jouvie, F., J. C. Bolomey, and G. Gaboriand, "Discussion of capabilities of microwave phased arrays for hyperthermia treatment of neck tumors," *IEEE Trans. on Microwave Theory and Techniques,* **MTT-34,** No. 5, 495–501, May 1986.

[51] Turner, P. F., "Hyperthermia and inhomogeneous tissue effects using an annular phased array," *IEEE Trans. on Microwave Theory and Techniques*, **MTT-32**, No. 8, 874–882, August 1984.

[52] Turner, P. F., "Regional hyperthermia with an annular phased array," *IEEE Trans. Biomedical Engineering*, **BME-31**, No. 1, 106–114, January 1984.

[53] Gibbs, F. A., M. D. Sapozink, K. S. Gates, and J. R. Stewart, "Regional hyperthermia with an annular phased array in the experimental treatment of cancer: report of work in progress with a technical emphasis," *IEEE Trans. on Biomedical Engineering,* **BME-31**, No. 1, 115–119, January 1984.

[54] Hagmann, M. J., R. L. Levin, and P. F. Turner, "A comparison of the annular phased array to helical coil application for limb and torso hyperthermia," *IEEE Trans. on Biomedical Engineering*, **BME-32**, No. 11, 916–927, November 1985.

[55] Turner, P. F., "Mini-annular phased array for limb hyperthermia," *IEEE Trans. on Microwave Theory and Techniques*, **MTT-34,** No. 5, 508–513, May 1986.

[56] Sathiaseelam, V., M. F. Iskander, G. C. W. Howard, and N. M. Blechen, "Theoretical analysis and clinical demonstration of the effects of power pattern control using the annular phased array hyperthermia system," *IEEE Trans. on Microwave Theory and Techniques,* **MTT-34**, No. 5, 517–519, May 1986.

[57] Iskander, M. F., P. F. Turner, J. B. DuBow, and J. Kao, " Two-

dimensional technique to calculate the EM power deposition pattern in the human body," *J. of Microwave Power*, **17**, No. 3, 1982.

[58] Spiegel, R. J., "A review of numerical models for predicting the energy deposition and resultant thermal response of humans exposed to electromagnetic fields," *IEEE Trans. on Microwave Theory and Techniques*, **MTT-32**, No. 8, 730–746, August 1984.

[59] Weil, C. M., "Absorption characteristics of multilayered sphere models exposed to UHF/microwave radiation," *IEEE Trans. on Biomedical Engineering*, **BME-22**, No. 6, 468–476, November 1975.

[60] Kritikos, H. N. and H. P. Schwan, "The distribution of heating potential inside lossy spheres," *IEEE Trans. on Biomedical Engineering*, **BME-22**, No. 6, 457–463, November 1975.

[61] Kong, J. A., *Electromagnetic Wave Theory*, Wiley, New York, 1986.

[62] Hildebrand, F. B., *Advanced Calculus for Applications*, Prentice-Hall, Englewood Cliffs, NJ, 1976.

[63] Abramowitz, M. and I. A. Stegun, *Handbook of Mathematical Functions with Formulas, Graphs, and Mathematical Tables*, US Government Printing Office, Washington, DC, NBS Applied Mathematics Series – 55, 1972.

[64] Stratton, J. A., *Electromagnetic Theory*, McGraw-Hill, New York, 1941.

[65] Andersen, J. B., Private communication.

[66] Christensen, D. A. and C. H. Durney, "Hyperthermia production for cancer therapy: a review of fundamentals and methods," *J. of Microwave Power*, **16**, No. 2, 89–105, 1981.

[67] Tsai, C. T., C. H. Durney, and D. A. Christensen, "Calculated power absorption for hyperthermia applications consisting of electric dipole arrays," *J. of Microwave Power*, **19**, No. 1, 1–13, 1984.

[68] Wait, J. R., "Focussed heating in cylindrical targets: part 1," *IEEE Trans. on Microwave Theory and Techniques*, **MTT-33**, No. 7, 647–649. July 1985.

[69] Strohbehn, J. W. and R. B. Roemer, "A survey of computer simulations of hyperthermia treaments," *IEEE Trans. on Biomedical Engineering,* **68**, No. 1, 136–149, January 1984.

[70] Sultan, M. F. and R. Mittra, "An iterative moment method for analyzing the electromagnetic field distribution inside inhomogenous lossy dielectric objects," *IEEE Trans. on Microwave Theory and Techniques,* **MTT-33**, No. 2, 163–168, February 1985.

[71] Holt, F. S., "Application of geometric optics to the design analysis of microwave antennas," *Tech. report AFCRL-67-0501*, Air Force Cambridge Research Laboratory, September 1967.

3

EFFECTIVE PROPAGATION CONSTANTS IN MEDIA WITH DENSELY DISTRIBUTED DIELECTRIC PARTICLES OF MULTIPLE SIZES AND PERMITTIVITIES

K. H. Ding and L. Tsang

3.1 Introduction

In a dense medium, the particles occupy an appreciable fractional volume. In a nontenuous medium, the dielectric properties of the particles are substantially different from that of the background medium. In recent years, studies of the propagation and scattering of waves in dense nontenuous media have been made with applications to geophysical terrain and composite materials [1–9]. Such media often consist of mixtures of particles with multiple sizes and permittivities that are embedded in a background medium. It has been shown both theoretically and experimentally that in a dense nontenuous medium, the assumption that the particles scatter independently is not valid. A more rigorous theory is needed to take into account correlated scattering between particles.

In section 3.2, we outline the governing multiple scattering equations for random discrete scatterers. Various approximations to calculate the first moment of the field are discussed. The treatment includes the important case of dense nontenuous medium with multiple species of particles. The multiple species refers to the fact that the medium is a mixture of particles with different sizes and permittivities. Special emphases are placed on the Quasicrystalline Approximation (QCA) and the Quasicrystalline Approximation with Coherent Potential (QCA–CP) [2,8].

In section 3.3, the pair distribution function is discussed. Specifically, the Percus–Yevick approximation is used to calculate the pair distribution functions in a medium with non-interpenetrable spheres of multiple sizes [10–16]. Numerical results of pair distribution functions are illustrated as functions of particle sizes and fractional volumes.

In section 3.4, the low frequency limit of the dispersion relations under the QCA and QCA–CP approximations is considered. Analytic closed form expressions are obtained for the complex effective propagation constants. Numerical results are illustrated as functions of particle sizes, fractional volumes, and dielectric properties.

In section 3.5, we consider the case of moderate size particles. The QCA equations are formulated in terms of the T-matrix formulism and utilizing vector spherical waves as basis functions [4,8]. The final equations of the dispersion relation for multi-species are solved numerically and illustrated as functions of the physical properties of the particles.

In section 3.6, we examine energy conservation. In multiple scattering problems of lossless media or slightly lossy media, one key objective is to calculate the second moment and the incoherent intensity of waves [17,18]. For these problems, the calculation of the first moment and the effective propagation constant serves as an intermediate step. Thus, approximation of the first moment must be done with due regard to energy conservation and consistency with the second moment. It is shown in section 3.6 that QCA–CP is energetically consistent with the ladder approximation of second moment. This is analogous to the nonlinear approximation of continuous random medium theory [19]. The QCA–CP has been used in conjunction with the ladder approximation modified with the pair distribution function to derive a set of dense medium radiative transfer equations that obey energy conservation and include multiple scattering of the incoherent fields [8,20–22]. In section 3.6, we also derive the intensity operator under ladder approximation of correlated scatterers for multiple species which will be useful for second moment calculations and which also obeys energy conservation.

The problem of multiple particle sizes and permittivities is an important one for both geophysical terrain and composite materials. The particles in natural geophysical terrain are usually not of identical sizes. They usually follow a drop size distribution which can be discretized and represented by a histogram. Thus, the approach given in this paper for the calculation of effective propagation constants for such cases of random discrete scatterers of multiple sizes gives a better description of the effects of correlated scattering.

3.2 Multiple Scattering Equations and Dispersion Relations

Consider a random medium consisting of N discrete scatterers of L different species, located at $\overline{r}_1, \overline{r}_2, \ldots, \overline{r}_N$ in a volume V of background medium with permittivity ϵ (Fig. 3.2.1). The different species refer to the fact that particles can be of different shapes, sizes, and permittivities. However, interpenetration of particles is not allowed. The different species are denoted by $s_j = 1, 2, \ldots, L$. Each species of particles can have a distinct size, shape, and permittivity ϵ_{s_j}.

To study the propagation of electromagnetic waves in such a medium, a multiple scattering theory based on dyadic Green's operator can be applied. The N-particle dyadic Green's operator $\overline{\overline{G}}$ satisfies

Figure 3.2.1 Discrete nontenuous media with multi-species particles with different sizes and permittivities.

the multiple scattering equations [2,8]

$$\overline{\overline{G}} = \overline{\overline{G}}_o + \overline{\overline{G}}_o \sum_{j=1}^{N} \overline{\overline{U}}_j^{s_j} \overline{\overline{G}} \tag{1}$$

where $\overline{\overline{G}}_o$ is the dyadic Green's operator of the background medium with propagation constant $k = \omega\sqrt{\mu\epsilon}$, ω is the angular frequency, and μ is the permeability. The scattering potential operator, $\overline{\overline{U}}_j^{s_j}$, associated with the j th scatterer of species s_j is such that

$$\begin{aligned}\left\langle \overline{r} \mid \overline{\overline{U}}_j^{s_j} \mid \overline{r}' \right\rangle &= U^{s_j}(\overline{r} - \overline{r}_j)\overline{\overline{I}}\delta(\overline{r} - \overline{r}') \\ &= U_j^{s_j}(\overline{r})\overline{\overline{I}}\delta(\overline{r} - \overline{r}')\end{aligned} \tag{2}$$

where $\overline{\overline{I}}$ is unit dyad and

$$U^{s_j}(\overline{r} - \overline{r}_j) = \begin{cases} 0 & \text{for } \overline{r} \text{ outside particle } j \\ k_{s_j}^2 - k^2 & \text{for } \overline{r} \text{ inside particle } j \end{cases} \tag{3}$$

with $k_{s_j} = \omega\sqrt{\mu\epsilon_{s_j}}$ being the wavenumber in the s_j type scatterers.

The particle distribution is described by the many-particle probability density function $p(\overline{r}_1, \overline{r}_2, \ldots, \overline{r}_N; s_1, s_2, \ldots, s_L)$. The first and second order statistics are given by

$$Np(\overline{r}_i; s_i) = n_{s_i} \tag{4}$$

$$(N-1)p(\overline{r}_j; s_j \mid \overline{r}_i; s_i) = n_{s_j} g_{s_i s_j}(\overline{r}_i, \overline{r}_j) \tag{5}$$

where $p(\mid)$ is the conditional probability function, n_{s_i} is the number density of s_i type particles, and $g_{s_i s_j}(\overline{r}_i, \overline{r}_j)$ is the pair distribution functions for two particles of species s_i and s_j. Using Bayes rule, it follows readily that $g_{s_i s_j}$ is symmetric so that $g_{s_i s_j}(\overline{r}_i, \overline{r}_j) = g_{s_j s_i}(\overline{r}_j, \overline{r}_i)$.

Generally, the Green's function in (1) depends on the locations and types of the N particles. The concept of configuration average [8] is used to obtain the average Green's function as a function of number densities and pair distribution functions. Quasicrystalline Approximation (QCA) and Quasicrystalline Approximation with Coherent Potential (QCA–CP) are applied to derive the dispersion relations and the results are obtained in the momentum representations.

a. *Quasicrystalline Approximation (QCA)*

Given the scattering potential operator $\overline{\overline{U}}_j^{s_j}$ for the jth scatterer of s_j species, the transition operator $\overline{\overline{T}}_j^{s_j}$ associated with this particle satisfies the Lippmann–Schwinger equation

$$\overline{\overline{T}}_j^{s_j} = \overline{\overline{U}}_j^{s_j} + \overline{\overline{U}}_j^{s_j} \overline{\overline{G}}_o \overline{\overline{T}}_j^{s_j} \tag{6}$$

which describes the scattering of wave from the jth particle in the absence of all other particles. In terms of this single particle transition operator $\overline{\overline{T}}_j^{s_j}$, a scattering operator $\overline{\overline{Q}}_j^{s_j}$ at $\overline{r}_j$ is defined as [8]

$$\overline{\overline{Q}}_j^{s_j} = \overline{\overline{T}}_j^{s_j} + \overline{\overline{T}}_j^{s_j} \overline{\overline{G}}_o \sum_{l \neq j}^{N} \overline{\overline{Q}}_l^{s_l} \tag{7}$$

which can be interpreted as scattering operator for the jth particle in the presence of other particles. In terms of $\overline{\overline{Q}}_j^{s_j}$, the multiple scattering equation (1) becomes

$$\overline{\overline{G}} = \overline{\overline{G}}_o + \overline{\overline{G}}_o \sum_{j=1}^{N} \overline{\overline{Q}}_j^{s_j} \overline{\overline{G}}_o \tag{8}$$

The configuration average of $\overline{\overline{G}}$ and $\overline{\overline{Q}}_j^{s_j}$ are

$$E\left(\overline{\overline{G}}\right) = \overline{\overline{G}}_o + N\overline{\overline{G}}_o E\left[E_j\left(\overline{\overline{Q}}_j^{s_j}\right)\right]\overline{\overline{G}}_o \tag{9}$$

$$E_l\left(\overline{\overline{Q}}_l^{s_l}\right) = \overline{\overline{T}}_l^{s_l} + (N-1)\overline{\overline{T}}_l^{s_l}\overline{\overline{G}}_o E_l\left[E_{lj}\left(\overline{\overline{Q}}_j^{s_j}\right)\right] \tag{10}$$

where E stands for the expectation value, E_j is the conditional average given the position and species of particle j, and E_{lj} is the conditional average of given the positions and species of particles j and l. Thus,

$$E\left[E_j\left(\overline{\overline{Q}}_j^{s_j}\right)\right] = \iint d\overline{r}_j ds_j p\left(\overline{r}_j; s_j\right) E_j\left(\overline{\overline{Q}}_j^{s_j}\right) \tag{11}$$

$$E_l\left[E_{lj}\left(\overline{\overline{Q}}_j^{s_j}\right)\right] = \iint d\overline{r}_j ds_j p\left(\overline{r}_j; s_j \mid \overline{r}_l; s_l\right) E_{lj}\left(\overline{\overline{Q}}_j^{s_j}\right) \tag{12}$$

Equations (9) and (10) indicate that the total average is given in terms of the conditional average with one particle fixed and the conditional average with one particle fixed is given in terms of the conditional average with two particles fixed. In a similar manner, we can express the conditional average with n particles fixed in terms of the conditional average with $n+1$ particles fixed. A hierarchy of equations is thus generated. Truncation at various stages leads to different approximations.

Under quasicrystalline approximation, truncation is made at the second stage of the hierarchy of equations

$$E_{lj}\left(\overline{\overline{Q}}_j^{s_j}\right) \approx E_j\left(\overline{\overline{Q}}_j^{s_j}\right) \tag{13}$$

Assuming discrete values of s_j, we can replace the integration with respect to s_j in (11) and (12) by the summation over s_j. From the properties of (4) and (5), $E\left(\overline{\overline{G}}\right)$ and $E_l\left(\overline{\overline{Q}}_l^{s_l}\right)$ can be expressed as

$$E\left(\overline{\overline{G}}\right) = \overline{\overline{G}}_o + \overline{\overline{G}}_o \sum_{s_j=1}^{L} n_{s_j} \int d\overline{r}_j E_j\left(\overline{\overline{Q}}_j^{s_j}\right)\overline{\overline{G}}_o \tag{14}$$

$$E_l\left(\overline{\overline{Q}}_l^{s_l}\right) = \overline{\overline{T}}_l^{s_l} + \overline{\overline{T}}_l^{s_l}\overline{\overline{G}}_o \sum_{s_j=1}^{L} n_{s_j} \int d\overline{r}_j g_{s_l s_j}\left(\overline{r}_l, \overline{r}_j\right) E_j\left(\overline{\overline{Q}}_j^{s_j}\right) \tag{15}$$

Assume the solution for $E_l\left(\overline{\overline{Q}}_l^{s_l}\right)$ in the momentum representation as [8]

$$\left\langle \overline{p} \mid E_l\left(\overline{\overline{Q}}_l^{s_l}\right) \mid \overline{p}' \right\rangle = e^{-i(\overline{p}-\overline{p}')\cdot\overline{r}_l}\, \overline{\overline{Q}}_p^{s_l}(\overline{p},\overline{p}') \tag{16}$$

Further let

$$\overline{\overline{C}}_p^{s_l}(\overline{p},\overline{p}') = \overline{\overline{Q}}_p^{s_l}(\overline{p},\overline{p}') \left[\overline{\overline{I}} + \overline{\overline{G}}_o(\overline{p}') \sum_{s_j=1}^{L} n_{s_j} \overline{\overline{Q}}_p^{s_j}(\overline{p}',\overline{p}')\right]^{-1} \tag{17}$$

The quantity $\overline{\overline{C}}_p^{s_l}(\overline{p},\overline{p}')$ is governed by an integral equation as follows

$$\overline{\overline{C}}_p^{s_l}(\overline{p},\overline{p}') = \overline{\overline{T}}_p^{s_l}(\overline{p},\overline{p}') + \sum_{s_j=1}^{L} n_{s_j} \times \int d\overline{p}'' \overline{\overline{T}}_p^{s_l}(\overline{p},\overline{p}'') \\ \overline{\overline{G}}_o(\overline{p}'')\, H_{s_l s_j}(\overline{p}''-\overline{p}')\, \overline{\overline{C}}_p^{s_j}(\overline{p}'',\overline{p}') \tag{18}$$

where $H_{s_l s_j}(\overline{p})$ is defined as

$$H_{s_l s_j}(\overline{p}) = \frac{1}{(2\pi)^3} \int d\overline{r} h_{s_l s_j}(\overline{r}) \exp(-i\overline{p}\cdot\overline{r}) \tag{19}$$

and

$$h_{s_l s_j}(\overline{r}) = g_{s_l s_j}(\overline{r}) - 1 \tag{20}$$

and $\overline{\overline{T}}_p^{s_j}(\overline{p},\overline{p}') = \left\langle \overline{p} \mid \overline{\overline{T}}^{s_j} \mid \overline{p}' \right\rangle$ is the momentum representation of $\overline{\overline{T}}^{s_j}$. For statistical homogeneous medium, the average dyadic Green's operator is diagonal in the momentum representation, $\left\langle \overline{p} \mid E\left(\overline{\overline{G}}\right) \mid \overline{p}' \right\rangle = \overline{\overline{G}}(\overline{p}) \langle \overline{p} \mid \overline{p}' \rangle$, and $\overline{\overline{G}}(\overline{p})$ is given by

$$\overline{\overline{G}}(\overline{p}) = \left[\overline{\overline{G}}_o^{-1}(\overline{p}) - \sum_{s_j=1}^{L} n_{s_j} \overline{\overline{C}}_p^{s_j}(\overline{p},\overline{p})\right]^{-1} \tag{21}$$

where $\overline{\overline{C}}_p^{s_j}(\overline{p},\overline{p}')$ satisfies the integral equation (18). By setting the determinant of the inverse of $\overline{\overline{G}}(\overline{p})$ equals to zero, the dispersion relation is obtained as

$$\det \left[\overline{\overline{G}}_o^{-1}(\overline{p}) - \sum_{s_j=1}^{L} n_{s_j} \overline{\overline{C}}_p^{s_j}(\overline{p},\overline{p})\right] = 0 \tag{22}$$

where det stands for determinant. The complex value of $\overline{p}$ given by solution of (22) is the effective propagation constant K. Thus, the equations of QCA for multiple species consist of solving (18) for $\overline{\overline{C}}_p^{s_j}(\overline{p},\overline{p}')$ and then substituting that in (22) to calculate the effective propagation constant K.

The dispersion equations of (18) and (22) can also be expressed in terms of mass operator. We define the operator $\overline{\overline{C}}_j^{s_j}$ that is a function of $\overline{r}_j$ and s_j such that

$$\left\langle \overline{p} \mid \overline{\overline{C}}_j^{s_j} \mid \overline{p}' \right\rangle = e^{-i(\overline{p}-\overline{p}')\cdot\overline{r}_j}\, \overline{\overline{C}}_p^{s_j}(\overline{p},\overline{p}') \tag{23}$$

Equation (18) can be put in operator form

$$\overline{\overline{C}}_l^{s_l} = \overline{\overline{T}}_l^{s_l} + \sum_{s_j=1}^{L} n_{s_j} \int d\overline{r}_j h_{s_l s_j}(\overline{r}_j - \overline{r}_l)\, \overline{\overline{T}}_l^{s_l}\, \overline{\overline{G}}_o\, \overline{\overline{C}}_j^{s_j} \tag{24}$$

Thus, $\overline{\overline{C}}_l^{s_l}$ can be interpreted as the scattering transition operator of particle l in the presence of other particles averaged over particle positions and properties under the QCA approximation. The operator equation equivalent to (21) is then

$$E\left(\overline{\overline{G}}\right) = \overline{\overline{G}}_o + \overline{\overline{G}}_o \sum_{s_j=1}^{L} n_{s_j} \int d\overline{r}_j\, \overline{\overline{C}}_j^{s_j} E\left(\overline{\overline{G}}\right) \tag{25}$$

Let

$$\overline{\overline{M}} = \sum_{s_j=1}^{L} n_{s_j} \int d\overline{r}_j\, \overline{\overline{C}}_j^{s_j} \tag{26}$$

be the dyadic mass operator for the medium with scatterers of L different species under the quasicrystalline approximation. Equation (25) can be rewritten as

$$E\left(\overline{\overline{G}}\right) = \overline{\overline{G}}_o + \overline{\overline{G}}_o \overline{\overline{M}} E\left(\overline{\overline{G}}\right) \tag{27}$$

In terms of the mass operator, the dispersion relation of (21) becomes

$$\overline{\overline{G}}(\overline{p}) = \left[\overline{\overline{G}}_o^{-1}(\overline{p}) - \overline{\overline{M}}(\overline{p},\overline{p})\right]^{-1} \tag{28}$$

with

$$\left\langle \overline{p} \mid \overline{\overline{M}} \mid \overline{p}' \right\rangle = \langle \overline{p} \mid \overline{p}' \rangle \sum_{s_j=1}^{L} n_{s_j} \overline{\overline{C}}_p^{s_j} (\overline{p}, \overline{p}) \tag{29}$$

b. Quasicrystalline Approximation with Coherent Potential (QCA–CP)

In the coherent potential approximation [8,23–26], the medium is described by a coherent potential operator $\overline{\overline{w}}(\overline{p})$ which is constant in space but may be a function of the momentum operator $\overline{p}$. By adding and subtracting an operator $n\overline{\overline{w}}(\overline{p})$, (1) can be expressed as

$$\overline{\overline{G}} = \left[\overline{\overline{G}}_o^{-1} - n\overline{\overline{w}}(\overline{p}) - \sum_{j=1}^{N} \left(\overline{\overline{U}}_j^{s_j} - \frac{\overline{\overline{w}}(\overline{p})}{V} \right) \right]^{-1} \tag{30}$$

where

$$n = \sum_{s_j=1}^{L} n_{s_j} \tag{31}$$

is the total number density of particles. Let $\overline{\overline{G}}_c^{-1} = \overline{\overline{G}}_o^{-1} - n\overline{\overline{w}}(\overline{p})$ be the the coherent potential Green's operator and $\tilde{\overline{\overline{U}}}_j^{s_j} = \overline{\overline{U}}_j^{s_j} - \overline{\overline{w}}(\overline{p})/V$ be the modified potential operator. In terms of $\overline{\overline{G}}_c$ and $\tilde{\overline{\overline{U}}}_j^{s_j}$, the multiple scattering equation (1) becomes

$$\overline{\overline{G}} = \overline{\overline{G}}_c + \overline{\overline{G}}_c \sum_{j=1}^{N} \tilde{\overline{\overline{U}}}_j^{s_j} \overline{\overline{G}} \tag{32}$$

The QCA–CP approximation is introduced by applying QCA to the scattering equation of (32) with the new potential $\tilde{\overline{\overline{U}}}_j^{s_j}$ and then imposing the self-consistent condition that the average Green's operator must be the same as the coherent potential Green's operator i.e. $E\left(\overline{\overline{G}}\right) = \overline{\overline{G}}_c$. With the $\overline{\overline{G}}_o$ in equation (6) replaced by $\overline{\overline{G}}_c$, a modified transition operator $\hat{\overline{\overline{t}}}_j^{s_j}$ is defined to satisfy

$$\hat{\overline{\overline{t}}}_j^{s_j} = \tilde{\overline{\overline{U}}}_j^{s_j} + \tilde{\overline{\overline{U}}}_j^{s_j} \overline{\overline{G}}_c \hat{\overline{\overline{t}}}_j^{s_j} \tag{33}$$

The result for the average Green's operator under QCA becomes

$$\overline{\overline{G}}(\overline{p}) = \overline{\overline{G}}_c(\overline{p}) = \left[\overline{\overline{G}}_o^{-1}(\overline{p}) - \sum_{s_j=1}^{L} n_{s_j} \widehat{\overline{\overline{C}}}_p^{s_j}(\overline{p},\overline{p}) \right]^{-1} \tag{34}$$

where $\widehat{\overline{\overline{C}}}_p^{s_j}(\overline{p},\overline{p}')$ satisfies the integral equation

$$\widehat{\overline{\overline{C}}}_p^{s_l}(\overline{p},\overline{p}') = \widehat{\overline{\overline{t}}}_p^{s_l}(\overline{p},\overline{p}') + \sum_{s_j=1}^{L} n_{s_j} \times \int d\overline{p}'' \widehat{\overline{\overline{t}}}_p^{s_l}(\overline{p},\overline{p}'') \overline{\overline{G}}_c(\overline{p}'') H_{s_l s_j}(\overline{p}'' - \overline{p}') \widehat{\overline{\overline{C}}}_p^{s_j}(\overline{p}'',\overline{p}') \tag{35}$$

and $\widehat{\overline{\overline{t}}}_p^{s_j}(\overline{p},\overline{p}') = \left\langle \overline{p} \,|\, \widehat{\overline{\overline{t}}}^{s_j} \,|\, \overline{p}' \right\rangle$ is the momentum representation of $\widehat{\overline{\overline{t}}}^{s_j}$. The dispersion relation for QCA–CP is

$$\det \left[\overline{\overline{G}}_o^{-1}(\overline{p}) - \sum_{s_j=1}^{L} n_{s_j} \widehat{\overline{\overline{C}}}_p^{s_j}(\overline{p},\overline{p}) \right] = 0 \tag{36}$$

Thus solution of QCA–CP is provided by first solving (32) and (34). Substituting $\widehat{\overline{\overline{C}}}_p^{s_j}(\overline{p},\overline{p}')$ in (35) and solving for $\overline{p}$ gives the complex effective propagation constant K.

Similarly, the dispersion relations of (35) and (36) can be expressed in terms of mass operator. Let the mass operator $\widehat{\overline{\overline{M}}}$ under QCA–CP be expressed as

$$\widehat{\overline{\overline{M}}} = \sum_{s_j=1}^{L} n_{s_j} \int d\overline{r}_j \widehat{\overline{\overline{C}}}_j^{s_j} \tag{37}$$

where

$$\widehat{\overline{\overline{C}}}_l^{s_l} = \widehat{\overline{\overline{t}}}_l^{s_l} + \sum_{s_j=1}^{L} n_{s_j} \int d\overline{r}_j h_{s_l s_j}(\overline{r}_j - \overline{r}_l) \widehat{\overline{\overline{t}}}_l^{s_l} \overline{\overline{G}}_c \widehat{\overline{\overline{C}}}_j^{s_j} \tag{38}$$

The average Green's operator under QCA–CP can be expressed as

$$E\left(\overline{\overline{G}}\right) = \overline{\overline{G}}_c = \overline{\overline{G}}_o + \overline{\overline{G}}_o \widehat{\overline{\overline{M}}} \overline{\overline{G}}_c \tag{39}$$

In this case,

$$\left\langle \overline{p} \mid \overline{\overline{\widehat{M}}} \mid \overline{p}' \right\rangle = \langle \overline{p} \mid \overline{p}' \rangle \sum_{s_j=1}^{L} n_{s_j} \overline{\overline{\widehat{C}}}_p^{s_j} (\overline{p},\overline{p}) \tag{40}$$

The mean Green's function then becomes

$$\overline{\overline{G}}(\overline{p}) = \left[\overline{\overline{G}}_o^{-1} (\overline{p}) - \overline{\overline{\widehat{M}}} (\overline{p},\overline{p}) \right]^{-1} \tag{41}$$

and the dispersion relation is

$$\det \left[\overline{\overline{G}}_o^{-1} (\overline{p}) - \overline{\overline{\widehat{M}}} (\overline{p},\overline{p}) \right] = 0 \tag{42}$$

Thus we can regard equations (26) and (37) as the mass operators under QCA and QCA–CP respectively.

3.3 Pair Distribution Functions for Media with Particles of Multiple Sizes

In applying QCA and QCA–CP to calculate the effective propagation constants, it is seen from (18), (19) and (35) that the results depend on the pair distribution functions for multi-species. The medium composed of N scatterers considered here is analogous to an ensemble of mixture of L types of particles in the study of statistical mechanics, by considering the dynamics and positions of the particles with regard to the interparticle forces. Studies have been made in obtaining the pair distribution functions using various approximate theories. One of the important results is based on the Percus–Yevick (PY) approximation.

The pair distribution function $g_{ij}(r)$ is proportional to the conditional probability of finding a particle of type j at a distance r from the origin given that there is a particle of type i at the origin. The total correlation function, $h_{ij}(r)$, is defined as

$$h_{ij}(r) = g_{ij}(r) - 1 \tag{43}$$

which describes the total influence of particle of species i on particle of species j. The direct correlation function, $c_{ij}(r)$, is related to $h_{ij}(r)$ by means of the generalized Ornstein–Zernike (OZ) relation [11,12]

$$h_{ij}(r) = c_{ij}(r) + \sum_{l=1}^{L} n_l \int d\overline{r}' c_{il}(r') h_{lj} (\mid \overline{r} - \overline{r}' \mid) \tag{44}$$

The physical interpretation of (44) is that the total correlation function can be decomposed into a sum of direct and indirect correlation functions, and the indirect influence of particle of species i on particle of species j is a result of particle of species i acting directly on a particle of species l at $\overline{r}'$, which in turn exerts total influence on particle of species j. The indirect influence is averaged over particle positions $\overline{r}'$ and particle species l and weighted by the number density n_l as indicated in (44).

In the following, we shall describe the results for non-interpenetrable spheres under the PY approximation. For non-interpenetrable spheres, the interparticle forces are zero except for the fact that two particles cannot interpenetrate each other. Let a_i and a_j be the particle radii of species i and j respectively. Then

$$g_{ij}(r) = 0 \qquad \text{for} \quad r < a_i + a_j \tag{45}$$

Under the PY approximation, it is assumed that for non-interpenetrable particles the direct correlation function $c_{ij}(r)$ is zero for separation larger than $a_i + a_j$. This physically means that any *net* interparticle influence between two particles for distance larger than $a_i + a_j$ is due to the presence of other particles and is attributed entirely to indirect influence. Thus,

$$c_{ij}(r) = 0 \qquad \text{for} \quad r > R_{ij} \tag{46}$$

where $R_{ij} = a_i + a_j$. Next, let

$$Y_{ij}(r) = \begin{cases} -c_{ij}(r) & \text{for} \quad r < R_{ij} \\ g_{ij}(r) & \text{for} \quad r \geq R_{ij} \end{cases} \tag{47}$$

Then using (43), (45)–(47), the OZ relation can be expressed entirely in terms of Y_{ij}

$$\begin{aligned} Y_{ij}(r) = 1 &+ \sum_{l=1}^{L} n_l \int_{r'<R_{lj};|\overline{r}-\overline{r}'|<R_{il}} d\overline{r}' Y_{il}(r') \\ &- \sum_{l=1}^{L} n_l \int_{r'<R_{lj};|\overline{r}-\overline{r}'|>R_{il}} d\overline{r}' \left[Y_{il}\left(\overline{r}-\overline{r}'\right) - 1 \right] Y_{lj}(r') \\ &\qquad i,j = 1,2,\ldots,L \end{aligned} \tag{48}$$

The solution of (48) for the case of single species has been solved by Wertheim, Thiele and Baxter. For the case of two species, the solution can be found in Lebowitz. For the case of general L species, the solution is obtained by Baxter based on a generalized Wiener–Hopf technique.

Let $\widetilde{H}_{ij}(\overline{p})$ and $\widetilde{C}_{ij}(\overline{p})$ be proportional to the three dimensional Fourier transform of $h_{ij}(\overline{r})$ and $c_{ij}(\overline{r})$ as follows

$$\widetilde{H}_{ij}(\overline{p}) = (n_i n_j)^{1/2} \int d\overline{r} e^{i\overline{p}\cdot\overline{r}} h_{ij}(\overline{r}) \tag{49}$$

$$\widetilde{C}_{ij}(\overline{p}) = (n_i n_j)^{1/2} \int d\overline{r} e^{i\overline{p}\cdot\overline{r}} c_{ij}(\overline{r}) \tag{50}$$

then in matrix form, the OZ relation becomes

$$\widetilde{H}(p) = \widetilde{C}(p) + \widetilde{C}(p)\widetilde{H}(p) \tag{51}$$

Because of the spherical particle assumption, the transform only depends on $p = |\overline{p}|$. In (51) all matrices are of dimensional $L \times L$. Based on the generalized Wiener–Hopf technique, the solution of the $L \times L$ matrix $\widetilde{C}(p)$ can be factorized as [14]

$$\widetilde{C}(p) = E - \widetilde{Q}^T(-p)\widetilde{Q}(p) \tag{52}$$

where E is the $L \times L$ unit matrix, and superscript T denotes transpose of the $L \times L$ matrix $\widetilde{Q}$

$$\widetilde{Q}_{ij}(p) = \delta_{ij} - \int_{S_{ij}}^{R_{ij}} dr e^{ipr} Q_{ij}(r) \tag{53}$$

δ_{ij} is the Kronecker delta, and

$$S_{ij} = a_i - a_j \tag{54}$$

$$R_{ij} = a_i + a_j \tag{55}$$

$$Q_{ij}(r) = 2\pi (n_i n_j)^{1/2} q_{ij}(r) \tag{56}$$

where $Q_{ij}(r)$ and $q_{ij}(r)$ are functions only nonzero over the range $S_{ij} \le r \le R_{ij}$. Note that S_{ij} can be negative. The solution of $q_{ij}(r)$

is [14]

$$q_{ij}(r) = A_i \frac{r^2}{2} + B_i r + D_{ij} \tag{57}$$

$$A_i = \frac{1 - \xi_3 + 6a_i \xi_2}{(1 - \xi_3)^2} \tag{58}$$

$$B_i = -\frac{6a_i^2 \xi_2}{(1 - \xi_3)^2} \tag{59}$$

$$D_{ij} = -A_i \frac{R_{ij}^2}{2} - B_i R_{ij} \tag{60}$$

$$\xi_\alpha = \frac{\pi}{6} \sum_{j=1}^{L} n_j (2a_j)^\alpha \tag{61}$$

$$\alpha = 1, 2, 3.$$

The process of computation is as follows. By using (54)–(61), $Q_{ij}(r)$ for $S_{ij} \le r \le R_{ij}$ can be computed. Then, $Q_{ij}(p)$ is calculated by using (53). Next, the matrix $\widetilde{C}(p)$ is evaluated by using (52).

Next the matrix equation of (51) is solved and $\widetilde{H}(p)$ is calculated. Then, $h_{ij}(r) = g_{ij}(r) - 1$ is computed by taking an inverse Fourier transform. In this manner, the pair distribution function $g_{ij}(r)$ is calculated and is numerically illustrated in Figs. 3.3.1–3.3.6.

In the QCA and QCA–CP equations $H_{s_i s_j}(p)$ as defined by (19) is needed. Comparing (19) and (49) gives

$$H_{ij}(\overline{p}) = \frac{\widetilde{H}_{ij}(-\overline{p})}{(2\pi)^3 (n_i n_j)^{1/2}} \tag{62}$$

In Fig. 3.3.1, the pair distribution functions for the case of particles of identical sizes are plotted for $f = 0.2$ and $f = 0.4$. We see that the conditional probability is zero for the separation less than one diameter of particle because the particles cannot interpenetrate each other. For higher concentration of particles, the pair distribution function shows more variations indicating that the particles have less freedom in positioning themselves.

In Fig. 3.3.2, the pair distribution functions for a mixture of two particle sizes are plotted with $a_2 = 2a_1$, $f_1 = 0.03$, and $f_2 = 0.01$. Because of small fractional volumes in this case, the pair distribution functions show only a small deviation from the results of hole correction

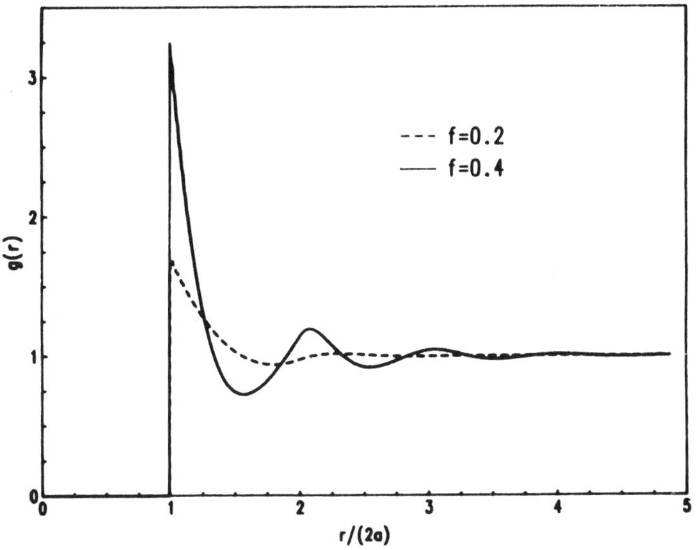

Figure 3.3.1 Percus–Yevick results of $g(r)$, for $f = 0.2$, and $f = 0.4$, vs $r/(2a)$.

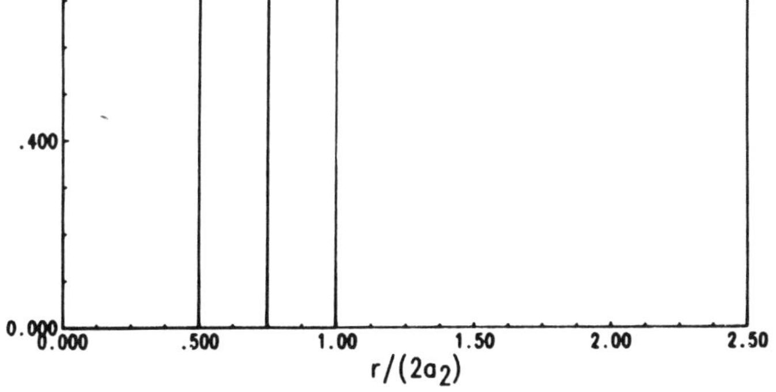

Figure 3.3.2 Percus–Yevick results of $g_{ij}(r)$, $i, j = 1, 2$, for $f_1 = 0.03$, and $f_2 = 0.01$, vs $r/(2a_2)$, $a_2 = 2a_1$.

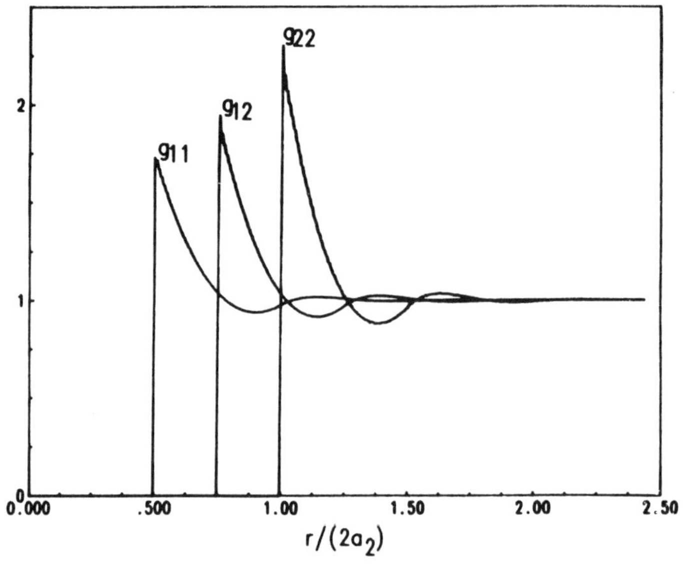

Figure 3.3.3 Percus–Yevick results of $g_{ij}(r)$, $i,j = 1,2$, for $f_1 = 0.2$, and $f_2 = 0.01$, vs $r/(2a_2)$, $a_2 = 2a_1$.

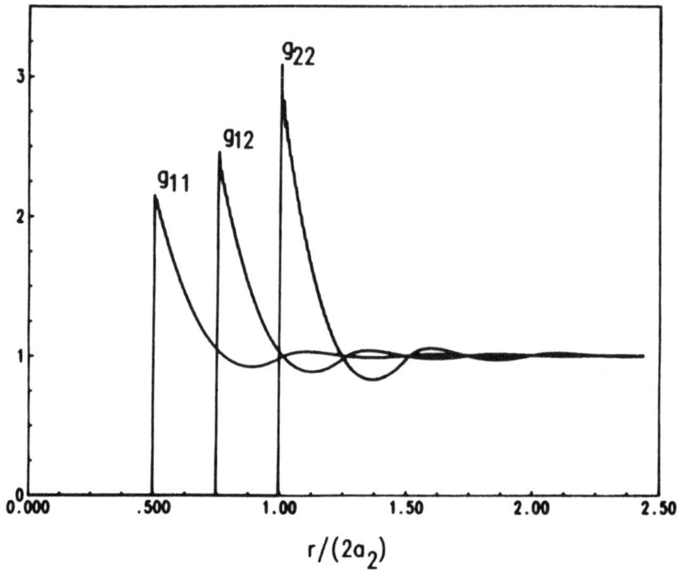

Figure 3.3.4 Percus–Yevick results of $g_{ij}(r)$, $i,j = 1,2$, for $f_1 = 0.2$, and $f_2 = 0.1$, vs $r/(2a_2)$, $a_2 = 2a_1$.

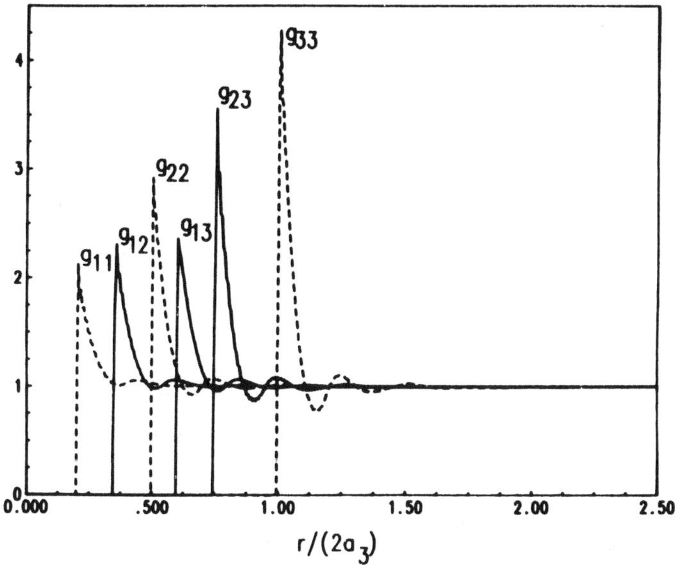

Figure 3.3.5 Percus–Yevick results of $g_{ij}(r)$, $i, j = 1, 2, 3$, for $f_1 = 0.15$, $f_2 = 0.1$, and $f_3 = 0.05$, vs $r/(2a_3)$, $a_3 = 5a_1$, $a_2 = 2.5a_1$.

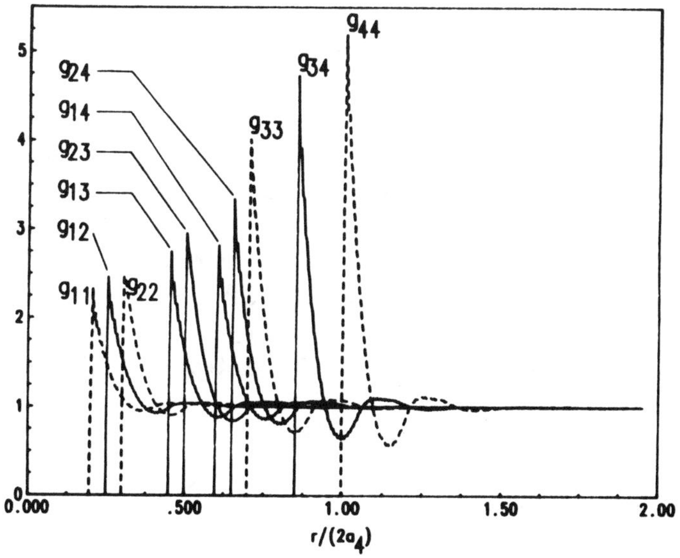

Figure 3.3.6 Percus–Yevick results of $g_{ij}(r)$, $i, j = 1, 2, 3, 4$, for $f_1 = 0.15$, $f_2 = 0.1$, $f_3 = 0.05$, and $f_4 = 0.02$, vs $r/(2a_4)$, $a_4 = 5a_1$, $a_3 = 3.5a_1$, $a_2 = 1.5a_1$.

(HC). In the HC approximation,

$$g_{ij}(r) = \begin{cases} 0 & r < a_i + a_j \\ 1 & r \geq a_i + a_j \end{cases} \tag{63}$$

When the pair distribution function is equal to 1, the two particle positions are independent. Thus, the hole correction approximation assumes that the particles are independent aside from the fact that they cannot interpenetrate each other. The deviation of pair distribution function from unity is a measure of the *freedom* of particle positions with larger deviation corresponding to a smaller freedom.

In Figs. 3.3.3 and 3.3.4, particle concentrations are higher so that the deviations of pair functions from unity are significant with less freedom for larger concentrations. The results also indicate that g_{22} is more different from unity than g_{11}. This is true in spite of the fact that the fractional volume of particle size 2 in both figures are small because the high concentration of particle size 1 affects the freedom of positions of particle size 2. Also, g_{22} is more different from unity because particles have less freedom in positions as particle size 2 becomes larger and larger. Fractional volume of particle size 2 in Fig. 3.3.4 is higher than that in Fig. 3.3.3 while fractional volume of particle size 1 is the same in both figures. Nevertheless, the increase in concentrations of particle size 2 affects that of particle size 1 as indicated in the figures. In Fig. 3.3.5, the results of a mixture of particles with three different sizes are illustrated with $a_3 > a_2 > a_1$. We see that g_{33} exhibit the least freedom while there is more freedom in g_{11} than in g_{22}. We also note that the value of g_{ij} generally lies between g_{ii} and g_{jj}. In Fig. 3.3.6, the results of the pair distribution functions for a medium containing particles of four different sizes are also illustrated. Because of the symmetry relation $g_{ij} = g_{ji}$, there are 10 pair distribution functions for this case as illustrated in the figure.

3.4 Effective Propagation Constants for Media with Small Particles

To calculate the effective propagation constants for QCA and QCA–CP, we need to solve the integral equations (18) and (35). In general, it requires numerical techniques to solve the integral equations for cases where particle sizes are comparable to or larger than wavelengths.

The numerical technique for calculating the effective propagation constant for moderate size particles under QCA will be discussed in section 3.5. In this section, we consider the case of small particles. Closed form expressions of the effective propagation constants can be derived in the low-frequency limit when particles are much smaller than wavelength. The results facilitate the understanding of effective attenuation rates and effective phase velocity for wave propagation in discrete random medium of multiple species.

The effective propagation constant is denoted by $K = K_r + iK_i$. The real part K_r is related to the effective phase velocity while he imaginary part K_i is related to the attenuation. Generally, $K_r \gg K_i$. We shall, in solving (18) and (35), (i) keep only the leading term of the real part and the leading term of the imaginary part and (ii) ignore terms of order less than $(ka)^3$ [2,8].

a. QCA

The momentum representation of $\overline{\overline{T}}^{s_j}$, $\overline{\overline{T}}_p^{s_j}(\overline{p},\overline{p}')$, can be calculated by using the mixed representation [2,8]. In the low-frequency limit

$$\overline{\overline{T}}_p^{s_j}(\overline{p},\overline{p}') = T_m^{s_j} v_{s_j} \overline{\overline{I}} \tag{64}$$

where $v_{s_j} = (4\pi/3)a_{s_j}^3$ is the volume of s_j type particle, and

$$T_m^{s_j} = 3k^2 y_{s_j}\left[1 + i\frac{2}{3}k^3 a_{s_j}^3 y_{s_j}\right] \tag{65}$$

$$y_{s_j} = \frac{\epsilon_{s_j} - \epsilon}{\epsilon_{s_j} + 2\epsilon} \tag{66}$$

Further, let $\overline{\overline{C}}_p^{s_j}(\overline{p},\overline{p}') = C_{s_j}\overline{\overline{I}}$ in the low-frequency limit. Using (64) in (18) gives the relation

$$C_{s_j} = T_m^{s_j} v_{s_j} + T_m^{s_j} v_{s_j} \sum_{s_l=1}^{L} n_{s_l} C_{s_l}\left[\frac{1}{3k^2} + i\frac{4\pi^2 k}{3} H_{s_j s_l}(\overline{p}=0)\right] \tag{67}$$

By keeping only the leading term in the real part and the leading term in the imaginary part, equation (67) can be solved for C_{s_j}. Substituting $\overline{\overline{C}}_p^{s_j}(\overline{p},\overline{p})$ into the dispersion relation (22), the effective propagation K^2 can be solved and is given by

$$K^2 = k^2 + \sum_{s_j=1}^{L} n_{s_j} C_{s_j} \tag{68}$$

After simplification, this gives

$$K^2 = k^2 + \frac{3k^2}{D} \sum_{s_j=1}^{L} f_{s_l} y_{s_l} \times \left\{ 1 + i\frac{2k^3}{3D} \left[a_{s_l}^3 y_{s_l} + \sum_{s_j=1}^{L} a_{s_j}^3 8\pi^3 n_{s_j} y_{s_j} H_{s_j s_l}(\overline{p}=0) \right] \right\} \tag{69}$$

where

$$D = 1 - \sum_{s_l=1}^{L} f_{s_l} y_{s_l} \tag{70}$$

and

$$f_{s_j} = n_{s_j} \frac{4\pi}{3} a_{s_j}^3 \tag{71}$$

is the fractional volume occupied by particles of type s_j.

b. QCA–CP

The low frequency QCA–CP solution can be obtained by solving (33), (35) and (36). The solution for (33) is

$$\hat{\overline{\overline{t}}}_p^{s_j}(\overline{p}, \overline{p}') = \hat{t}_m^{s_j} v_{s_j} \overline{\overline{I}} \tag{72}$$

where

$$\hat{t}_m^{s_j} = 3K^2 \hat{y}_{s_j} \left[1 + i\frac{2}{3} K^3 a_{s_j}^3 \hat{y}_{s_j} \right] \tag{73}$$

$$\hat{y}_{s_j} = \frac{k_{s_j}^2 - k^2}{3K^2 + \left(k_{s_j}^2 - k^2 \right)} \tag{74}$$

Solving (35) and (36) gives a nonlinear equation for the effective propagation constant as follows

$$\hat{\overline{\overline{C}}}_p^{s_j}(\overline{p}, \overline{p}') = \hat{C}_{s_j} \overline{\overline{I}} \tag{75}$$

$$\hat{C}_{s_j} = \hat{t}_m^{s_j} v_{s_j} + \hat{t}_m^{s_j} v_{s_j} \sum_{s_l=1}^{L} n_{s_l} \hat{C}_{s_l} \left[\frac{1}{3K^2} + i\frac{4\pi^2 K}{3} H_{s_j s_l}(\overline{p}=0) \right] \tag{76}$$

$$
\begin{aligned}
K^2 &= k^2 + \sum_{s_j=1}^{L} n_{s_j} \widehat{C}_{s_j} \\
&= k^2 + \frac{3K^2}{\widehat{D}} \sum_{s_l=1}^{L} f_{s_l} \widehat{y}_{s_l} \\
&\times \left\{ 1 + i \frac{2K^3}{3\widehat{D}} \left[a_{s_l}^3 \widehat{y}_{s_l} + \sum_{s_j=1}^{L} a_{s_j}^3 8\pi^3 n_{s_j} \widehat{y}_{s_j} H_{s_j s_l} (\overline{p} = 0) \right] \right\}
\end{aligned}
\tag{77}
$$

$$
\widehat{D} = 1 - \sum_{s_l=1}^{L} f_{s_l} \widehat{y}_{s_l} \tag{78}
$$

Note that $\widehat{y}_{s_j}$ depends on K^2 so that (77) is a nonlinear equation of K^2. Solution of (77) for K^2 can be conveniently computed in the following manner. First, the terms in the square bracket in (77) are ignored. The remaining terms can be cast into a polynomial equation for K^2 which can be solved readily. Experience indicates that only one solution of K^2 has the correct physical meaning of effective propagation constant. This solution is used as an initial approximation and inserted into the right hand side of (77). The solution quickly converges after a few iterations.

Thus solutions (69) and (77) give K^2 under QCA and QCA–CP respectively. The real part of K is the angular frequency divided by the effective phase velocity. The imaginary part of K gives effective attenuation rate due to both absorption and scattering.

In Figs. 3.4.1 to 3.4.10, by solving the equations (69) and (77), we illustrate the normalized attenuation rate $2K_i/k$ and the effective dielectric constant $(K^2/k^2)_r$ where subscripts r and i denote real and imaginary parts respectively. The parameters chosen represent typical values found in microwave and millimeter wave remote sensing in snow. Dry snow is a mixture of ice particles and air with fractional volume of ice between 0.2 and 0.4. The permittivity of ice particles is $3.2\epsilon_o$. The ice particles are not of identical sizes and are usually described by a particle size distribution ranging from 0.01 mm to 2 mm. Wet snow is a mixture of air, ice particles and water droplet with sizes of water droplets comparable to that of ice particles [27,28].

In Figs. 3.4.1 and 3.4.2, the case of particles of identical sizes is considered. The attenuation rate $2K_i/k$ and the effective dielectric constant $(K^2/k^2)_r$ are plotted as a function of f for $ka = 0.1047$

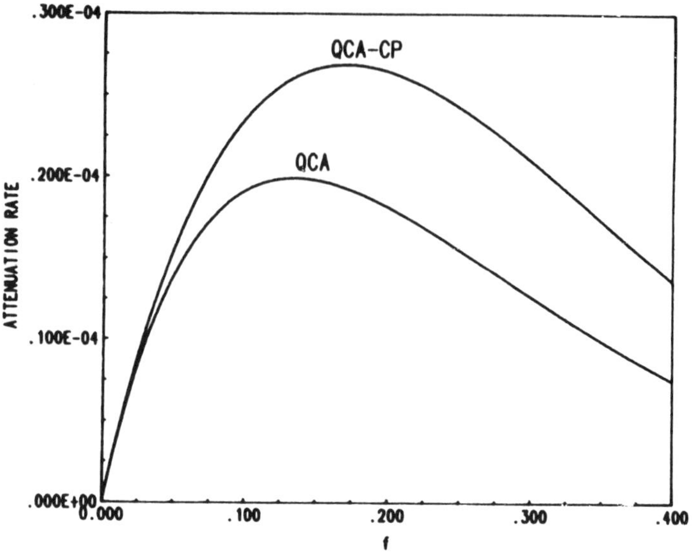

Figure 3.4.1 Attenuation rate $2K_i/k$ as a function of f, for $ka = 0.1047$, and $\epsilon = 3.2\epsilon$, $\epsilon = \epsilon_o$ is the background permittivity. The results of Quasicrystalline Approximation (QCA) and Quasicrystalline Approximation with Coherent Potential (QCA–CP) are compared.

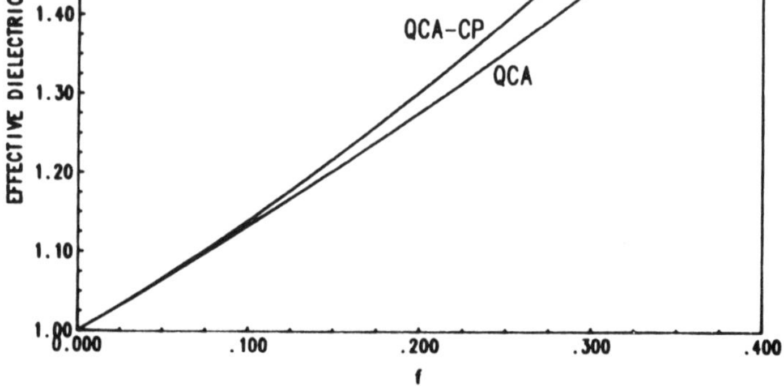

Figure 3.4.2 Effective dielectric constant $(K^2/k^2)_r$ as a function of f, for $ka = 0.1047$, and $\epsilon = 3.2\epsilon$, $\epsilon = \epsilon_o$ is the background permittivity. The results of QCA and QCA–CP are compared.

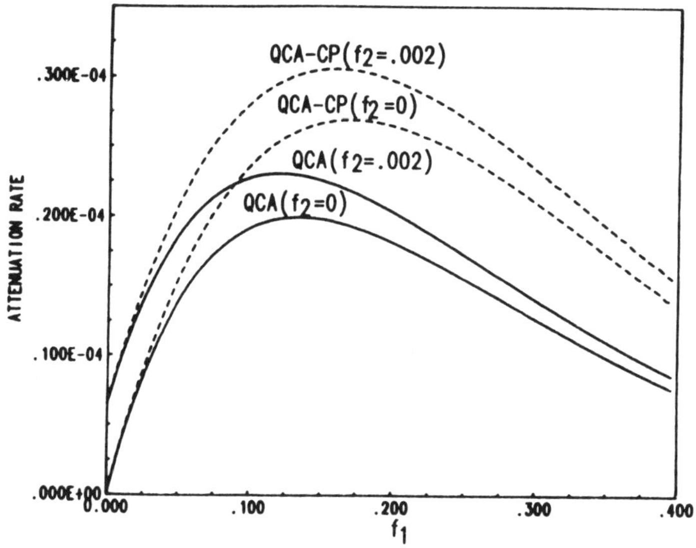

Figure 3.4.3 Attenuation rate $2K_i/k$ as a function of f_1, for (1) $f_2 = 0$ and (2) $f_2 = 0.002$, where $ka_1 = 0.1047$, $ka_2 = 0.2094$, and $\epsilon_1 = \epsilon_2 = 3.2\epsilon$, $\epsilon = \epsilon_o$ is the background permittivity. The results of Quasicrystalline Approximation (QCA) and Quasicrystalline Approximation with Coherent Potential (QCA–CP) are compared.

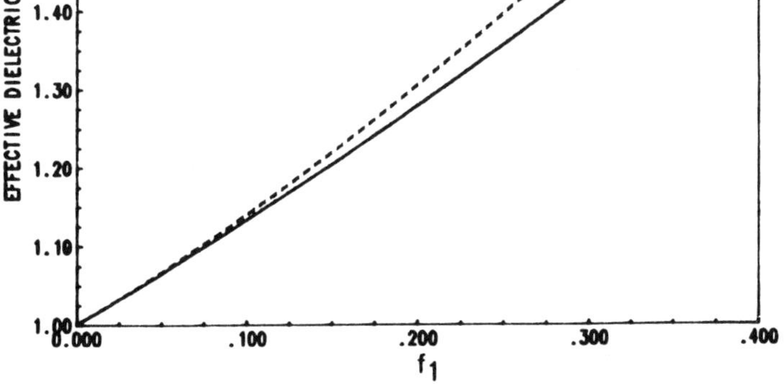

Figure 3.4.4 Effective dielectric constant $(K^2/k^2)_r$ as a function of f_1, for (1) $f_2 = 0$ and (2) $f_2 = 0.002$, where $ka_1 = 0.1047$, $ka_2 = 0.2094$, and $\epsilon_1 = \epsilon_2 = 3.2\epsilon$, $\epsilon = \epsilon_o$ is the background permittivity. The results of QCA and QCA–CP are compared.

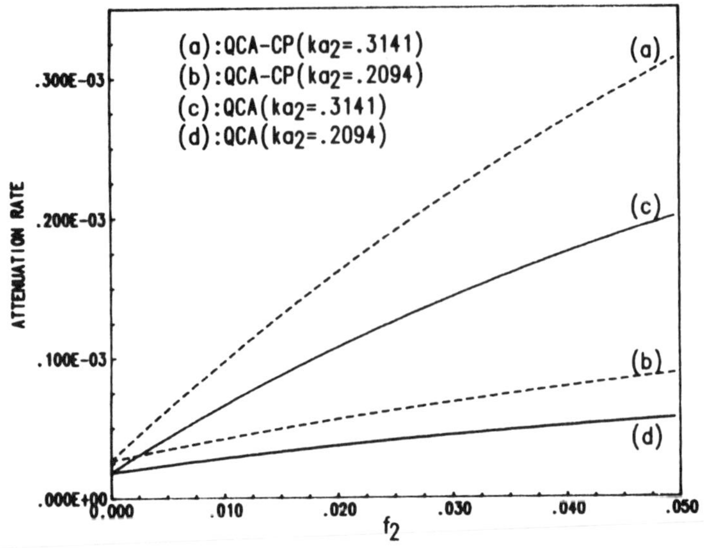

Figure 3.4.5 **Attenuation rate $2K_i/k$ as a function of f_2, for (1) $ka_2 =$ 0.2094 and (2) $ka_2 = 0.3141$, where $f_1 = 0.2$, $ka_1 = 0.1047$, $\epsilon_1 = \epsilon_2 = 3.2\epsilon$, $\epsilon = \epsilon_o$ is the background permittivity.**

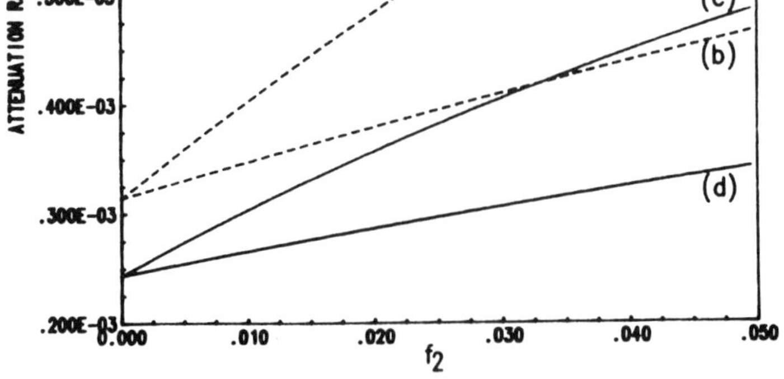

Figure 3.4.6 **Attenuation rate $2K_i/k$ as a function of f_2, for (1) $ka_2 =$ 0.2094 and (2) $ka_2 = 0.3141$, where $f_1 = 0.2$, $ka_1 = 0.1047$, $\epsilon_1 = \epsilon_2 =$ $3.2(1 + i0.001)\epsilon$, $\epsilon = \epsilon_o$ is the background permittivity.**

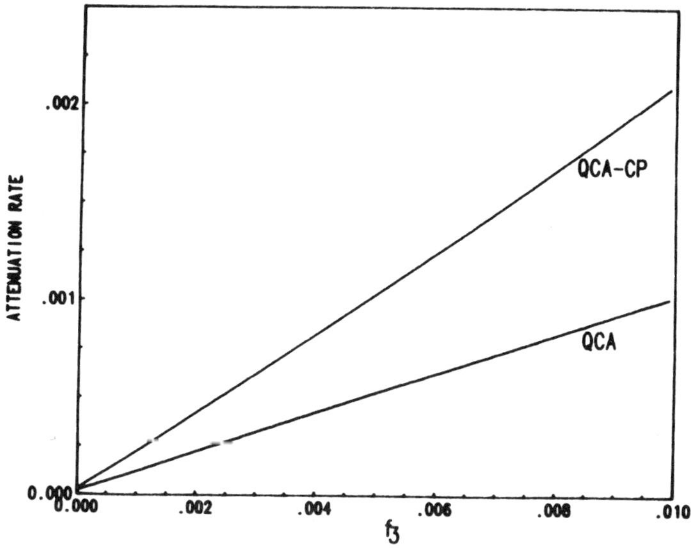

Figure 3.4.7 Attenuation rate $2K_i/k$ as a function of f_3, for $f_1 = 0.2$, and $f_2 = 0.002$, where $ka_1 = 0.1047$, $ka_2 = 0.2094$, $ka_3 = 0.3141$, and $\epsilon_1 = \epsilon_2 = 3.2\epsilon$, $\epsilon_3 = 66.2(1 + i0.55)\epsilon$, $\epsilon = \epsilon_o$ is the background permittivity.

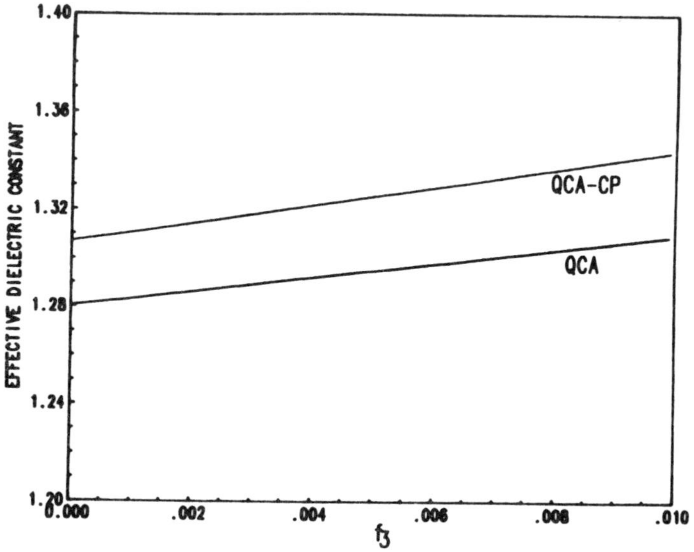

Figure 3.4.8 Effective dielectric constant $(K^2/k^2)_r$ as a function of f_3, for $f_1 = 0.2$, and $f_2 = 0.002$, where $ka_1 = 0.1047$, $ka_2 = 0.2094$, $ka_3 = 0.3141$, and $\epsilon_1 = \epsilon_2 = 3.2\epsilon$, $\epsilon_3 = 66.2(1 + i0.55)\epsilon$, $\epsilon = \epsilon_o$ is the background permittivity.

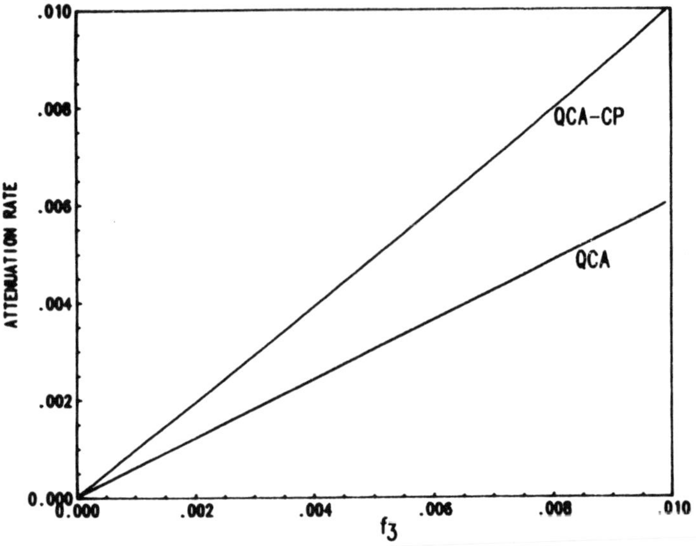

Figure 3.4.9 Attenuation rate $2K_i/k$ as a function of f_3, for $f_1 = 0.2$, and $f_2 = 0.002$, where $ka_1 = 0.1047$, $ka_2 = 0.2094$, $ka_3 = 0.3141$, and $\epsilon_1 = \epsilon_2 = 3.2\epsilon$, $\epsilon_3 = 6.205(1 + i1.224)\epsilon$, $\epsilon = \epsilon_o$ is the background permittivity.

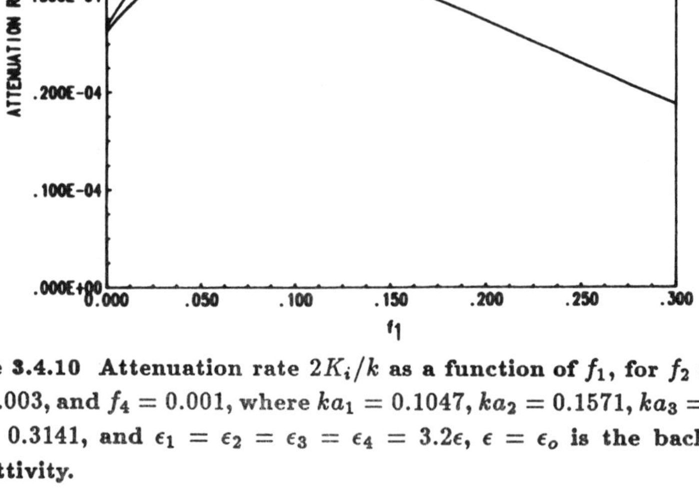

Figure 3.4.10 Attenuation rate $2K_i/k$ as a function of f_1, for $f_2 = 0.005$, $f_3 = 0.003$, and $f_4 = 0.001$, where $ka_1 = 0.1047$, $ka_2 = 0.1571$, $ka_3 = 0.2094$, $ka_4 = 0.3141$, and $\epsilon_1 = \epsilon_2 = \epsilon_3 = \epsilon_4 = 3.2\epsilon$, $\epsilon = \epsilon_o$ is the background permittivity.

and for real particle permittivity $\epsilon_s = 3.2\epsilon_o$. In this case there is no absorption. Attenuation is due entirely to scattering. We note that the attenuation rate first increase with f, then rises to a peak and decreases when f further increases. This indicates that the medium appears more homogeneous as the particle concentration further increases beyond a certain value.

In Figs. 3.4.3 and 3.4.4, the attenuation rate $2K_i/k$ and the effective dielectric constant $(K^2/k^2)_r$ are illustrated respectively as a function of f_1 for $f_2 = 0$ and $f_2 = 0.002$. The two particle sizes are $ka_1 = 0.1047$ and $ka_2 = 0.2094$. The results based on QCA and QCA–CP are illustrated. In spite of the much smaller fractional volume ($f_2 = 0.002$), larger particles have a larger weighting factor in scattering and can contribute significantly to the scattering attenuation rates. This indicates that it is generally not valid to use the average particle size to calculate the scattering effects. On the other hand, Fig. 3.4.4 shows that the larger particles with a very small fractional volume have relatively a small effect on the effective dielectric constant. The results in Fig. 3.4.3 also indicate that QCA–CP generally predicts a larger scattering attenuation rate than QCA. This is apparent by comparing equations (6) and (33). The scattering operator for a single scatterer for QCA is embedded in medium with propagator $\overline{\overline{G}}_o$. For QCA–CP, the single particle transition operator $\hat{\overline{\overline{t}}}^{s}_{j}$ is embedded in an effective medium with propagator $\overline{\overline{G}}_c$.

In Fig. 3.4.5, the attenuation rates are plotted as a function of f_2 with $f_1 = 0.2$. Because the concentration of f_2 is small, the attenuation rate increases almost linearly with f_2. In Fig. 3.4.6, the attenuation rates are illustrated for the case of slightly lossy particles with $\epsilon_1 = \epsilon_2 = 3.2(1+i0.001)\epsilon_o$. The particle sizes in the figure are such that the attenuation rate due to absorption is comparable to that due to scattering. In Figs. 3.4.7–3.4.9, we illustrate the case when the medium is a mixture of three particle species with different sizes and permittivities. In Figs. 3.4.7 and 3.4.8, particles of species 3 are largest in size and have permittivity equal to $\epsilon_3 = 66.2(1+i0.55)\epsilon_o$ corresponding to that of water at 5 GHz. In Fig. 3.4.9, $\epsilon_3 = 6.205(1+i1.224)\epsilon_o$ which corresponds to that of water at 90 GHz. Thus the model can be used to represent wet snow. We note that both the effective dielectric constant and the attenuation rate increases rapidly with f_3 which corresponds to wetness content of wet snow. In Fig. 3.4.10, we illustrate the case when there are four different particle sizes.

c. *Dielectric Mixing Formula Ignoring Scattering Attenuation Rates*

Classical dielectric mixing formulae aim at calculating the effective permittivity of a mixture of components. For these cases, the particle size is much smaller than wavelength so that scattering attenuation rates can be ignored. Such dielectric mixing formula can be obtained from the results of effective propagation constants by setting particle size a_{s_j} to zero and number density n_{s_j} to infinity such that fractional volume f_{s_j} of (71) remains finite. The effective permittivity ϵ_{eff} is given by

$$\frac{\epsilon_{eff}}{\epsilon} = \frac{K^2}{k^2} \tag{79}$$

d. *Mixing Formula for QCA*

From (69), we get the mixing formula

$$\frac{\epsilon_{eff}}{\epsilon} = 1 + \frac{3\sum_{s_j=1}^{L} f_{s_j} y_{s_j}}{1 - \sum_{s_j=1}^{L} f_{s_j} y_{s_j}} \tag{80}$$

Using expression for y_{s_j} as given by (66), the QCA result of (80) can be rearranged to give

$$\frac{\epsilon_{eff} - \epsilon}{\epsilon_{eff} + 2\epsilon} = \sum_{s_j=1}^{L} f_{s_j} \frac{\epsilon_{s_j} - \epsilon}{\epsilon_{s_j} + 2\epsilon} \tag{81}$$

which is the same as the classical Rayleigh mixing formula [29].

e. *Mixing Formula for QCA–CP*

The mixing formula for QCA–CP, from (77) and (74), is

$$\epsilon_{eff} = \epsilon + \frac{3\epsilon_{eff} \sum_{s_j=1}^{L} f_{s_j} \dfrac{\epsilon_{s_j} - \epsilon}{3\epsilon_{eff} + \epsilon_{s_j} - \epsilon}}{1 - \sum_{s_j=1}^{L} f_{s_j} \dfrac{\epsilon_{s_j} - \epsilon}{3\epsilon_{eff} + \epsilon_{s_j} - \epsilon}} \tag{82}$$

3.5 Effective Propagation Constants for Media with Moderate Size Particles

Generally, for cases where particle sizes are comparable to or larger than wavelengths, it requires numerical technique to solve for the effective propagation constants. Such cases will be illustrated using the quasicrystalline approximation. We also find it convenient to use the T-matrix formulism with vector spherical wave functions as basis to formulate the multiple scattering equation and dispersion relation [4,8].

Let $\overline{E}^{inc}(\overline{r})$ be the incident electric field in a medium with multi-species of particles and $\overline{E}_j^S(\overline{r})$ be the scattered field form the jth particle. The total field $\overline{E}(\overline{r})$ at point $\overline{r}$ is the sum of the incident field and the scattered fields from all particles,

$$\overline{E}(\overline{r}) = \overline{E}^{inc}(\overline{r}) + \sum_{j=1}^{N} \overline{E}_j^S(\overline{r}) \tag{83}$$

The scattered field $\overline{E}_j^S(\overline{r})$ is related to the jth particle exciting field $\overline{E}_j^E(\overline{r})$ by

$$\overline{E}_j^S(\overline{r}) = \overline{\overline{T}}_j^{s_j}\overline{E}_j^E(\overline{r}) \tag{84}$$

where $\overline{\overline{T}}_j^{s_j}$ is the transition operator for the particle j of the s_j species. The exciting field for the jth particle can also be expressed as the total field less its own scattered field

$$\begin{aligned}\overline{E}_j^E(\overline{r}) &= \overline{E}^{inc}(\overline{r}) + \sum_{l\neq j}^{N} \overline{E}_l^S(\overline{r}) \\ &= \overline{E}^{inc}(\overline{r}) + \sum_{l\neq j}^{N} \overline{\overline{T}}_l^{s_l}\overline{E}_l^E(\overline{r})\end{aligned} \tag{85}$$

with $j = 1, \ldots, N$.

Under the quasicrystalline approximation, the integral equation for the conditional average of the jth particle exciting field, $\left\langle \overline{E}_j^E(\overline{r}) \right\rangle_j$, is

$$\left\langle \overline{E}_j^E(\overline{r}) \right\rangle_j = \overline{E}^{inc}(\overline{r}) + \sum_{s_l=1}^{L} n_{s_l} \int d\overline{r}_l\, g_{s_j s_l}(\overline{r}_l, \overline{r}_j)\overline{\overline{T}}_l^{s_l}\left\langle \overline{E}_l^E(\overline{r}) \right\rangle_l \tag{86}$$

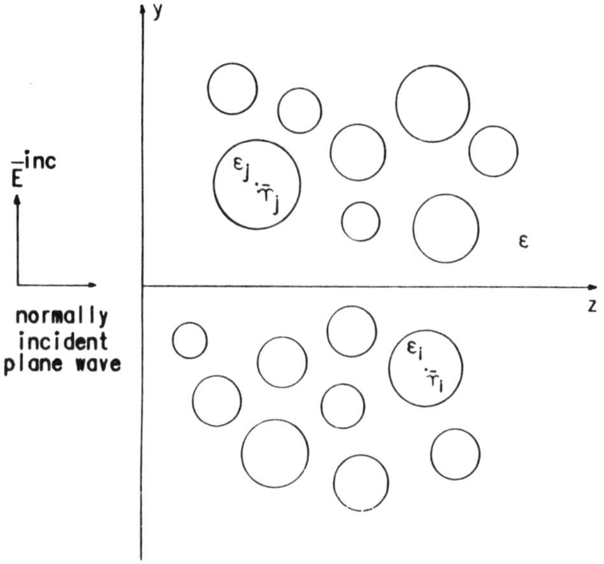

Figure 3.5.1 Plane electromagnetic wave normally incident on a half space of multiple species of particles with different sizes and permittivities.

where $\overline{\overline{T}}_l^{s_l} \left\langle \overline{E}_l^E(\overline{r}) \right\rangle_l$ is the field scattered by a scatterer of s_l species at $\overline{r}_l$ when excited by the field $\left\langle \overline{E}_l^E(\overline{r}) \right\rangle_l$.

For a plane electromagnetic wave normally incident on a half-space of spherical scatterers (Fig. 3.5.1), the incident field is $\overline{E}^{inc}(\overline{r}) = \hat{y}e^{ikz}$ and can be expanded in vector spherical waves

$$\overline{E}_{inc}(\overline{r}) = e^{ikz_j} \sum_{n=1}^{\infty} \frac{i^n}{2} \sqrt{4\pi(2n+1)} \left[Rg\overline{M}_{1n}(k\overline{r r_j}) - Rg\overline{M}_{-1n}(k\overline{r r_j}) + Rg\overline{N}_{1n}(k\overline{r r_j}) + Rg\overline{N}_{-1n}(k\overline{r r_j}) \right] \quad (87)$$

where $Rg\overline{M}_{mn}(k\overline{r r_j})$ and $Rg\overline{N}_{mn}(k\overline{r r_j})$ are regular vector spherical wave functions [8,30]

$$Rg\overline{M}_{mn}(k\overline{r}) = \sqrt{\frac{(2n+1)(n-m)!}{4\pi n(n+1)(n+m)!}} \overline{\nabla} \times \left[\overline{r} j_n(kr) P_n^m(\cos\theta) e^{im\phi} \right] \quad (88)$$

$$Rg\overline{N}_{mn}(k\overline{r}) = \frac{1}{k} \overline{\nabla} \times Rg\overline{M}_{mn}(k\overline{r}) \quad (89)$$

and $j_n(kr)$ is the spherical Bessel function and $P_n^m(\cos\theta)$ is the associated Legendre polynomial. The symbol $\overline{rr_j}$ is used to denote the vector pointing from $\overline{r}_j$ to $\overline{r}$. To solve (86), we express the exciting field and the scattered field in terms of vector spherical waves. At a point $\overline{r}$ in the vicinity of the jth particle, let [4,8]

$$\left\langle \overline{E}_j^E(\overline{r})\right\rangle_j = \sum_{\nu=1}^{\infty}\sum_{\mu=-\nu}^{\nu}\left[a_{\mu\nu}^{s_j}(z_j)Rg\overline{M}_{\mu\nu}(k\overline{rr_j}) + b_{\mu\nu}^{s_j}(z_j)Rg\overline{N}_{\mu\nu}(k\overline{rr_j})\right] \tag{90}$$

Then, for scattering by spheres

$$\overline{\overline{T}}_l^{s_l}\left\langle \overline{E}_l^E(\overline{r})\right\rangle_l = \sum_{n=1}^{\infty}\sum_{m=-n}^{n}\left[T_n^{(M)s_l}a_{mn}^{s_l}(z_l)\overline{M}_{mn}(k\overline{rr_l}) + T_n^{(N)s_l}b_{mn}^{s_l}(z_l)\overline{N}_{mn}(k\overline{rr_l})\right] \tag{91}$$

where $\overline{M}_{mn}(k\overline{rr_j})$ and $\overline{N}_{mn}(k\overline{rr_j})$ are vector spherical functions with the spherical Bessel function j_n in (88) and (89) replaced by spherical Hankel function of the first kind h_n, and $a_{\mu\nu}^{s_j}, b_{\mu\nu}^{s_j}, a_{mn}^{s_l}, b_{mn}^{s_l}$ are the unknown expansion coefficients. The scattering coefficients, $T_n^{(M)s_l}$ and $T_n^{(N)s_l}$ for spheres with radius a_{s_l} and permittivity ϵ_{s_l}, are those of Mie scattering [4,8]

$$T_n^{(M)s_l} = -\frac{[\rho_{s_l}j_n(\rho_{s_l})]'\,j_n(\zeta_{s_l}) - [\zeta_{s_l}j_n(\zeta_{s_l})]'\,j_n(\rho_{s_l})}{[\rho_{s_l}h_n(\rho_{s_l})]'\,j_n(\zeta_{s_l}) - [\zeta_{s_l}j_n(\zeta_{s_l})]'\,h_n(\rho_{s_l})} \tag{92}$$

$$T_n^{(N)s_l} = -\frac{[\rho_{s_l}j_n(\rho_{s_l})]'\,\zeta_{s_l}^2\,j_n(\zeta_{s_l}) - [\zeta_{s_l}j_n(\zeta_{s_l})]'\,\rho_{s_l}^2\,j_n(\rho_{s_l})}{[\rho_{s_l}h_n(\rho_{s_l})]'\,\zeta_{s_l}^2\,j_n(\zeta_{s_l}) - [\zeta_{s_l}j_n(\zeta_{s_l})]'\,\rho_{s_l}^2\,h_n(\rho_{s_l})} \tag{93}$$

where $\rho_{s_l} = ka_{s_l}$ and $\zeta_{s_l} = k_{s_l}a_{s_l}$. We can further make use of the vector translational addition theorem to expand the spherical vector waves $\overline{M}_{mn}(k\overline{rr_l})$ and $\overline{N}_{mn}(k\overline{rr_l})$ about $\overline{r}_j$ as center. Since $|\,\overline{r} - \overline{r}_j\,| < |\,\overline{r}_j - \overline{r}_l\,|$, the vector translational addition theorem is [8,30–33]

$$\overline{M}_{mn}(k\overline{rr_l}) = \sum_{\nu=1}^{\infty}\sum_{\mu=-\nu}^{\nu}\left[A_{\mu\nu mn}(k\overline{r_jr_l})\,Rg\overline{M}_{\mu\nu}(k\overline{rr_j}) + B_{\mu\nu mn}(k\overline{r_jr_l})\,Rg\overline{N}_{\mu\nu}(k\overline{rr_j})\right] \tag{94}$$

$$\overline{N}_{mn}(k\overline{rr_l}) = \sum_{\nu=1}^{\infty}\sum_{\mu=-\nu}^{\nu}\left[B_{\mu\nu mn}(k\overline{r_jr_l})\,Rg\overline{M}_{\mu\nu}(k\overline{rr_j}) + A_{\mu\nu mn}(k\overline{r_jr_l})\,Rg\overline{N}_{\mu\nu}(k\overline{rr_j})\right] \tag{95}$$

where

$$A_{\mu\nu mn}\left(k\overline{r_j r_l}\right) = \frac{\gamma_{mn}}{\gamma_{\mu\nu}}(-1)^{\mu}\sum_{p=0}^{\infty} a(m,n \mid -\mu,\nu \mid p)a(n,\nu,p) \times h_p\left(k \mid \overline{r_j r_l} \mid\right) P_p^{m-\mu}\left(\cos\theta_{\overline{r_j r_l}}\right) e^{i(m-\mu)\phi_{\overline{r_j r_l}}} \tag{96}$$

$$B_{\mu\nu mn}\left(k\overline{r_j r_l}\right) = \frac{\gamma_{mn}}{\gamma_{\mu\nu}}(-1)^{\mu+1}\sum_{p=0}^{\infty} a(m,n \mid -\mu,\nu \mid p,p-1)b(n,\nu,p) \times h_p\left(k \mid \overline{r_j r_l} \mid\right) P_p^{m-\mu}\left(\cos\theta_{\overline{r_j r_l}}\right) e^{i(m-\mu)\phi_{\overline{r_j r_l}}} \tag{97}$$

$$\gamma_{mn} = \sqrt{\frac{(2n+1)(n-m)!}{4\pi n(n+1)(n+m)!}} \tag{98}$$

$$a(m,n \mid -\mu,\nu \mid p) = (-1)^{m+\mu}(2p+1)\sqrt{\frac{(n+m)!(\nu+\mu)!(p-m-\mu)!}{(n-m)!(\nu-\mu)!(p+m+\mu)!}} \times \begin{pmatrix} n & \nu & p \\ m & \mu & -(m+\mu) \end{pmatrix} \begin{pmatrix} n & \nu & p \\ 0 & 0 & 0 \end{pmatrix} \tag{99}$$

$$a(m,n \mid -\mu,\nu \mid p,q) = (-1)^{m+\mu}(2p+1)\sqrt{\frac{(n+m)!(\nu+\mu)!(p-m-\mu)!}{(n-m)!(\nu-\mu)!(p+m+\mu)!}} \times \begin{pmatrix} n & \nu & p \\ m & \mu & -(m+\mu) \end{pmatrix} \begin{pmatrix} n & \nu & q \\ 0 & 0 & 0 \end{pmatrix} \tag{100}$$

$$a(n,\nu,p) = \frac{(2\nu+1)}{2\nu(\nu+1)} i^{\nu-n+p}\left[\nu(\nu+1)+n(n+1)-p(p+1)\right] \tag{101}$$

$$b(n,\nu,p) = -\frac{(2\nu+1)}{2\nu(\nu+1)} i^{\nu-n+p}\left[(n+\nu+p+1)(\nu+p-n) \times (n+p-\nu)(n+\nu-p+1)\right]^{1/2} \tag{102}$$

and

$$\begin{pmatrix} j_1 & j_2 & j_3 \\ m_1 & m_2 & -(m_1+m_2) \end{pmatrix}$$

is the Wigner $3j$ symbol [33].

Equations (87), (90), (91), and (94)–(97) can be substituted into (86). To solve (86), we assume the trial solution

$$a_{\mu\nu}^{s_j}(z_j) = \alpha_{\mu\nu}^{s_j} e^{iKz_j} \tag{103}$$

$$b_{\mu\nu}^{s_j}(z_j) = \beta_{\mu\nu}^{s_j} e^{iKz_j} \tag{104}$$

where K is the effective propagation constant. Equating terms with the same $Rg\overline{M}_{\mu\nu}(k\overline{r}\overline{r_j})$ and the same $Rg\overline{N}_{\mu\nu}(k\overline{r}\overline{r_j})$ give the following two equations:

$$\begin{aligned}\alpha_{\mu\nu}^{s_j} e^{iKz_j} =& e^{ikz_j} \frac{i^\nu}{2} \sqrt{4\pi(2\nu+1)} \left(\delta_{\mu 1} - \delta_{\mu(-1)}\right) \\ &+ e^{iKz_j} \sum_{s_l=1}^{L} \sum_{n=1}^{\infty} \sum_{p=0}^{\infty} n_{s_l} (-1)^\mu \frac{\gamma_{\mu n}}{\gamma_{\mu\nu}} (-1)^p I_p^{s_j s_l} \\ &\times \left\{ T_n^{(M)s_l} \alpha_{\mu n}^{s_l} a(\mu, n \mid -\mu, \nu \mid p) a(n,\nu,p) \right. \\ &\left. - T_n^{(N)s_l} \beta_{\mu n}^{s_l} a(\mu, n \mid -\mu, \nu \mid p, p-1) b(n,\nu,p) \right\} \end{aligned} \tag{105}$$

$$\begin{aligned}\beta_{\mu\nu}^{s_j} e^{iKz_j} =& e^{ikz_j} \frac{i^\nu}{2} \sqrt{4\pi(2\nu+1)} \left(\delta_{\mu 1} + \delta_{\mu(-1)}\right) \\ &+ e^{iKz_j} \sum_{s_l=1}^{L} \sum_{n=1}^{\infty} \sum_{p=0}^{\infty} n_{s_l} (-1)^{\mu+1} \frac{\gamma_{\mu n}}{\gamma_{\mu\nu}} (-1)^p I_p^{s_j s_l} \\ &\times \left\{ T_n^{(M)s_l} \alpha_{\mu n}^{s_l} a(\mu, n \mid -\mu, \nu \mid p, p-1) b(n,\nu,p) \right. \\ &\left. - T_n^{(N)s_l} \beta_{\mu n}^{s_l} a(\mu, n \mid -\mu, \nu \mid p) a(n,\nu,p) \right\} \end{aligned} \tag{106}$$

where δ_{mn} is the Kronecker delta function and

$$I_p^{s_j s_l} = \int_{z_l \geq 0} d\overline{r}_l g_{s_j s_l}(\overline{r}_l - \overline{r}_j) e^{iK(z_l - z_j)} h_p(k \mid \overline{r_l r_j} \mid) P_p\left(\cos\theta_{\overline{r_l r_j}}\right) \tag{107}$$

The volume of integration of (107) consists of the half space $z_l \geq 0$ less a sphere of radius $R_{s_j s_l} = a_{s_j} + a_{s_l}$ centered at the point $\overline{r}_j$. This is due to the fact that particles do not interpenetrate each other so that $g_{s_j s_l}(\overline{r}) = 0$ for $r < R_{s_j s_l}$. We rewrite $g_{s_j s_l}(\overline{r}_l - \overline{r}_j)$ as follows

$$g_{s_j s_l}(\overline{r}_l - \overline{r}_j) = 1 + \left[g_{s_j s_l}(\overline{r}_l - \overline{r}_j) - 1\right] \tag{108}$$

so that the integral $I_p^{s_j s_l}$ in (107) can be decomposed in the following manner

$$I_p^{s_j s_l} = I_1 + I_2 \tag{109}$$

$$
\begin{aligned}
I_1 &= \int_{|\bar{r}_l - \bar{r}_j| > R_{s_j s_l}} d\bar{r}_l \, e^{iK(z_l - z_j)} \, h_p\left(k \mid \overline{r_l r_j} \mid\right) P_p\left(\cos\theta_{\overline{r_l r_j}}\right) \\
&= \frac{2\pi \, i^{p+1} \, e^{ikz_j}}{k^2(K-k)} e^{-iKz_j} + 4\pi i^p L_p\left(k, K \mid R_{s_j s_l}\right) \qquad (110)
\end{aligned}
$$

$$
\begin{aligned}
I_2 &= \int_{|\bar{r}_l - \bar{r}_j| > R_{s_j s_l}} d\bar{r}_l \left[g_{s_j s_l}\left(\bar{r}_l - \bar{r}_j\right) - 1\right] e^{iK(z_l - z_j)} \\
&\qquad \times h_p\left(k \mid \overline{r_l r_j} \mid\right) P_p\left(\cos\theta_{\overline{r_l r_j}}\right) \\
&= 4\pi \, i^p \, M_p\left(k, K \mid R_{s_j s_l}\right) \qquad (111)
\end{aligned}
$$

where

$$
\begin{aligned}
L_p\left(k, K \mid R_{s_j s_l}\right) &= -\frac{R^2_{s_j s_l}}{K^2 - k^2} \left[k h_p'(kR_{s_j s_l}) j_p(KR_{s_j s_l}) \right. \\
&\qquad \left. - K h_p(kR_{s_j s_l}) j_p'(KR_{s_j s_l})\right] \qquad (112)
\end{aligned}
$$

$$
M_p\left(k, K \mid R_{s_j s_l}\right) = \int_{R_{s_j s_l}}^{\infty} dr r^2 \left[g_{s_j s_l}(r) - 1\right] h_p(kr) j_p(Kr) \qquad (113)
$$

Substituting (110) and (111) into (105) and (106) gives two types of terms in (105) and (106). One type of terms has a $\exp(ikz_j)$ dependence corresponding to waves travelling with the propagation constant of the incident wave. The terms with propagation constant k should balance each other giving the generalized Ewald–Oseen extinction theorem [8,34]. The physical interpretation is that the medium generates a wave that extinguishes the original incident wave.

$$
\begin{aligned}
&\frac{i^\nu}{2}\sqrt{4\pi(2\nu+1)}\left(\delta_{\mu 1} - \delta_{\mu(-1)}\right) \\
&\quad + \sum_{s_l=1}^{L} \sum_{n=1}^{\infty} \sum_{p=0}^{\infty} n_{s_l} (-1)^\mu (-i)^p \frac{\gamma_{\mu n}}{\gamma_{\mu\nu}} \frac{2\pi i}{k^2(K-k)} \\
&\quad \times \left\{ T_n^{(M)s_l} \alpha_{\mu n}^{s_l} a(\mu, n \mid -\mu, \nu \mid p) a(n, \nu, p) \right. \\
&\quad \left. - T_n^{(N)s_l} \beta_{\mu n}^{s_l} a(\mu, n \mid -\mu, \nu \mid p, p-1) b(n, \nu, p) \right\} = 0 \qquad (114)
\end{aligned}
$$

$$
\begin{aligned}
&\frac{i^\nu}{2}\sqrt{4\pi(2\nu+1)}\left(\delta_{\mu 1} + \delta_{\mu(-1)}\right) \\
&\quad + \sum_{s_l=1}^{L} \sum_{n=1}^{\infty} \sum_{p=0}^{\infty} n_{s_l} (-1)^{\mu+1} (-i)^p \frac{\gamma_{\mu n}}{\gamma_{\mu\nu}} \frac{2\pi i}{k^2(K-k)}
\end{aligned}
$$

$$\times \left\{ T_n^{(M)s_l} \alpha_{\mu n}^{s_l} a(\mu, n \mid -\mu, \nu \mid p, p-1)\, b(n, \nu, p) - T_n^{(N)s_l} \beta_{\mu n}^{s_l} a(\mu, n \mid -\mu, \nu \mid p) a(n, \nu, p) \right\} = 0 \tag{115}$$

The other type of terms has an $\exp(iKz_j)$ dependence corresponding to waves travelling with the propagation constant of the effective medium. Balancing the terms with propagation constant K gives the generalized Lorentz–Lorenz law [8,34].

$$\begin{aligned}\alpha_{\mu\nu}^{s_j} =& 4\pi \sum_{s_l=1}^{L} \sum_{n=1}^{\infty} \sum_{p=0}^{\infty} n_{s_l} (-1)^{\mu} (-i)^p \frac{\gamma_{\mu n}}{\gamma_{\mu\nu}} \\ & \left[L_p\left(k, K \mid R_{s_j s_l}\right) + M_p\left(k, K \mid R_{s_j s_l}\right) \right] \\ & \times \left\{ T_n^{(M)s_l} \alpha_{\mu n}^{s_l} a(\mu, n \mid -\mu, \nu \mid p)\, a(n, \nu, p) \right. \\ & \left. - T_n^{(N)s_l} \beta_{\mu n}^{s_l} a(\mu, n \mid -\mu, \nu \mid p, p-1)\, b(n, \nu, p) \right\} \end{aligned} \tag{116}$$

$$\begin{aligned}\beta_{\mu\nu}^{s_j} =& 4\pi \sum_{s_l=1}^{L} \sum_{n=1}^{\infty} \sum_{p=0}^{\infty} n_{s_l} (-1)^{\mu+1} (-i)^p \frac{\gamma_{\mu n}}{\gamma_{\mu\nu}} \\ & \left[L_p\left(k, K \mid R_{s_j s_l}\right) + M_p\left(k, K \mid R_{s_j s_l}\right) \right] \\ & \times \left\{ T_n^{(M)s_l} \alpha_{\mu n}^{s_l} a(\mu, n \mid -\mu, \nu \mid p, p-1)\, b(n, \nu, p) \right. \\ & \left. - T_n^{(N)s_l} \beta_{\mu n}^{s_l} a(\mu, n \mid -\mu, \nu \mid p)\, a(n, \nu, p) \right\} \end{aligned} \tag{117}$$

The generalized Lorentz–Lorenz law provides the necessary dispersion relation for the effective propagation constant K.

On examining the generalized Ewald–Oseen extinction theorem and using the relations [8],

$$a(-\mu, n \mid \mu, \nu \mid p) = \frac{(n-\mu)!(\nu+\mu)!}{(n+\mu)!(\nu-\mu)!} a(\mu, n \mid -\mu, \nu \mid p) \tag{118}$$

$$a(-\mu, n \mid \mu, \nu \mid p, p-1) = -\frac{(n-\mu)!(\nu+\mu)!}{(n+\mu)!(\nu-\mu)!} a(\mu, n \mid -\mu, \nu \mid p, p-1) \tag{119}$$

it follows that $\alpha_{-1\nu}^{s_l} = -\alpha_{1\nu}^{s_l}$ and $\beta_{-1\nu}^{s_l} = \beta_{1\nu}^{s_l}$. Furthermore, by making use of the following summation identities

$$-\sum_{p=0}^{\infty} (-i)^p a(1, n \mid -1, \nu \mid p, p-1) b(n, \nu, p)$$

$$= \sum_{p=0}^{\infty} (-i)^p a(1, n \mid -1, \nu \mid p) a(n, \nu, p)$$

$$= -i^{\nu - n} \frac{(2\nu + 1) n(n+1)}{2\nu(\nu + 1)} \tag{120}$$

and setting

$$\alpha_{1n}^{s_l} = X_{1n}^{s_l} \frac{i^n}{2} \sqrt{4\pi(2n+1)} \tag{121}$$

$$\beta_{1n}^{s_l} = Y_{1n}^{s_l} \frac{i^n}{2} \sqrt{4\pi(2n+1)} \tag{122}$$

equations (114) and (115) of the generalized Ewald–Oseen extinction theorem can be simplified to a single equation

$$K - k = -\frac{\pi i}{k^2} \sum_{s_l=1}^{L} \sum_{n=1}^{\infty} n_{s_l} \left[T_n^{(M)s_l} X_{1n}^{s_l} + T_n^{(N)s_l} Y_{1n}^{s_l} \right] (2n+1) \tag{123}$$

Equations (116) and (117) of the generalized Lorentz–Lorenz law are also simplified to

$$\begin{aligned} X_{1\nu}^{s_j} = & -2\pi \sum_{s_l=1}^{L} \sum_{n=1}^{\infty} \sum_{p=0}^{\infty} n_{s_l} (2n+1) \\ & \left[L_p \left(k, K \mid R_{s_j s_l} \right) + M_p \left(k, K \mid R_{s_j s_l} \right) \right] \\ & \times \left\{ T_n^{(M)s_l} X_{1n}^{s_l} a(1, n \mid -1, \nu \mid p) A(n, \nu, p) \right. \\ & \left. + T_n^{(N)s_l} Y_{1n}^{s_l} a(1, n \mid -1, \nu \mid p, p-1) B(n, \nu, p) \right\} \end{aligned} \tag{124}$$

$$\begin{aligned} Y_{1\nu}^{s_j} = & -2\pi \sum_{s_l=1}^{L} \sum_{n=1}^{\infty} \sum_{p=0}^{\infty} n_{s_l} (2n+1) \\ & \left[L_p \left(k, K \mid R_{s_j s_l} \right) + M_p \left(k, K \mid R_{s_j s_l} \right) \right] \\ & \times \left\{ T_n^{(M)s_l} X_{1n}^{s_l} a(1, n \mid -1, \nu \mid p, p-1) B(n, \nu, p) \right. \\ & \left. + T_n^{(N)s_l} Y_{1n}^{s_l} a(1, n \mid -1, \nu \mid p) A(n, \nu, p) \right\} \end{aligned} \tag{125}$$

where

$$A(n, \nu, p) = \frac{\nu(\nu+1) + n(n+1) - p(p+1)}{n(n+1)} \tag{126}$$

$$B(n, \nu, p) = \frac{\sqrt{(n + \nu + p + 1)(\nu + p - n)(n + p - \nu)(n + \nu - p + 1)}}{n(n+1)} \tag{127}$$

Equations (124) and (125) form a system of simultaneous homogeneous equations for the unknowns $X_{1n}^{s_l}$ and $Y_{1n}^{s_l}$. For a nontrivial solution, the determinant of the coefficients must vanish. The requirement for the vanishing of the determinant gives an equation for the effective wavenumber K. Thus, for given particle sizes a_{s_l}, number densities n_{s_j}, and permittivities ϵ_{s_l}, $s_l = 1, 2, \ldots, L$, we calculate the Mie scattering coefficients $T_n^{(M)s_l}$ and $T_n^{(N)s_l}$, the pair distribution functions $g_{s_j s_l}(\overline{r})$, and the expressions $L_p\left(k, K \mid R_{s_j s_l}\right)$ and $M_p\left(k, K \mid R_{s_j s_l}\right)$. The effective propagation constant K is then calculated by searching it in the complex plane such that the determinant of equations (124)–(125) will vanish.

In the following, we summarize the numerical algorithm that we used to compute the effective propagation constant K. For the values of ka up to 2.5, and L species of particles, the determinant of the coefficients of $X_{1\nu}^{s_j}$ and $Y_{1\nu}^{s_j}$ is computed numerically by retaining a maximum of $12 \cdot L$ simultaneous homogeneous equations for $X_{1\nu}^{s_l}$ and $Y_{1\nu}^{s_l}$, $\nu = 1, 2, \ldots, 6; s_l = 1, 2, \ldots, L$. The Wigner $3j$ symbols are generated by a computer code. The elements of $M_p\left(k, K \mid R_{s_j s_l}\right)$ for $p = 0, 1, \ldots, 12$ were computed by numerically evaluating the integral in (113) for r between $R_{s_j s_l}$ and $5R_{s_j s_l}$, the value of $g_{s_j s_l}(r) - 1$ is assumed to be zero for r larger than $5R_{s_j s_l}$.

For the given values of ka_{s_j} and f_{s_j}, the roots of the determinant were searched in the complex K plane $(K_r + iK_i)$ using Muller's method. We have used two good initial guesses. One initial guess is provided by the low frequency solution for the effective propagation constant under quasicrystalline approximation of (69). The other good initial guess is the result under Foldy's approximation obtained by replacing $X_{1n}^{s_j}$ and $Y_{1n}^{s_j}$ by 1 in (123)

$$K^{(F)} = k - \frac{\pi i}{k^2} \sum_{s_l=1}^{L} \sum_{n=1}^{\infty} n_{s_l} \left[T_n^{(M)s_l} + T_n^{(N)s_l}\right] (2n+1) \tag{128}$$

where $K^{(F)}$ denotes the effective propagation constant under Foldy's approximation. These two guesses could be used systematically to obtain quick convergence of roots at increasing higher values of ka_{s_j}. In the following, we shall present the results for the normalized phase velocity k/K_r and the effective loss tangent $2K_i/K_r$. The effective loss tangent accounts for attenuation due to both scattering and absorption.

In Figs. 3.5.2–3.5.5, we illustrate the effective propagation constant for the case when the particles are of identical sizes. In Figs. 3.5.2 and 3.5.3, the normalized phase velocity and the effective loss tangent are plotted as a function of ka for $f = 0.1$ and $f = 0.2$. The results for the case of lossless particles, $\epsilon = 3.2\epsilon_o$, and the case of lossy particles, $\epsilon = (3.2 + i0.15)\epsilon_o$, are also illustrated. We note that the phase velocity first decreases with increasing particle size and oscillates as particle size increases further. The oscillation is a characteristic of resonant scattering for large particles. Comparing the loss tangents for the two cases of lossy and lossless particles, we note that, for small particles, absorption dominates over scattering as the effective loss tangent for lossy case is much larger than that for lossless case. For ka up to 2.5, the loss tangents are comparable in magnitude in the two cases. In Figs. 3.5.4 and 3.5.5, we show respectively phase velocity and loss tangent versus f for $ka = 1.5$ and $ka = 2$ with $\epsilon = 3.2\epsilon_o$. We note that the phase velocity decreases as f increases since the phase velocity for the scatterers is lower. As a function of f, the effective loss tangent first increases, then rises to a peak and decreases as f further increases. This indicates that the medium appears more homogeneous as f further increases beyond a certain value.

In Figs. 3.5.6–3.5.9, we illustrate the case when the particles are of two different sizes. In Figs. 3.5.6 and 3.5.7, the normalized phase velocity and the effective loss tangent are plotted as a function of ka_2 for $f_2 = 0.1$ and $f_2 = 0.2$ with $ka_1 = 0.5$ and $f_1 = 0.1$. The results for the case of lossless particles, $\epsilon_1 = \epsilon_2 = 3.2\epsilon_o$, and the case of lossy particles, $\epsilon_1 = \epsilon_2 = (3.2 + i0.15)\epsilon_o$, are illustrated. In Figs. 3.5.8 and 3.5.9, we show respectively phase velocity and loss tangent vs f_2 for $ka_2 = 1.5$ and $ka_2 = 2$ with $ka_1 = 0.5$, $f_1 = 0.1$, and $\epsilon_1 = \epsilon_2 = 3.2\epsilon_o$. We see that the location of maximum attenuation occurs at a larger f_2 for the case of larger ka. This indicates that independent scattering has a wider regime of validity for larger particles.

In Figs. 3.5.10–3.5.13, we illustrate the case when there are four species of particles represented by four different sizes. In Figs. 3.5.10 and 3.5.11, the normalized phase velocity and the effective loss tangent are plotted as a function of ka_4 for $f_4 = 0.05$ and $f_4 = 0.1$ with $ka_1 = 0.2$, $ka_2 = 0.3$, $ka_3 = 0.5$ and $f_1 = 0.1$, $f_2 = 0.15$, and $f_3 = 0.05$. We also compare the results for the lossless particle case, $\epsilon_1 = \epsilon_2 = \epsilon_3 = \epsilon_4 = 3.2\epsilon_o$ and the lossy particle case, $\epsilon_1 = \epsilon_2 = \epsilon_3 = \epsilon_4 = (3.2 + i0.15)\epsilon_o$. Figs. 3.5.12 and 3.5.13 plot respectively the phase

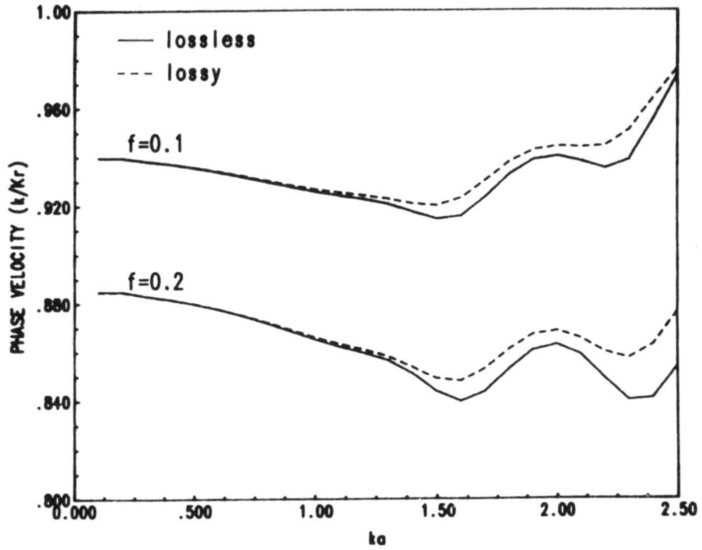

Figure 3.5.2 Normalized phase velocity k/K_r as a function of ka for (1)$f = 0.1$ and (2)$f = 0.2$. The results of lossless particles, $\epsilon = 3.2\epsilon_o$, and of lossy particles, $\epsilon = (3.2 + i0.15)\epsilon_o$, are illustrated.

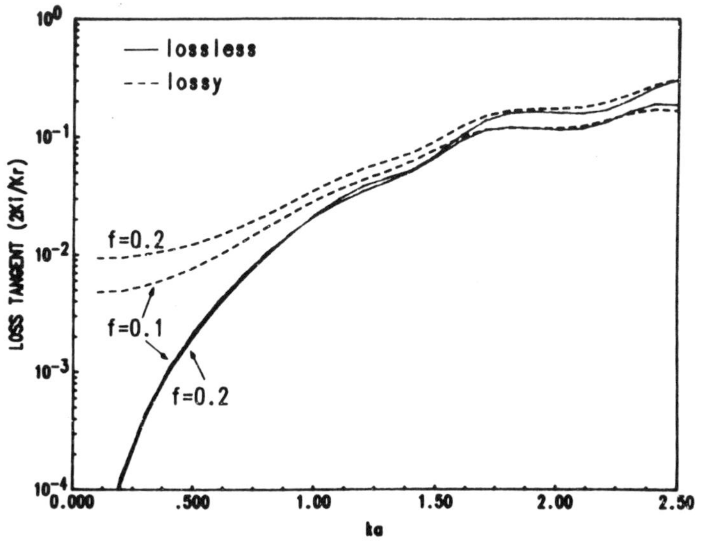

Figure 3.5.3 Effective loss tangent $2K_i/K_r$ as a function of ka for (1)$f =$ 0.1 and (2)$f = 0.2$. The results of lossless particles, $\epsilon = 3.2\epsilon_o$, and of lossy particles, $\epsilon = (3.2 + i0.15)\epsilon_o$, are illustrated.

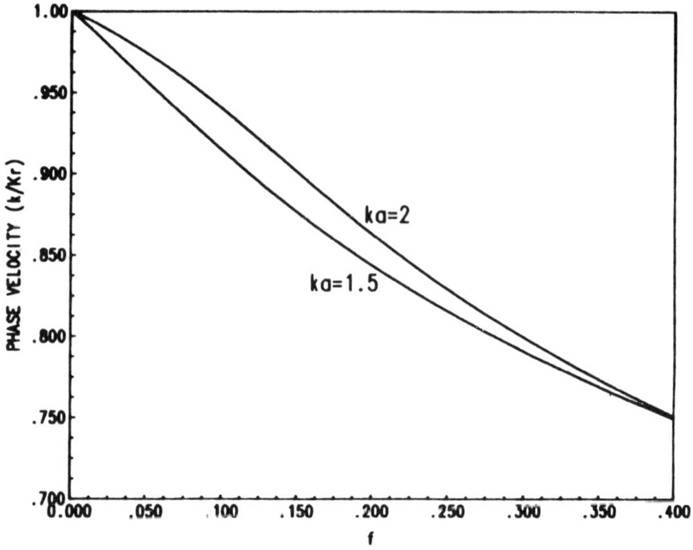

Figure 3.5.4 Normalized phase velocity k/K_r as a function of f for (1)$ka = 1.5$ and (2)$ka = 2$, and $\epsilon = 3.2\epsilon_o$.

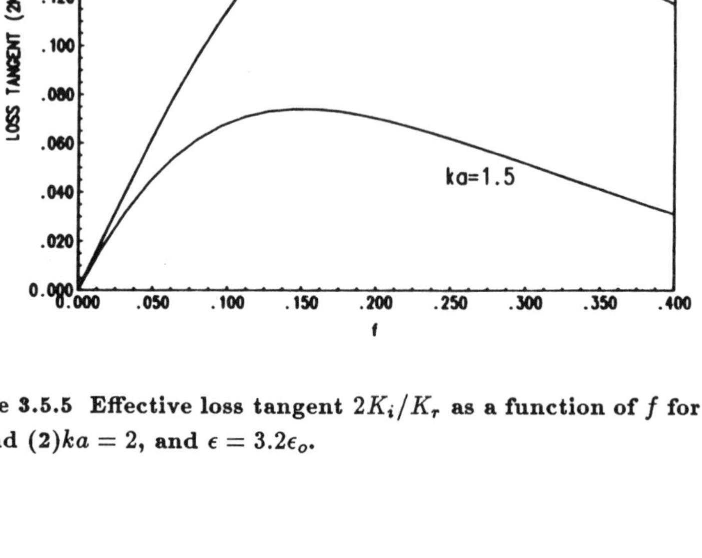

Figure 3.5.5 Effective loss tangent $2K_i/K_r$ as a function of f for (1)$ka = 1.5$ and (2)$ka = 2$, and $\epsilon = 3.2\epsilon_o$.

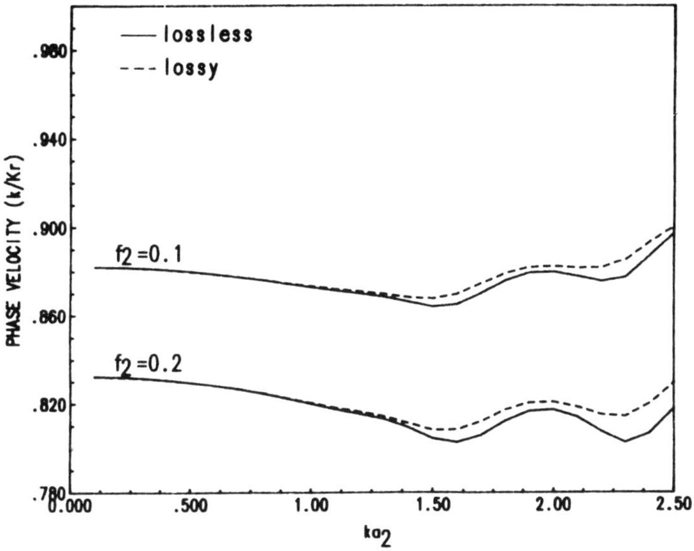

Figure 3.5.6 Normalized phase velocity k/K_r as a function of ka_2 for u $(1) f_2 = 0.1$ and $(2) f_2 = 0.2$, where $ka_1 = 0.5$ and $f_1 = 0.1$. The results of lossless particles, $\epsilon_1 = \epsilon_2 = 3.2\epsilon_o$, and of lossy particles, $\epsilon_1 = \epsilon_2 = (3.2 + i0.15)\epsilon_o$, are illustrated.

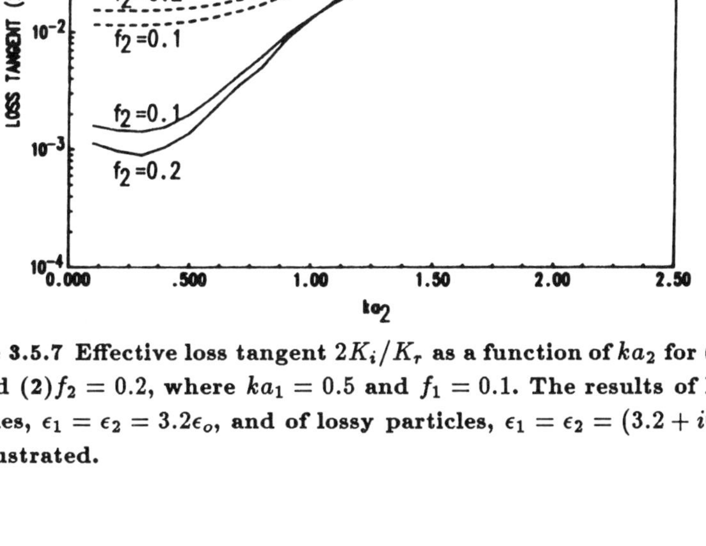

Figure 3.5.7 Effective loss tangent $2K_i/K_r$ as a function of ka_2 for $(1) f_2 = 0.1$ and $(2) f_2 = 0.2$, where $ka_1 = 0.5$ and $f_1 = 0.1$. The results of lossless particles, $\epsilon_1 = \epsilon_2 = 3.2\epsilon_o$, and of lossy particles, $\epsilon_1 = \epsilon_2 = (3.2 + i0.15)\epsilon_o$, are illustrated.

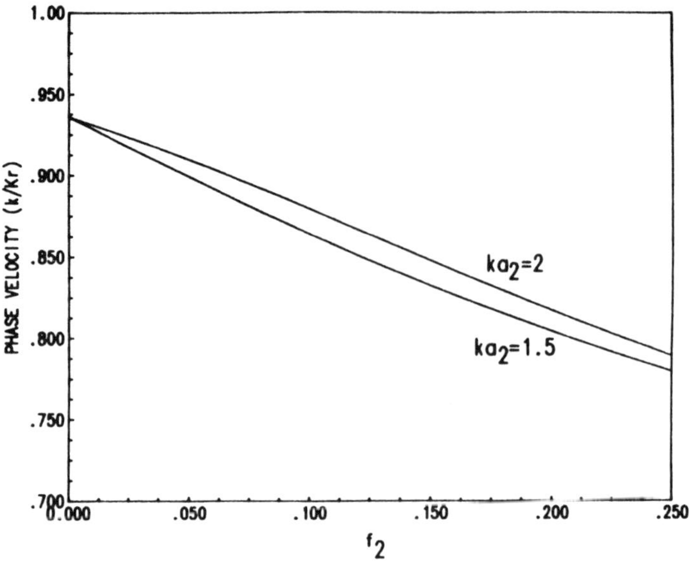

Figure 3.5.8 Normalized phase velocity k/K_r as a function of f_2 for $(1)ka_2 = 1.5$ and $(2)ka_2 = 2$, where $ka_1 = 0.5$ and $f_1 = 0.1$ and $\epsilon_1 = \epsilon_2 = 3.2\epsilon_o$.

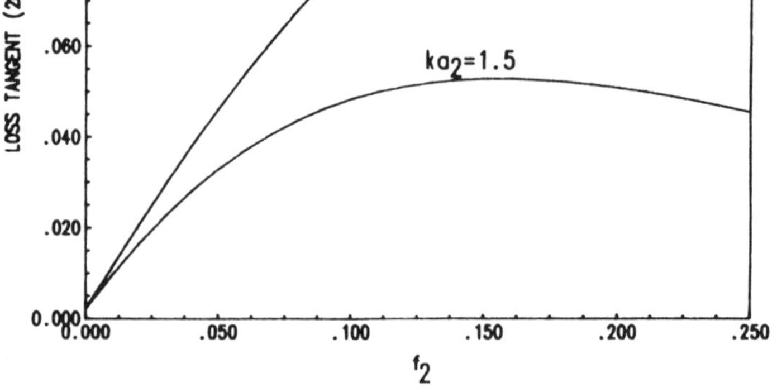

Figure 3.5.9 Effective loss tangent $2K_i/K_r$ as a function of f_2 for $(1)ka_2 = 1.5$ and $(2)ka_2 = 2$, where $ka_1 = 0.5$ and $f_1 = 0.1$ and $\epsilon_1 = \epsilon_2 = 3.2\epsilon_o$.

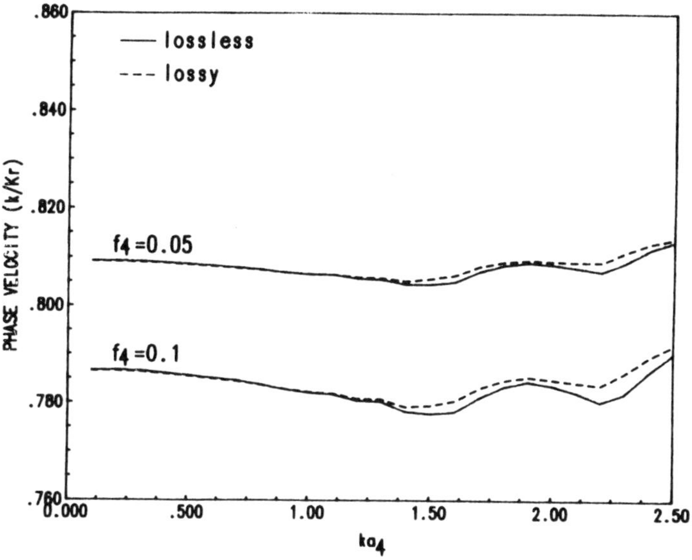

Figure 3.5.10 Normalized phase velocity k/K_r as a function of ka_4 for (1)$f_4 = 0.05$ and (2)$f_4 = 0.1$, where $ka_1 = 0.2$, $ka_2 = 0.3$, $ka_3 = 0.5$ and $f_1 = 0.1$, $f_2 = 0.15$, $f_3 = 0.05$. The results of lossless particles, $\epsilon_1 = \epsilon_2 = \epsilon_3 = \epsilon_4 = 3.2\epsilon_o$, and of lossy particles, $\epsilon_1 = \epsilon_2 = \epsilon_3 = \epsilon_4 = (3.2 + i0.15)\epsilon_o$, are illustrated.

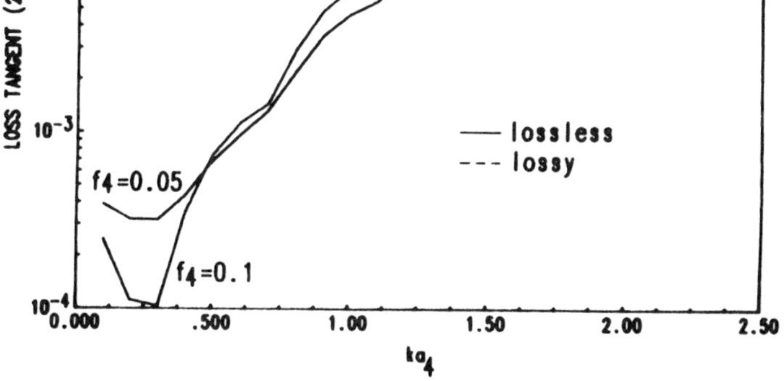

Figure 3.5.11 Effective loss tangent $2K_i/K_r$ as a function of ka_4 for (1)$f_4 = 0.05$ and (2)$f_4 = 0.1$, where $ka_1 = 0.2$, $ka_2 = 0.3$, $ka_3 = 0.5$ and $f_1 = 0.1$, $f_2 = 0.15$, $f_3 = 0.05$. The results of lossless particles, $\epsilon_1 = \epsilon_2 = \epsilon_3 = \epsilon_4 = 3.2\epsilon_o$, and of lossy particles, $\epsilon_1 = \epsilon_2 = \epsilon_3 = \epsilon_4 = (3.2 + i0.15)\epsilon_o$, are illustrated.

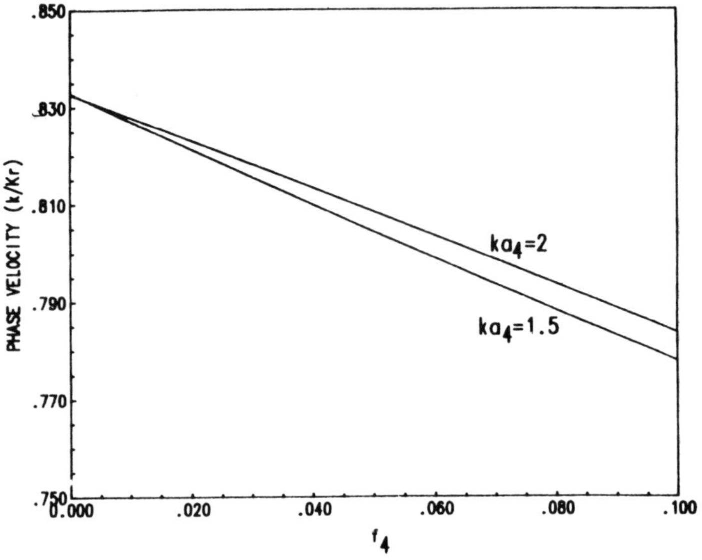

Figure 3.5.12 Normalized phase velocity k/K_r as a function of f_4 for $(1)ka_4 = 1.5$ and $(2)ka_4 = 2$, where $ka_1 = 0.2$, $ka_2 = 0.3$, $ka_3 = 0.5$, $f_1 = 0.1$, $f_2 = 0.15$, $f_3 = 0.05$, and $\epsilon_1 = \epsilon_2 = \epsilon_3 = \epsilon_4 = 3.2\epsilon_o$.

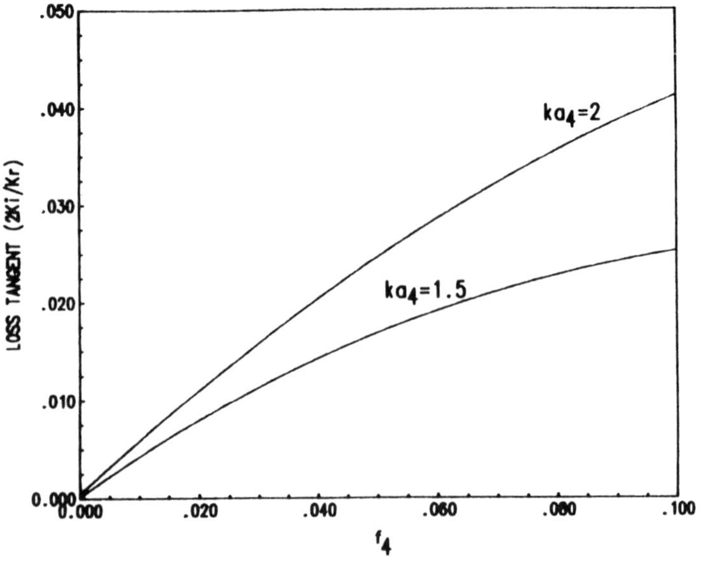

Figure 3.5.13 Effective loss tangent $2K_i/K_r$ as a function of f_4 for $(1)ka_4 = 1.5$ and $(2)ka_4 = 2$, where $ka_1 = 0.2$, $ka_2 = 0.3$, $ka_3 = 0.5$, $f_1 = 0.1$, $f_2 = 0.15$, $f_3 = 0.05$, and $\epsilon_1 = \epsilon_2 = \epsilon_3 = \epsilon_4 = 3.2\epsilon_o$.

velocity and the loss tangent as a function of f_4 for $ka_4 = 1.5$ and $ka_4 = 2$ with $ka_1 = 0.2$, $ka_2 = 0.3$, $ka_3 = 0.5$, $f_1 = 0.1$, $f_2 = 0.15$, $f_3 = 0.05$, and $\epsilon_1 = \epsilon_2 = \epsilon_3 = \epsilon_4 = 3.2\epsilon_o$.

We have shown the numerical results for the cases of one, two and four different species. The numerical algorithm can be extended to more number of species.

3.6 Energy Conservation and Ladder Approximation for Dense Media with Multiple Species

In multiple scattering of waves in lossless media or slightly lossy media, one important problem is to calculate the second moment of the incoherent fields (such a problem is often neglected for media of moderate loss or for cases of extremely long wavelengths). A common approximation to the Bethe–Salpeter equation of the second moment is the ladder approximation. The second moment equation also utilizes the first moment Green's function and the first moment of the field. Thus, the result of the second moment and the question of energy conservation depends on the choices of approximations for the first moment equation as well as that for the second moment equation. A test of energy conservation is provided by the integrated optical relation [8].

In the following, we shall show that the ladder approximation and the quasicrystalline approximation with coherent potential satisfy the integrated optical relation exactly. In the process, we have also derived the ladder approximation for dense medium with multiple species of particles taking into account the effect of correlated scatterers.

The time-averaged Poynting's vector $\overline{S}$ is defined as

$$\overline{S} = \frac{1}{4i\omega\mu}\left\{-\overline{E}\times\left(\overline{\nabla}\times\overline{E}^*\right)+\overline{E}^*\times\left(\overline{\nabla}\times\overline{E}\right)\right\} \tag{129}$$

We also use angular bracket to denote ensemble average. The integrated optical relation states that, for lossless background and particles, the integration over space of the divergence of the time averaged Poynting's vector must be zero.

$$\int d\overline{r}\,\langle\overline{\nabla}\cdot\overline{S}\rangle = 0 \tag{130}$$

The vector wave equation governing the electric field is

$$\overline{E}(\overline{r}) = \overline{E}^{inc}(\overline{r}) + \sum_{l=1}^{N} \int d\overline{r}' \overline{\overline{G}}_o(\overline{r}, \overline{r}') U_l^{s\prime}(\overline{r}') \overline{E}(\overline{r}') \tag{131}$$

Let the operator $\overline{\overline{W}}$ be defined by

$$\overline{\overline{W}}(\overline{r}) \cdot \overline{E}(\overline{r}) = \overline{\nabla} \times \overline{\nabla} \times \overline{E}(\overline{r}) - k^2 \overline{E}(\overline{r}) \tag{132}$$

so that

$$\overline{\overline{W}}(\overline{r}) \cdot \overline{E}(\overline{r}) = \sum_{l=1}^{N} U_l^{s\prime}(\overline{r}) \overline{E}(\overline{r}) \tag{133}$$

The integrated optical relation of (130), by using (129), (133), and the fact that $U_l^{s\prime}$ is real for lossless scatterers, assumes the form

$$4i\omega\mu \int d\overline{r} \langle \overline{\nabla} \cdot \overline{S}(\overline{r}) \rangle = \int d\overline{r} \sum_{\alpha,\beta} \lim_{\overline{r}' \to \overline{r}} \Big\{ W_{\alpha\beta}(\overline{r}') \langle E_\alpha(\overline{r}) E_\beta^*(\overline{r}') \rangle - W_{\alpha\beta}(\overline{r}) \langle E_\beta(\overline{r}) E_\alpha^*(\overline{r}') \rangle \Big\} \tag{134}$$

In (134), the Greek index subscripts $\alpha, \beta = 1, 2, 3$ are used to denote the Cartesian field components.

The coherent field $\langle \overline{E}(\overline{r}) \rangle$ and the coherent Green's function both obey the Dyson's equation [8,17]

$$\overline{\nabla} \times \overline{\nabla} \times \langle \overline{E}(\overline{r}) \rangle - k^2 \langle \overline{E}(\overline{r}) \rangle = \int d\overline{r}' \overline{\overline{M}}(\overline{r}, \overline{r}') \cdot \langle \overline{E}(\overline{r}') \rangle \tag{135}$$

$$\overline{\nabla} \times \overline{\nabla} \times \left\langle \overline{\overline{G}}(\overline{r}, \overline{r}') \right\rangle - k^2 \left\langle \overline{\overline{G}}(\overline{r}, \overline{r}') \right\rangle = \overline{\overline{I}} \delta(\overline{r} - \overline{r}') + \int d\overline{r}'' \overline{\overline{M}}(\overline{r}, \overline{r}'') \cdot \left\langle \overline{\overline{G}}(\overline{r}'', \overline{r}') \right\rangle \tag{136}$$

where $\overline{\overline{M}}$ is the mass operator. Equations (135) and (136) are generally true and the exact mass operator contains an infinite number of terms. Comparisons with (28) and (41) of section 3.2 show that the QCA and QCA–CP can be cast in the form of (135) and (136). Thus, (26) and

(37) give the approximations for the mass operator under QCA and QCA–CP respectively.

Generally, the second moment of the field obeys the Bethe–Salpeter equation [8,17]

$$
\begin{aligned}
\langle E_\alpha(\bar r)\, E^*_\beta(\bar r')\rangle &= \langle E_\alpha(\bar r)\rangle \langle E^*_\beta(\bar r')\rangle \\
&+ \sum_{\alpha',\beta'} \sum_{\alpha'',\beta''} \int d\bar r_1 \int d\bar r_2 \int d\bar r_1' \int d\bar r_2' \\
&\cdot \langle G_{\alpha\alpha'}(\bar r, \bar r_1)\rangle \langle G^*_{\beta\beta'}(\bar r', \bar r_1')\rangle \\
&\cdot I_{\alpha'\alpha'';\beta'\beta''}(\bar r_1, \bar r_2; \bar r_1', \bar r_2') \langle E_{\alpha''}(\bar r_2)\, E^*_{\beta''}(\bar r_2')\rangle
\end{aligned}
\tag{137}
$$

From (135)–(137), it follows that

$$
\begin{aligned}
&\sum_\beta W_{\alpha\beta}(\bar r') \langle E_\alpha(\bar r)\, E^*_\beta(\bar r')\rangle = \\
&\quad \langle E_\alpha(\bar r)\rangle \sum_\beta \int d\bar r_2 M^*_{\alpha\beta}(\bar r', \bar r_2) \langle E^*_\beta(\bar r_2)\rangle \\
&\quad + \sum_{\alpha'} \sum_{\alpha'',\beta''} \int d\bar r_1 \int d\bar r_2 \int d\bar r_2' \langle G_{\alpha\alpha'}(\bar r, \bar r_1)\rangle \\
&\quad \cdot I_{\alpha'\alpha'';\alpha\beta''}(\bar r_1, \bar r_2; \bar r', \bar r_2') \langle E_{\alpha''}(\bar r_2)\, E^*_{\beta''}(\bar r_2')\rangle \\
&\quad + \sum_\beta \sum_{\alpha',\beta'} \sum_{\alpha'',\beta''} \int d\bar r_1 \int d\bar r_2 \int d\bar r_1' \int d\bar r_2' \langle G_{\alpha\alpha'}(\bar r, \bar r_1)\rangle \\
&\quad \cdot \int d\bar r_2'' M^*_{\alpha\beta}(\bar r', \bar r_2'') \langle G^*_{\beta\beta'}(\bar r_2'', \bar r_1')\rangle \\
&\quad \cdot I_{\alpha'\alpha'';\beta'\beta''}(\bar r_1, \bar r_2; \bar r_1', \bar r_2') \langle E_{\alpha''}(\bar r_2)\, E^*_{\beta''}(\bar r_2')\rangle
\end{aligned}
\tag{138}
$$

Next apply the Bethe–Salpeter equation of (137) to the second term on the right hand side of (138). This gives

$$
\begin{aligned}
&\sum_\beta W_{\alpha\beta}(\bar r') \langle E_\alpha(\bar r)\, E^*_\beta(\bar r')\rangle = \\
&\quad \langle E_\alpha(\bar r)\rangle \sum_\beta \int d\bar r_2 M^*_{\alpha\beta}(\bar r', \bar r_2) \langle E^*_\beta(\bar r_2)\rangle \\
&\quad + \sum_{\alpha'} \sum_{\alpha'',\beta''} \int d\bar r_1 \int d\bar r_2 \int d\bar r_2' \langle G_{\alpha\alpha'}(\bar r, \bar r_1)\rangle
\end{aligned}
$$

$$
\begin{aligned}
&\cdot I_{\alpha'\alpha'';\alpha\beta''}\left(\bar{r}_1,\bar{r}_2;\bar{r}',\bar{r}_2'\right)\left\langle E_{\alpha''}\left(\bar{r}_2\right)\right\rangle\left\langle E^*_{\beta''}\left(\bar{r}_2'\right)\right\rangle\\
&+\sum_{\alpha'}\sum_{\alpha'',\beta''}\int d\bar{r}_1\int d\bar{r}_2\int d\bar{r}_2'\left\langle G_{\alpha\alpha'}\left(\bar{r},\bar{r}_1\right)\right\rangle I_{\alpha'\alpha'';\alpha\beta''}\left(\bar{r}_1,\bar{r}_2;\bar{r}',\bar{r}_2'\right)\\
&\cdot\sum_{\alpha_1,\beta_1}\sum_{\alpha_2,\beta_2}\int d\bar{r}_3\int d\bar{r}_4\int d\bar{r}_3'\int d\bar{r}_4'\left\langle G_{\alpha''\alpha_1}\left(\bar{r}_2,\bar{r}_3\right)\right\rangle\\
&\cdot\left\langle G^*_{\beta''\beta_1}\left(\bar{r}_2',\bar{r}_3'\right)\right\rangle I_{\alpha_1\alpha_2;\beta_1\beta_2}\left(\bar{r}_3,\bar{r}_4;\bar{r}_3',\bar{r}_4'\right)\left\langle E_{\alpha_2}\left(\bar{r}_4\right)E^*_{\beta_2}\left(\bar{r}_4'\right)\right\rangle\\
&+\sum_{\beta}\sum_{\alpha',\beta'}\sum_{\alpha'',\beta''}\int d\bar{r}_1\int d\bar{r}_2\int d\bar{r}_1'\int d\bar{r}_2'\left\langle G_{\alpha\alpha'}\left(\bar{r},\bar{r}_1\right)\right\rangle\\
&\cdot\int d\bar{r}_2''M^*_{\alpha\beta}\left(\bar{r}',\bar{r}_2''\right)\left\langle G^*_{\beta\beta'}\left(\bar{r}_2'',\bar{r}_1'\right)\right\rangle\\
&\cdot I_{\alpha'\alpha'';\beta'\beta''}\left(\bar{r}_1,\bar{r}_2;\bar{r}_1',\bar{r}_2'\right)\left\langle E_{\alpha''}\left(\bar{r}_2\right)E^*_{\beta''}\left(\bar{r}_2'\right)\right\rangle
\end{aligned}
\tag{139}
$$

Similar manipulation can be applied to $\sum_{\beta} W_{\alpha\beta}(\bar{r})\left\langle E^*_{\alpha}(\bar{r}')E_{\beta}(\bar{r})\right\rangle$. The result expression and (139) are then substituted into the right hand side of equation (134). The limit $\bar{r}'\longrightarrow\bar{r}$ is taken and the integration is carried out over $d\bar{r}$. There are eight terms. By interchanging dummy indices and arguments, the eight terms can be rearranged, factorized and cast into the following form

$$
\begin{aligned}
&4i\omega\mu\int d\bar{r}\left\langle\overline{\nabla}\cdot\overline{S}(\bar{r})\right\rangle=\\
&\int d\bar{r}_2\int d\bar{r}_2'\sum_{\alpha'',\beta''}\left\{M^*_{\alpha''\beta''}\left(\bar{r}_2,\bar{r}_2'\right)-M_{\beta''\alpha''}\left(\bar{r}_2',\bar{r}_2\right)\right.\\
&+\int d\bar{r}\int d\bar{r}_1\sum_{\alpha,\alpha''}\left[\left\langle G_{\alpha\alpha'}\left(\bar{r},\bar{r}_1\right)\right\rangle-\left\langle G^*_{\alpha'\alpha}\left(\bar{r}_1,\bar{r}\right)\right\rangle\right]\\
&\quad\left.\cdot I_{\alpha'\alpha'';\alpha\beta''}\left(\bar{r}_1,\bar{r}_2;\bar{r},\bar{r}_2'\right)\right\}\left\langle E_{\alpha''}\left(\bar{r}_2\right)\right\rangle\left\langle E^*_{\beta''}\left(\bar{r}_2'\right)\right\rangle\\
&+\sum_{\alpha_1\beta_1}\sum_{\alpha_2,\beta_2}\sum_{\alpha'',\beta''}\int d\bar{r}_2\int d\bar{r}_3\int d\bar{r}_4\int d\bar{r}_3'\int d\bar{r}_4'\int d\bar{r}_2'\\
&\quad\cdot\left\langle G_{\alpha''\alpha_1}\left(\bar{r}_2,\bar{r}_3\right)\right\rangle\left\langle G^*_{\beta''\beta_1}\left(\bar{r}_2',\bar{r}_3'\right)\right\rangle\\
&\quad\cdot I_{\alpha_1\alpha_2;\beta_1\beta_2}\left(\bar{r}_3,\bar{r}_4;\bar{r}_3',\bar{r}_4'\right)\left\langle E_{\alpha_2}\left(\bar{r}_4\right)E^*_{\beta_2}\left(\bar{r}_4'\right)\right\rangle\\
&\quad\cdot\left\{M^*_{\alpha''\beta''}\left(\bar{r}_2,\bar{r}_2'\right)-M_{\beta''\alpha''}\left(\bar{r}_2',\bar{r}_2\right)+\int d\bar{r}_1\sum_{\alpha,\alpha'}\int d\bar{r}\right.
\end{aligned}
$$

$$\left.\cdot\left[\langle G_{\alpha\alpha'}(\overline{r},\overline{r}_1)\rangle - \langle G^*_{\alpha'\alpha}(\overline{r}_1,\overline{r})\rangle\right] I_{\alpha'\alpha'';\alpha\beta''}(\overline{r}_1,\overline{r}_2;\overline{r},\overline{r}'_2)\right\} \quad (140)$$

From (140), it follows that the integrated optical relation is satisfied exactly if

$$M^*_{\alpha\beta}(\overline{r},\overline{r}') - M_{\beta\alpha}(\overline{r}',\overline{r}) + \sum_{\alpha',\beta'}\int d\overline{r}_1 \int d\overline{r}_2$$

$$\cdot\left[\langle G_{\alpha'\beta'}(\overline{r}_1,\overline{r}_2)\rangle - \langle G^*_{\beta'\alpha'}(\overline{r}_2,\overline{r}_1)\rangle\right] I_{\beta'\alpha;\alpha'\beta}(\overline{r}_2,\overline{r};\overline{r}_1,\overline{r}') = 0 \quad (141)$$

In the following we shall show that the condition is obeyed by QCA–CP and the ladder approximation that is modified for dense medium. From (37), the mass operator under QCA–CP is $\overline{\overline{\widehat{M}}}$ which

$$\overline{\overline{\widehat{M}}} = \sum_{s_j=1}^{L} n_{s_j}\int d\overline{r}_j \overline{\overline{\widehat{C}}}_j^{s_j} \quad (142)$$

Let superscript $\dagger$ denote Hermitian adjoint such that

$$\left(\overline{\overline{A}}^{\dagger}\right)_{\alpha''\beta''}(\overline{r}_2,\overline{r}'_2) = \overline{\overline{A}}^*_{\beta''\alpha''}(\overline{r}'_2,\overline{r}_2) \quad (143)$$

It follows from (33) and (38) that

$$\overline{\overline{\widehat{C}}}_j^{s_j} = \overline{\overline{U}}_j^{s_j} + \sum_{s_l=1}^{L}\int d\overline{r}_l q_{s_j s_l}(\overline{r}_j - \overline{r}_l)\overline{\overline{U}}_j^{s_j}\overline{\overline{G}}_c\overline{\overline{\widehat{C}}}_l^{s_l} \quad (144)$$

where

$$q_{s_j s_l}(\overline{r}_j - \overline{r}_l) = \delta_{s_j s_l}\delta(\overline{r}_j - \overline{r}_l) + n_{s_l}h_{s_j s_l}(\overline{r}_j - \overline{r}_l) \quad (145)$$

From (142), (144), and

$$\overline{\overline{U}}_j^{s_j\dagger} = \overline{\overline{U}}_j^{s_j} \quad (146)$$

for lossless scatterers, it follows that

$$\overline{\overline{\widehat{M}}}^{\dagger} - \overline{\overline{\widehat{M}}} = \sum_{s_j=1}^{L}\sum_{s_l=1}^{L} n_{s_j}\int d\overline{r}_j \int d\overline{r}_l\, q_{s_j s_l}(\overline{r}_j - \overline{r}_l)$$

$$\cdot\left[\overline{\overline{\widehat{C}}}_l^{s_l\dagger}\overline{\overline{G}}_c^{\dagger}\overline{\overline{U}}_j^{s_j} - \overline{\overline{U}}_j^{s_j}\overline{\overline{G}}_c\overline{\overline{\widehat{C}}}_l^{s_l}\right] \quad (147)$$

Next, use $\overline{\overline{U}}_j^{s_j}$ from (144) to substitute in the first $\overline{\overline{U}}_j^{s_j}$ in (147). For the second $\overline{\overline{U}}_j^{s_j}$ in (147), use

$$\overline{\overline{U}}_j^{s_j} = \widehat{\overline{\overline{C}}}_j^{s_j\dagger} - \sum_{s_m=1}^{L} \int d\overline{r}_m q_{s_j s_m}(\overline{r}_j - \overline{r}_m)\widehat{\overline{\overline{C}}}_m^{s_m\dagger}\overline{\overline{G}}_c^{\dagger}\overline{\overline{U}}_j^{s_j} \tag{148}$$

which is a consequence of (144) and (146). Thus, we get four terms in the right hand side of (147). Two of the terms cancel on exchange of dummy indices and arguments. The result is

$$\widehat{\overline{\overline{M}}}^{\dagger} - \widehat{\overline{\overline{M}}} = \sum_{s_j=1}^{L}\sum_{s_l=1}^{L} n_{s_j}\int d\overline{r}_j \int d\overline{r}_l\, q_{s_j s_l}(\overline{r}_j - \overline{r}_l) \cdot \left[\widehat{\overline{\overline{C}}}_l^{s_l\dagger}\overline{\overline{G}}_c^{\dagger}\widehat{\overline{\overline{C}}}_j^{s_j} - \widehat{\overline{\overline{C}}}_j^{s_j\dagger}\overline{\overline{G}}_c\widehat{\overline{\overline{C}}}_l^{s_l}\right] \tag{149}$$

Since $h_{s_j s_l}(\overline{r}_j - \overline{r}_l)$ is symmetric with respect to s_j and s_l as well as $\overline{r}_j$ and $\overline{r}_l$, it follows that $n_{s_j}q_{s_j s_l}(\overline{r}_j - \overline{r}_l) = n_{s_l}q_{s_l s_j}(\overline{r}_j - \overline{r}_l)$. Thus (149) is of the same form as the requirement of energy conservation of (141) provided that the intensity operator is chosen as

$$I_{\alpha\beta;\alpha'\beta'}(\overline{r},\overline{r}_1;\overline{r}',\overline{r}_1') = \sum_{s_j=1}^{L}\sum_{s_l=1}^{L}\int d\overline{r}_j \int d\overline{r}_l \left[n_{s_j}\delta_{s_j s_l}\delta(\overline{r}_j - \overline{r}_l) + n_{s_j}n_{s_l}h_{s_j s_l}(\overline{r}_j - \overline{r}_l)\right]\widehat{C}^{s_l}_{l\alpha\beta}(\overline{r},\overline{r}_1)\widehat{C}^{s_j *}_{j\alpha'\beta'}(\overline{r}',\overline{r}_1') \tag{150}$$

The intensity operator represents a sum of contribution form the same scatterer as represented by the first term in (150) and from different scatterers weighted by $n_{s_j}n_{s_l}h_{s_j s_l}(\overline{r}_j - \overline{r}_l)$ as represented by the second term in (150). As noted in the previous sections, $\widehat{\overline{\overline{C}}}_j^{s_j}$ represents transition scattering matrix of the j th scatterer in the presence of other scatterers.

It is interesting to compare (150) with the usually ladder approximation. In the usual ladder approximation, the second term in (150) is absent and secondly the $\widehat{\overline{\overline{C}}}_l$ in (150) is replaced by $\overline{\overline{T}}_l$ which is the transition matrix of the l th particle in the absence of other scatterers. Obviously, the two cases reduce to each other for sparse concentration of particles when $n_{s_j} n_{s_l} h_{s_j s_l}$ can be neglected and $\overline{\overline{\widehat{C}}}_l$ is approximately equal to $\overline{\overline{T}}_l$. Equation (150) is also the intensity operator under ladder approximation for multiple species which will be useful for second moment calculations.

3.7 Conclusion

In this chapter, we have studied the problem of effective propagation constants in media with densely distributed dielectric particles of multiple sizes and permittivities. The problem is complicated by the fact that for high concentrations of particles, the particles do not scatter independently. In this chapter, we have employed QCA and QCA–CP approximations to calculate the effective propagation constants incorporating the effect of correlated scatterers. The imaginary part of the complex effective propagation constant gives the attenuation rate of the coherent wave due to both scattering and absorption. The particles in natural geophysical terrain usually follow a drop size distribution which can be discretized and represented by a histogram. The approach described in this paper can find useful applications to such problems of random discrete scatterers of multiple sizes. It also finds applications to problems of random discrete scatterers of multiple permittivities in geophysical terrain and composite materials.

The results of the effective propagation constants under QCA and QCA–CP depend on the pair distribution functions of particle positions. In this chapter, the pair distribution functions of non-interpenetrable spheres of multiple sizes are calculated by the Percus–Yevick approximation.

For the case of small particles, analytic closed form expressions are derived for the effective propagation constants under QCA and QCA–CP. The effective propagation constants for moderate size particles under QCA are solved by numerical techniques.

For problems of lossless particles or slightly lossy particles, it is also important to calculate the second moment of the incoherent fields. A common approximation for the Bethe–Salpeter second moment equation is the ladder approximation which can include the effect of correlated scattering through the pair distribution function as discussed in section 3.6. It is also seen in section 3.6 that the QCA–CP is consistent with the ladder approximation in that the integrated optical relation is satisfied. The ladder approximation has been used in conjunction with QCA–CP to derive a set of dense media radiative transfer equations that includes the effect of correlated scatterers.

Finally, we will mention a few other approaches that have been used to calculate the effective propagation constants in dense media. The strong permittivity fluctuation approach uses the bilocal approximation of continuous random medium theory and takes into account the singularity of the dyadic Green's function [8,35–37]. The effective medium approximation consisting of a set of coupled integral equations for the effective T-matrix and effective distribution functions has also been used [9,38–41]. The QCA approximation has been used to consider the case of nonspherical particles [42]. Monte Carlo calculations have also been made on the pair distribution function of aligned identical spheroidal particles [43].

References

[1] Twersky, V., "Coherent scalar field in pair-correlated random distributions of aligned scatterers," *J. Math. Phys.,* **18,** 2468–2486, 1977.

[2] Tsang, L. and J. A. Kong, "Multiple scattering of electromagnetic waves by random distribution of discrete scatterers with coherent potential and quantum mechanical formulism," *J. Appl. Phys.,* **15,** 3465–3485, 1980.

[3] Tsang, L., J. A. Kong, and T. Habashy, "Multiple scattering of acoustic waves by random distribution of discrete spherical scatterers with the quasi-crystalline and Percus-Yevick approximation," *J. Acoust. Soc. Am.,* **71,** 552–558, 1982.

[4] Tsang, L. and J. A. Kong, "Effective propagation constants for coherent electromagnetic wave propagation in media embedded with dielectric scatterers," *J. Appl. Phys.,* **53,** 7162–7173, 1982.

[5] Ishimaru, A. and Y. Kuga, "Attenuation constant of a coherent field in a dense distribution of particles,," *J. Opt. Soc. Am.,* **72,** 1317–1320, 1982.

[6] Tsang, L. and J. A. Kong, "Scattering of electromagnetic waves

from a half space of densely distributed dielectric scatterers," *Radio Science,* **18**, 1260–1272, 1983.

[7] Varadan, V. K., V. N. Bringi, V. V. Varadan, and A. Ishimaru, "Multiple scattering theory for waves in discrete random media and comparison with experiments," *Radio Science,* **18**, 321–327, 1983.

[8] Tsang, L., J. A. Kong, and R. T. Shin, *Theory of Microwave Remote Sensing,* Wiley-Interscience, New York, 1985.

[9] Zhu, P. A., A. K. Fung, and K. W. Wong, "Effective propagation constants in dense random media under effective medium approximation," *Radio Science,* **22**, 234–250, 1987.

[10] Wertheim, M. S., "Exact solution of the Percus-Yevick integral equation for hard spheres," *Phys. Rev. Lett.,* **20**, 321–323, 1963.

[11] Lebowitz, J. L., "Exact solution of generalized Percus-Yevick equation for a mixture of hard spheres," *Phys. Rev.,* **133**, A895–899, 1964.

[12] Wertheim, M. S., "Analytic solution of the Percus-Yevick equation," *J. Math. Phys.,* **5**, 643–651, 1964.

[13] Baxter, R. J., "Ornstein-Zernike relation for a disordered fluid," *Aust. J. Phys.,* **21**, 563–569, 1968.

[14] Baxter, R. J., "Ornstein-Zernike relation and Percus-Yevick approximation for fluid mixtures," *J. Chem. Phys.,* **52**, 4559–4562, 1970.

[15] Leonard, P. J., D. Henderson, and J. A. Barker, "Calculation of the radial distribution function of hard-sphere mixtures in the Percus-Yevick approximation," *Molec. Phys.,* **21**, 107–111, 1971.

[16] Waseda, Y., *The Structure of Non-Crystalline Materials, Liquids and Amorphous Solids,* McGraw-Hill, New York, 1980.

[17] Frisch, V., "Wave propagation in random medium," in *Probabilistic Methods in Applied Mathematics,* A. T. Bharucha-Reid, ed., Academic Press, New York, 1968.

[18] Ishimaru, A., *Wave Propagation and Scattering in Random Media,* **1** , **2**, Academic Press, New York, 1978.

[19] Tatarskii, V. I., *The Effects of Turbulent Atmosphere on Wave Propagation,* National Technical Information Service, Springfield,

Va., 1971.

[20] Tsang, L. and A. Ishimaru, "Radiative wave and cyclical transfer equation for dense nontenuous media," *J. Opt. Soc. Am. A,* **2**, 2187–2194, 1985.

[21] Tsang, L. and A. Ishimaru, "Radiative wave equations for vector electromagnetic propagation in dense nontenuous media," *J. of Electromagnetic Waves and Applications,* **1**, 59–72, 1987.

[22] Tsang, L., "Passive remote sensing of dense nontenuous media," *J. of Electromagnetic Waves and Applications,* **1**, 159–173, 1987.

[23] Soven, P., "Coherent-potential model of substitutional disordered alloys," *Phys. Rev.,* **156**, 809–813, 1967.

[24] Gyorffy, B. L., "Electronic states in liquid metals: A generalization of the coherent–potential approximation for a system with short-range order," *Phys. Rev. B.,* **1**, 3290–3299, 1970.

[25] Korringa, J. and R. L. Mills, "Coherent potential approximation for random systems with short range correlations," *Phys. Rev. B.,* **5**, 1654–1655, 1972.

[26] Kohler, W. E. and G. C. Papanicolaou, "Some applications of the coherent potential approximation," in *Multiple Scattering and Waves in Random Media,* P. L. Chow, W. K. Kohler and G. C. Papanicolaou, eds., North-Holland, New York, 199–223, 1981.

[27] Schanda, E., "Microwave modelling of snow and soil," *J. of Electromagnetic Waves and Applications,* **1**, 1–24, 1987.

[28] Hallikainen, M. T., F. T. Ulaby, and T. E. Van Deventer, "Extinction behavior of dry snow in the 18-90 GHz range," *IEEE Trans. Geosci. Remote Sensing,* **GE-25**, 737–745, 1987.

[29] Rayleigh, L., "On the influence of obstacles arranged in rectangular order on the properties of a medium," *Phil. Mag.,* **34**, 481–502, 1892.

[30] Tai, C. T., *Dyadic Green's Function in Electromagnetic Theory,* International Textbook, Scranton, PA., 1971.

[31] Cruzan, O. R., "Translational addition theorems for spherical vector wave functions," TR–906, Diamond Ordinance Fuse Laboratories, Department of the Army, Washington DC., 1961.

[32] Cruzan, O. R., "Translational addition theorems for spherical vec-

tor wave functions," *Quart. J. Appl. Math.,* **20,** 33–40, 1962.

[33] Edmonds, A. R., *Angular Momentum in Quantum Mechanics,* Princeton University, Princeton, NJ., 1957.

[34] Born, M. and E. Wolf, *Principles of Optics,* fifth edition, Pergamon, New York, 1975.

[35] Tsang, L. and J. A. Kong, "Scattering of electromagnetic waves from random media with strong permittivity fluctuations," *Radio Science,* **16,** 303–320, 1981.

[36] Tsang, L., J. A. Kong, and R. W. Newton, "Application of strong fluctuation random medium theory to scattering of electromagnetic waves from a half space of dielectric mixture," *IEEE Trans. Ant. Prop.,* **AP-30,** 292–302, 1982.

[37] Stogryn, A., "Correlation functions for random granular media in strong fluctuation theory," *IEEE Trans. Geosci. Remote Sensing,* **GE-22,** 150–154, 1984.

[38] Roth, L., "Effective medium approximation for liquid metals," *Phys. Rev. B.,* **4,** 2476–2484, 1974.

[39] Singh, V. A., "Analytic unitary preserving approximations for the electronic structure of amorphous systems," *Phys. Rev. B.,* **24,** 4852–4854, 1981.

[40] Schwartz, L. and T. J. Plona, "Ultrasonic propagation in closed-packaged disordered suspensions," *J. Appl. Phys.,* **55,** 3971–3977, 1984.

[41] Davis, V. A. and L. Schwartz, "Electromagnetic propagation in close-packed disordered suspensions," *Phys. Rev. B.,* **31,** 5155–5165, 1985.

[42] Tsang, L., "Scattering of electromagnetic waves from a half space of nonspherical particles," *Radio Science,* **19,** 1450–1460, 1984.

[43] Varadan, V. V., V. K. Varadan, Y. Ma, and W. A. Steele, "Effects of nonspherical statistics on EM wave propagation in discrete random media," *Radio Science,* **22,** 491–498, 1987.

4

SCATTERING FROM RANDOMLY PERTURBED PERIODIC AND QUASIPERIODIC SURFACES

H. A. Yueh, R. T. Shin, and J. A. Kong

4.1 Introduction

Scattering of electromagnetic waves from a randomly perturbed periodic surface is of interest in the active remote sensing of plowed field and ocean surface. The variations of the radar scattering coefficients due to the change in the look direction relative to the row direction have been well documented [Batlivala and Ulaby, 1976; Ulaby and Bare, 1979; Fenner *et al.*, 1980]. In the past, the problem of electromagnetic wave scattering from periodic [Waterman, 1975; Jordan and Lang, 1979; Chuang and Kong, 1982] or random [Rice, 1951; Beckmann and Spizzichino, 1963; Stogryn, 1967; Valenzuela, 1967; Sancer, 1969; Leader, 1971; Tsang and Kong, 1980a] rough surfaces has been extensively studied. The problem of scattering by randomly perturbed surface has been studied by assuming that the periodic surface causes a tilting effect [Ulaby *et al.*, 1982]. In this approach the scattering coefficients of the random rough surfaces obtained using the Kirchhoff approximation or small perturbation method is averaged over the change in local incidence angle due to the periodic surface. This approach has also been used to solve the scattering from a composite random rough surface with small and large variations [Semenov, 1966; Wu and Fung, 1972].

In section 4.2, we use the Extended Boundary Condition (EBC) method to formulate the scattering of electromagnetic waves from a randomly perturbed dielectric periodic surface, and apply the small perturbation method (SPM) to solve the surface currents and scattered fields. The surface currents and the scattered fields are expanded and solved up to the second order. The zeroth order problem which is the scattering from periodic surface is solved exactly. This solution is then used in the small perturbation method to solve for the higher order scattered field from a randomly perturbed periodic surface. The random perturbation is modeled as a Gaussian random process. The theoretical results are illustrated by calculating the bistatic and backscattering coefficients and comparing these with the results obtained using the Kirchhoff approximation.

In section 4.3, we use the Kirchhoff approximation to study the scattering of electromagnetic waves from a randomly perturbed quasiperiodic surface in order to more realistically model the plowed fields. In the plowed fields there are some random variations on the period and amplitude of the sinusoidal variation as we move from one row to the next. This variation can be modelled by introducing the narrow-band Gaussian random process on top of the basic sinusoidal variation, which will cause the surface to be quasiperiodic. Therefore, we characterize the rough surface as a composite surface with a Gaussian random variation, a sinusoidal variation and a narrow-band Gaussian random variation around the same spatial frequency. The physical optics inte-

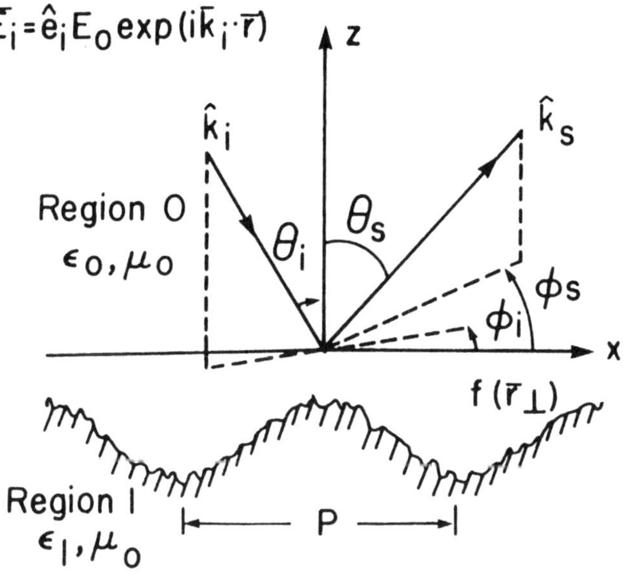

Figure 4.2.1 Geometric configuration of the problem.

gral obtained with the Kirchhoff approximation is evaluated to obtain the coherent and incoherent bistatic scattering coefficients. In the geometrical optics limit, the stationary phase method is used to further simplify the results. In this limit it can be shown that the bistatic scattering coefficients are proportional to the probability of the occurrence of the slopes which will specularly reflect the incident wave into the observation direction. The theoretical results are illustrated for the various cases by plotting the backscattering cross sections as a function of the angle of incidence with the incident wave vector either parallel or perpendicular to the row direction. The appearances of peaks will be explained in terms of the scattering patterns for sinusoidal surfaces.

4.2 Scattering by a Randomly Perturbed Periodic Surface

a. Formulation

Consider a plane wave incident upon a periodic surface with roughness (Fig. 4.2.1). The electric field of the incident wave is given by

$$\overline{E}_i = \hat{e}_i E_0 \exp(i\overline{k}_i \cdot \overline{r}) \tag{1}$$

where

$$\overline{k}_i = \overline{k}_{\perp i} + \hat{z}k_{zi} = \hat{x}k_{xi} + \hat{y}k_{yi} + \hat{z}k_{zi} \tag{2}$$

denotes the incident wave vector, and $\hat{e}_i$ is the unit vector in the polarization direction of the incident electric field. The rough surface is characterized by a height function $z = f(x, y)$, which is given by

$$f(x, y) = H \cos(\frac{2\pi}{p}x) + \xi(x, y) \tag{3}$$

where $\xi(x, y)$ is a small random perturbation on the top of a sinusoidal surface and is modelled as a Gaussian random process with zero mean,

$$< \xi(x, y) >= 0 \tag{4}$$

From Huygens' principle the electric field $\overline{E}(\overline{r})$ in the free space and the transmitted field $\overline{E}_1(\overline{r})$ in the dielectric medium satisfy the following equations [Tsang et al., 1985; Kong, 1986]:

$$\overline{E}_i(\overline{r}) + \int_{S'} dS' \left\{ i\omega\mu_0 \overline{\overline{G}}(\overline{r}, \overline{r}') \cdot [\hat{n} \times \overline{H}(\overline{r}')] + \nabla \times \overline{\overline{G}}(\overline{r}, \overline{r}') \cdot [\hat{n} \times \overline{E}(\overline{r}')] \right\}$$

$$= \begin{cases} \overline{E}(\overline{r}) & z > f(\overline{r}_\perp) \quad (5a) \\ 0 & z < f(\overline{r}_\perp) \quad (5b) \end{cases}$$

$$- \int_{S'} dS' \left\{ i\omega\mu_1 \overline{\overline{G}}_1(\overline{r}, \overline{r}') \cdot [\hat{n} \times \overline{H}_1(\overline{r}')] + \nabla \times \overline{\overline{G}}_1(\overline{r}, \overline{r}') \cdot [\hat{n} \times \overline{E}_1] \right\}$$

$$= \begin{cases} 0 & z > f(\overline{r}_\perp) \quad (6a) \\ \overline{E}_1(\overline{r}) & z < f(\overline{r}_\perp) \quad (6b) \end{cases}$$

where $\overline{\overline{G}}(\overline{r}, \overline{r}')$ and $\overline{\overline{G}}_1(\overline{r}, \overline{r}')$ are the dyadic Green's functions for free space and the homogeneous dielectric of region 1 [Tsang et al., 1985], respectively, and the surface integration is to be carried out over the rough surface S'. The unit vector normal to the rough surface $\hat{n}$ points into the free space

$$\hat{n} = \frac{-\alpha\hat{x} - \beta\hat{y} + \hat{z}}{\sqrt{1 + \alpha^2 + \beta^2}} \tag{7}$$

where α and β are the local slopes in the x and y directions,

$$\alpha = \frac{\partial f(x', y')}{\partial x'} \tag{8a}$$

$$\beta = \frac{\partial f(x', y')}{\partial y'} \tag{8b}$$

Since tangential fields are continuous, we can define surface field unknowns as

$$dS' \sqrt{\frac{\mu_o}{\epsilon_o}} \hat{n} \times \overline{H}(\overline{r}') = d\overline{r}'_\perp \overline{a}(\overline{r}'_\perp) = dS' \sqrt{\frac{\mu_o}{\epsilon_o}} \hat{n} \times \overline{H}_1(\overline{r}') \tag{9a}$$

$$dS' \hat{n} \times \overline{E}(\overline{r}') = d\overline{r}'_\perp \overline{b}(\overline{r}'_\perp) = dS' \hat{n} \times \overline{E}_1(\overline{r}') \tag{9b}$$

where $d\overline{r}'_\perp$ is the projection of dS' on the x - y plane, and the surface fields are tangential to the surface. Thus

$$\hat{n}(\overline{r}'_\perp) \cdot \overline{a}(\overline{r}'_\perp) = 0 \tag{10a}$$

$$\hat{n}(\overline{r}'_\perp) \cdot \overline{b}(\overline{r}'_\perp) = 0 \tag{10b}$$

Let f_{min} and f_{max} be, respectively, the minimum and maximum values of the surface profiles $f(\overline{r})$. Evaluating (5b) for $z < f_{min}$ and (6a) for $z > f_{max}$, we obtain

$$\begin{aligned} \overline{E}_i(\overline{r}) = & \frac{1}{8\pi^2} \int d\overline{k}_\perp e^{i\overline{k}_\perp \cdot \overline{r}_\perp} e^{-ik_z z} \frac{k}{k_z} \int d\overline{r}'_\perp e^{-i\overline{k}_\perp \cdot \overline{r}'_\perp} e^{ik_z f(\overline{r}'_\perp)} \\ & \times \Big\{ [\hat{h}(-k_z)\hat{h}(-k_z) + \hat{v}(-k_z)\hat{v}(-k_z)] \cdot \overline{a}(\overline{r}'_\perp) \\ & + [-\hat{v}(-k_z)\hat{h}(-k_z) + \hat{h}(-k_z)\hat{v}(-k_z)] \cdot \overline{b}(\overline{r}'_\perp) \Big\} \end{aligned} \tag{11a}$$

$$\begin{aligned} 0 = & \frac{1}{8\pi^2} \int d\overline{k}_\perp e^{i\overline{k}_\perp \cdot \overline{r}_\perp} e^{ik_{1z} z} \frac{k_1}{k_{1z}} \int d\overline{r}'_\perp e^{-i\overline{k}_\perp \cdot \overline{r}'_\perp} e^{-ik_{1z} f(\overline{r}'_\perp)} \\ & \times \Big\{ \frac{k}{k_1} [\hat{h}_1(k_{1z})\hat{h}_1(k_{1z}) + \hat{v}_1(k_{1z})\hat{v}_1(k_{1z})] \cdot \overline{a}(\overline{r}'_\perp) \\ & + [-\hat{v}_1(k_{1z})\hat{h}_1(k_{1z}) + \hat{h}_1(k_{1z})\hat{v}_1(k_{1z})] \cdot \overline{b}(\overline{r}'_\perp) \Big\} \end{aligned} \tag{11b}$$

where

$$\overline{k} = \overline{k}_\perp + \hat{z}k_z = \hat{x}k_x + \hat{y}k_y + \hat{z}k_z\,, \quad k = \omega\sqrt{\mu_o\epsilon_o} \tag{12a}$$

$$\overline{k}_1 = \overline{k}_\perp + \hat{z}k_{1z} = \hat{x}k_x + \hat{y}k_y + \hat{z}k_{1z}\,, \quad k_1 = \omega\sqrt{\mu_o\epsilon_1} \tag{12b}$$

$$k_\rho = \sqrt{k_x^2 + k_y^2}, \quad k_z = \sqrt{k^2 - k_\rho^2}, \quad k_{1z} = \sqrt{k_1^2 - k_\rho^2} \tag{12c}$$

$$\hat{h}(k_z) = \frac{\hat{k} \times \hat{z}}{|\hat{k} \times \hat{z}|} = \frac{1}{k_\rho}(\hat{x}k_y - \hat{y}k_x) \tag{12d}$$

$$\hat{v}(k_z) = \frac{1}{k}\hat{h} \times \overline{k} = \frac{-k_z}{kk_\rho}(\hat{x}k_x + \hat{y}k_y) + \frac{k_\rho}{k}\hat{z} \tag{12e}$$

$$\hat{h}_1(k_{1z}) = \frac{\hat{k}_1 \times \hat{z}}{|\hat{k}_1 \times \hat{z}|} = \frac{1}{k_\rho}(\hat{x}k_y - \hat{y}k_x) \tag{12f}$$

$$\hat{v}_1(k_{1z}) = \frac{1}{k_1}\hat{h}_1 \times \overline{k}_1 = \frac{-k_{1z}}{k_1k_\rho}(\hat{x}k_x + \hat{y}k_y) + \frac{k_\rho}{k_1}\hat{z} \tag{12g}$$

The above equations are the Extended Boundary Conditions, and are used to solve for the surface fields along with (10). Using (7), (10) can be written as

$$a_z(\overline{r}'_\perp) = (\hat{x}\frac{\partial f(\overline{r}'_\perp)}{\partial x'} + \hat{y}\frac{\partial f(\overline{r}'_\perp)}{\partial y'}) \cdot \overline{a}_\perp(\overline{r}'_\perp) \tag{13a}$$

$$b_z(\overline{r}'_\perp) = (\hat{x}\frac{\partial f(\overline{r}'_\perp)}{\partial x'} + \hat{y}\frac{\partial f(\overline{r}'_\perp)}{\partial y'}) \cdot \overline{b}_\perp(\overline{r}'_\perp) \tag{13b}$$

where a_z and b_z are the z components of $\overline{a}$ and $\overline{b}$, respectively.

Once the surface fields are obtained, then the scattered field $\overline{E}_s(\overline{r})$ in region 0 can be derived by evaluating (5a) for $z > f_{max}$. Thus, we obtain

$$\begin{aligned}\overline{E}_s(\overline{r}) = &-\frac{1}{8\pi^2}\int d\overline{k}_\perp e^{i\overline{k}_\perp\cdot\overline{r}_\perp} e^{ik_z z}\frac{k}{k_z}\int d\overline{r}'_\perp e^{-i\overline{k}_\perp\cdot\overline{r}'_\perp} e^{-ik_z f(\overline{r}'_\perp)}\\ &\times\Big\{[\hat{h}(k_z)\hat{h}(k_z) + \hat{v}(k_z)\hat{v}(k_z)]\cdot\overline{a}(\overline{r}'_\perp)\\ &+[-\hat{v}(k_z)\hat{h}(k_z) + \hat{h}(k_z)\hat{v}(k_z)]\cdot\overline{b}(\overline{r}'_\perp)\Big\}\end{aligned} \tag{14}$$

Equations (11), (13) and (14) are exact, and the approach is to solve for the surface fields using (11) and (13) and then to solve for the scattered fields using (14).

To solve for the surface fields by perturbation method, let

$$\overline{a}(\overline{r}'_\perp) = \sum_{m=0}^{\infty} \frac{\overline{a}^{(m)}(\overline{r}'_\perp)}{m!} \tag{15a}$$

$$\overline{b}(\overline{r}'_\perp) = \sum_{m=0}^{\infty} \frac{\overline{b}^{(m)}(\overline{r}'_\perp)}{m!} \tag{15b}$$

where $\overline{a}^{(m)}$ and $\overline{b}^{(m)}$ are the mth order terms of $\overline{a}$ and $\overline{b}$, respectively. We also have

$$\exp[\pm ik_z f(\overline{r}'_\perp)] = \exp\left[\pm ik_z H\cos(\frac{2\pi x}{p})\right] \sum_{m=0}^{\infty} \frac{[\pm ik_z \xi(\overline{r}'_\perp)]^m}{m!} \tag{16a}$$

$$\exp[\pm ik_{1z} f(\overline{r}'_\perp)] = \exp\left[\pm ik_{1z} H\cos(\frac{2\pi x}{p})\right] \sum_{m=0}^{\infty} \frac{[\pm ik_{1z} \xi(\overline{r}'_\perp)]^m}{m!} \tag{16b}$$

In the small perturbation method (SPM), ξ and its derivatives are regarded as small parameters. Thus we assume

$$k_z\xi(\overline{r}'_\perp),\ k_{1z}\xi(\overline{r}'_\perp),\ \frac{\partial\xi}{\partial x'},\ \frac{\partial\xi}{\partial y'} \ll 1 \tag{17}$$

Substituting (15) into (13) and equating the same order terms, we have

$$a_z^{(0)}(\overline{r}'_\perp) = \frac{-2\pi H}{P}\sin\left(\frac{2\pi}{P}x\right) a_x^{(0)} \tag{18a}$$

$$b_z^{(0)}(\overline{r}'_\perp) = \frac{-2\pi H}{P}\sin\left(\frac{2\pi}{P}x\right) b_x^{(0)} \tag{18b}$$

$$\begin{aligned} a_z^{(m)}(\overline{r}'_\perp) =& \frac{-2\pi H}{P}\sin\left(\frac{2\pi}{P}x\right) a_x^{(m)} \\ &+ m\left(\hat{x}\frac{\partial\xi(\overline{r}'_\perp)}{\partial x'} + \hat{y}\frac{\partial\xi(\overline{r}'_\perp)}{\partial y'}\right)\cdot \overline{a}_\perp^{(m-1)}(\overline{r}'_\perp) \end{aligned} \tag{18c}$$

$$\begin{aligned} b_z^{(m)}(\overline{r}'_\perp) =& \frac{-2\pi H}{P}\sin\left(\frac{2\pi}{P}x\right) b_x^{(m)} \\ &+ m\left(\hat{x}\frac{\partial\xi(\overline{r}'_\perp)}{\partial x'} + \hat{y}\frac{\partial\xi(\overline{r}'_\perp)}{\partial y'}\right)\cdot \overline{b}_\perp^{(m-1)}(\overline{r}'_\perp) \end{aligned} \tag{18d}$$

Similarly, substituting (15) and (16) into (11) and equating the same order terms, we can calculate the surface fields to different orders. Then, from (14), the scattered fields can be calculated to each order. In the following, we solve for the surface fields and scattered fields up to the second-order. The zeroth-order solutions are just the Bragg diffracted fields from a periodic surface. The first-order solution gives the lowest order incoherent scattered intensities. When the incident direction is perpendicular to the row direction of the periodic surface, the first-order solution does not give the depolarization effect in the backscattering direction. The second-order solution gives the depolarized scattered fields even when the incident direction is perpendicular to the row direction and also the lowest order correction to the coherent fields.

Zeroth-order solution

Substituting only the lowest order terms in the expansion of (15) and (16) into (11), the zeroth-order equations are given by

$$\begin{aligned}\overline{E}_i(\overline{r}) =& \frac{1}{8\pi^2}\int d\overline{k}_\perp e^{i\overline{k}_\perp\cdot\overline{r}_\perp} e^{-ik_z z}\frac{k}{k_z}\int d\overline{r}'_\perp e^{-i\overline{k}_\perp\cdot\overline{r}'_\perp}\exp\left[ik_z H\cos(\frac{2\pi x'}{p})\right]\\ &\times\Big\{[\hat{h}(-k_z)\hat{h}(-k_z)+\hat{v}(-k_z)\hat{v}(-k_z)]\cdot\overline{a}^{(0)}(\overline{r}'_\perp)\\ &+[-\hat{v}(-k_z)\hat{h}(-k_z)+\hat{h}(-k_z)\hat{v}(-k_z)]\cdot\overline{b}^{(0)}(\overline{r}'_\perp)\Big\}\end{aligned} \tag{19a}$$

$$\begin{aligned}0 =& \frac{1}{8\pi^2}\int d\overline{k}_\perp e^{i\overline{k}_\perp\cdot\overline{r}_\perp} e^{ik_{1z}z}\frac{k_1}{k_{1z}}\int d\overline{r}'_\perp e^{-i\overline{k}_\perp\cdot\overline{r}'_\perp}\exp\left[-ik_{1z}H\cos(\frac{2\pi x'}{p})\right]\\ &\times\Big\{\frac{k}{k_1}[\hat{h}_1(k_{1z})\hat{h}_1(k_{1z})+\hat{v}_1(k_{1z})\hat{v}_1(k_{1z})]\cdot\overline{a}^{(0)}(\overline{r}'_\perp)\\ &+[-\hat{v}_1(k_{1z})\hat{h}_1(k_{1z})+\hat{h}_1(k_{1z})\hat{v}_1(k_{1z})]\cdot\overline{b}^{(0)}(\overline{r}'_\perp)\Big\}\end{aligned} \tag{19b}$$

To solve the above equations, the incident electric field is written as

$$\begin{aligned}\overline{E}_i(\overline{r}) &= \hat{e}_i E_o e^{i\overline{k}_{\perp i}\cdot\overline{r}_\perp - ik_{zi}z}\\ &= \frac{\hat{e}_i}{4\pi^2}\int d\overline{k}_\perp e^{i\overline{k}_\perp\cdot\overline{r}_\perp - ik_z z}\int d\overline{r}'_\perp e^{i\overline{k}_{\perp i}\cdot\overline{r}'_\perp - i\overline{k}_\perp\cdot\overline{r}'_\perp}\end{aligned} \tag{20}$$

Since the zeroth-order solutions are exactly the scattered fields from a periodic surface, the zeroth-order surface currents can then be expanded by the Floquet modes [Chuang and Kong, 1981 and 1982]. Let

$$a_x^{(0)} = \sum_{n=-\infty}^{\infty} \alpha_{xn}^{(0)} e^{i\overline{k}_{\perp i}\cdot\overline{r}_\perp} e^{i\frac{2n\pi}{P}x} \tag{21a}$$

$$a_y^{(0)} = \sum_{n=-\infty}^{\infty} \alpha_{yn}^{(0)} e^{i\overline{k}_{\perp i}\cdot\overline{r}_\perp} e^{i\frac{2n\pi}{P}x} \tag{21b}$$

$$b_x^{(0)} = \sum_{n=-\infty}^{\infty} \beta_{xn}^{(0)} e^{i\overline{k}_{\perp i}\cdot\overline{r}_\perp} e^{i\frac{2n\pi}{P}x} \tag{21c}$$

$$b_y^{(0)} = \sum_{n=-\infty}^{\infty} \beta_{yn}^{(0)} e^{i\overline{k}_{\perp i}\cdot\overline{r}_\perp} e^{i\frac{2n\pi}{P}x} \tag{21d}$$

where $\overline{r}_\perp = x\hat{x} + y\hat{y}$, and $a_x^{(0)}$, $a_y^{(0)}$, $b_x^{(0)}$, and $b_y^{(0)}$ are the corresponding x and y components of $\overline{a}^{(0)}$ and $\overline{b}^{(0)}$. Substituting (20) and (21) into (19), we have the following matrix equation:

$$M^{(0)} \begin{bmatrix} \alpha_x^{(0)} \\ \alpha_y^{(0)} \\ \beta_x^{(0)} \\ \beta_y^{(0)} \end{bmatrix} = \begin{bmatrix} \hat{h}(-k_{zi}) \cdot \hat{e}_i \delta_m \\ \hat{v}(-k_{zi}) \cdot \hat{e}_i \delta_m \\ 0 \\ 0 \end{bmatrix} \tag{22}$$

where $\alpha_x^{(0)}$, $\alpha_y^{(0)}$, $\beta_x^{(0)}$ and $\beta_y^{(0)}$ are the column vectors of $\alpha_{xn}^{(0)}$, $\alpha_{yn}^{(0)}$, $\beta_{xn}^{(0)}$ and $\beta_{yn}^{(0)}$, respectively. The matrix equation (22) is given explicitly in Appendix A.

Once (22) is solved, the zeroth-order scattered field $\overline{E}_s^{(0)}(\overline{r})$ is given by

$$\overline{E}_s^{(0)}(\overline{r}) = \sum_{m=-\infty}^{\infty} e^{i\overline{k}_{\perp mi}\cdot\overline{r}_\perp} e^{ik_{zmi}z}[\hat{h}(k_{zmi})E_{hsm}^{(0)} + \hat{v}(k_{zmi})E_{vsm}^{(0)}] \tag{23a}$$

where

$$\begin{bmatrix} E_{hs}^{(0)} \end{bmatrix} = H_s^{(0)} \begin{bmatrix} \alpha_x^{(0)} \\ \alpha_y^{(0)} \\ \beta_x^{(0)} \\ \beta_y^{(0)} \end{bmatrix} = H_s^{(0)} M^{(0)-1} \begin{bmatrix} \hat{h}(-k_{zi}) \cdot \hat{e}_i \delta_m \\ \hat{v}(-k_{zi}) \cdot \hat{e}_i \delta_m \\ 0 \\ 0 \end{bmatrix} \tag{23b}$$

$$\left[E_{vs}^{(0)}\right] = V_s^{(0)} \begin{bmatrix} \alpha_x^{(0)} \\ \alpha_y^{(0)} \\ \beta_x^{(0)} \\ \beta_y^{(0)} \end{bmatrix} = V_s^{(0)} M^{(o)-1} \begin{bmatrix} \hat{h}(-k_{zi}) \cdot \hat{e}_i \delta_m \\ \hat{v}(-k_{zi}) \cdot \hat{e}_i \delta_m \\ 0 \\ 0 \end{bmatrix} \tag{23c}$$

$$\overline{k}_{\perp mi} = \overline{k}_{\perp i} + \hat{x}\frac{2\pi m}{p}, \quad k_{zmi} = \sqrt{k^2 - k_{\perp mi}^2} \tag{23d}$$

and $E_{hs}^{(0)}$ and $E_{vs}^{(0)}$ are the column vectors formed by $E_{hsm}^{(0)}$ and $E_{vsm}^{(0)}$, respectively. The explicit forms of $H_s^{(0)}$, $V_s^{(0)}$, $E_{hsm}^{(0)}$ and $E_{vsm}^{(0)}$ are given in Appendix A.

First-order solution

The first-order equation can be obtained by keeping Eqs. (15) and (16) up to the first order. Substituting the first-order expansion into (11), the following first-order equation can be obtained:

$$\begin{aligned} 0 = \frac{1}{8\pi^2} \int d\overline{k}_\perp e^{i\overline{k}_\perp \cdot \overline{r}_\perp} e^{-ik_z z} \frac{k}{k_z} \int d\overline{r}'_\perp e^{-i\overline{k}_\perp \cdot \overline{r}'_\perp} \exp\left[ik_z H \cos(\frac{2\pi}{p}x')\right] \\ \times \Big\{ [\hat{h}(-k_z)\hat{h}(-k_z) + \hat{v}(-k_z)\hat{v}(-k_z)] \\ \cdot [\overline{a}^{(1)}(\overline{r}'_\perp) + ik_z \xi(\overline{r}'_\perp) \overline{a}^{(0)}(\overline{r}'_\perp)] \\ + [-\hat{v}(-k_z)\hat{h}(-k_z) + \hat{h}(-k_z)\hat{v}(-k_z)] \\ \cdot [\overline{b}^{(1)}(\overline{r}'_\perp) + ik_z \xi(\overline{r}'_\perp) \overline{b}^{(0)}(\overline{r}'_\perp)] \Big\} \end{aligned} \tag{24a}$$

$$\begin{aligned} 0 = \frac{1}{8\pi^2} \int d\overline{k}_\perp e^{i\overline{k}_\perp \cdot \overline{r}_\perp} e^{ik_{1z} z} \frac{k_1}{k_{1z}} \int d\overline{r}'_\perp e^{-i\overline{k}_\perp \cdot \overline{r}'_\perp} \exp\left[-ik_{1z} H \cos(\frac{2\pi}{p}x')\right] \\ \times \Big\{ \frac{k}{k_1} [\hat{h}_1(k_{1z})\hat{h}_1(k_{1z}) + \hat{v}_1(k_{1z})\hat{v}_1(k_{1z})] \\ \cdot [\overline{a}^{(1)}(\overline{r}'_\perp) - ik_{1z} \xi(\overline{r}'_\perp) \overline{a}^{(0)}(\overline{r}'_\perp)] \\ + [-\hat{v}_1(k_{1z})\hat{h}_1(k_{1z}) + \hat{h}_1(k_{1z})\hat{v}_1(k_{1z})] \\ \cdot [\overline{b}^{(1)}(\overline{r}'_\perp) - ik_{1z} \xi(\overline{r}'_\perp) \overline{b}^{(0)}(\overline{r}'_\perp)] \Big\} \end{aligned} \tag{24b}$$

To solve the first-order equations in the spatial frequency domain, we take the Fourier transform of the first-order surface currents $\overline{a}^{(1)}$, $\overline{b}^{(1)}$

and the roughness function $\xi(x,y)$.

$$A_x^{(1)}(\overline{k}_\perp) = \frac{1}{(2\pi)^2}\int d\overline{r}'_\perp a_x^{(1)}(\overline{r}'_\perp)e^{-i\overline{k}_{\perp i}\cdot\overline{r}'_\perp}e^{-i\overline{k}_\perp\cdot\overline{r}'_\perp} \tag{25a}$$

$$A_y^{(1)}(\overline{k}_\perp) = \frac{1}{(2\pi)^2}\int d\overline{r}'_\perp a_y^{(1)}(\overline{r}'_\perp)e^{-i\overline{k}_{\perp i}\cdot\overline{r}'_\perp}e^{-i\overline{k}_\perp\cdot\overline{r}'_\perp} \tag{25b}$$

$$B_x^{(1)}(\overline{k}_\perp) = \frac{1}{(2\pi)^2}\int d\overline{r}'_\perp b_x^{(1)}(\overline{r}'_\perp)e^{-i\overline{k}_{\perp i}\cdot\overline{r}'_\perp}e^{-i\overline{k}_\perp\cdot\overline{r}'_\perp} \tag{25c}$$

$$B_y^{(1)}(\overline{k}_\perp) = \frac{1}{(2\pi)^2}\int d\overline{r}'_\perp b_y^{(1)}(\overline{r}'_\perp)e^{-i\overline{k}_{\perp i}\cdot\overline{r}'_\perp}e^{-i\overline{k}_\perp\cdot\overline{r}'_\perp} \tag{25d}$$

$$F(\overline{k}_\perp) = \frac{1}{(2\pi)^2}\int d\overline{r}'_\perp \xi(\overline{r}'_\perp)e^{-i\overline{k}_\perp\cdot\overline{r}'_\perp} \tag{26}$$

Substituting (18), (25) and (26) into (24), we have the following matrix equation (The matrix equation is given explicitly in Appendix B.):

$$M^{(1)}\begin{bmatrix} A_x^{(1)} \\ A_y^{(1)} \\ B_x^{(1)} \\ B_y^{(1)} \end{bmatrix} = R^{(1)}\left[F(\overline{k}_\perp - \overline{k}_{\perp mi})\right] \tag{27}$$

where $A_x^{(1)}$, $A_y^{(1)}$, $B_x^{(1)}$, and $B_y^{(1)}$ are the column vectors formed by $A_x^{(1)}(\overline{k}_\perp - \overline{k}_{\perp mi})$, $A_y^{(1)}(\overline{k}_\perp - \overline{k}_{\perp mi})$, $B_x^{(1)}(\overline{k}_\perp - \overline{k}_{\perp mi})$ and $B_y^{(1)}(\overline{k}_\perp - \overline{k}_{\perp mi})$, respectively.

The matrix $R^{(1)}$ is dependent on the zeroth-order solution $\alpha_{xn}^{(0)}$, $\alpha_{yn}^{(0)}$, $\beta_{xn}^{(0)}$ and $\beta_{yn}^{(0)}$. Once the zeroth-order solution is solved, the matrix $R^{(1)}$ can be determined. The above matrix equation accounts for the coupling between the first-order surface fields and the roughness function with the set of frequency components

$$\overline{k}_\perp - \overline{k}_{\perp mi} = \overline{k}_\perp - \overline{k}_{\perp i} - \hat{x}\frac{2m\pi}{p} \qquad m = -\infty, ..., \infty \tag{28}$$

The above coupling equations are due to the Bragg scattering mechanism of the large scale periodic surface. Thus, the first-order solution can be solved in terms of the zeroth-order solution and the Fourier components of the small random perturbation.

Keeping (14) to first order, we have the first-order scattered field.

$$\overline{E}_s^{(1)}(\overline{r}) = \int d\overline{k}_\perp e^{i\overline{k}_\perp\cdot\overline{r}_\perp}e^{ik_z z}\left\{\hat{h}(k_z)\sum_m E_{hsm}^{(1)} + \hat{v}(k_z)\sum_m E_{vsm}^{(1)}\right\} \tag{29}$$

where $E^{(1)}_{hsm}$ and $E^{(1)}_{vsm}$ are linearly dependent on the Fourier components $A^{(1)}_x(\overline{k}_\perp - \overline{k}_{\perp mi})$, $A^{(1)}_y(\overline{k}_\perp - \overline{k}_{\perp mi})$, $B^{(1)}_x(\overline{k}_\perp - \overline{k}_{\perp mi})$, $B^{(1)}_y(\overline{k}_\perp - \overline{k}_{\perp mi})$, and $F(\overline{k}_\perp - \overline{k}_{\perp mi})$.

$$\left[E^{(1)}_{hs} \right] = H^{(1)}_s \begin{bmatrix} A^{(1)}_x \\ A^{(1)}_y \\ B^{(1)}_x \\ B^{(1)}_y \end{bmatrix} + H^{(1)}_f \left[F(\overline{k}_\perp - \overline{k}_{\perp mi}) \right]$$

$$= \left(H^{(1)}_s M^{(1)-1} R^{(1)} + H^{(1)}_f \right) \left[F(\overline{k}_\perp - \overline{k}_{\perp mi}) \right] \qquad (30a)$$

$$\left[E^{(1)}_{vs} \right] = V^{(1)}_s \begin{bmatrix} A^{(1)}_x \\ A^{(1)}_y \\ B^{(1)}_x \\ B^{(1)}_y \end{bmatrix} + V^{(1)}_f \left[F(\overline{k}_\perp - \overline{k}_{\perp mi}) \right]$$

$$= \left(V^{(1)}_s M^{(1)-1} R^{(1)} + V^{(1)}_f \right) \left[F(\overline{k}_\perp - \overline{k}_{\perp mi}) \right] \qquad (30b)$$

where $E^{(1)}_{hs}$ and $E^{(1)}_{vs}$ are the column matrices formed by $E^{(1)}_{hsm}$ and $E^{(1)}_{vsm}$, respectively, and $H^{(1)}_f$ and $V^{(1)}_f$ are diagonal matrices [Appendix B]. Then we can obtain the following equations by summing up all the components of $E^{(1)}_{hs}$ and $E^{(1)}_{vs}$.

$$\sum_m E^{(1)}_{hsm} = \sum_m A^{(1)}_{hsm} F(\overline{k}_\perp - \overline{k}_{\perp mi}) \qquad (31a)$$

$$\sum_m E^{(1)}_{vsm} = \sum_m A^{(1)}_{vsm} F(\overline{k}_\perp - \overline{k}_{\perp mi}) \qquad (31b)$$

If the roughness is modelled as a Gaussian random process with zero mean (i.e., $< F(\overline{k}_\perp - \overline{k}_{\perp mi}) >= 0$), the ensemble average of the first-order scattered field is zero,

$$< \overline{E}^{(1)}_s(\overline{r}) >= 0 \qquad (32)$$

This means that the first-order scattered field is an incoherent field.

Second-order solution

By keeping the expansion of (15) and (16) up to the second order and substituting the second-order expansion into (11), the following second-order equations can be obtained:

$$0 = \frac{1}{8\pi^2}\int d\overline{k}_\perp e^{i\overline{k}_\perp\cdot\overline{r}_\perp} e^{-ik_z z}\frac{k}{k_z}\int d\overline{r}'_\perp e^{-i\overline{k}_\perp\cdot\overline{r}'_\perp}\exp\left[ik_z H\cos(\frac{2\pi}{p}x')\right]$$
$$\times\Bigg\{[\hat{h}(-k_z)\hat{h}(-k_z)+\hat{v}(-k_z)\hat{v}(-k_z)]$$
$$\cdot\left[\frac{1}{2}\overline{a}^{(2)}(\overline{r}'_\perp)+ik_z\xi(\overline{r}'_\perp)\overline{a}^{(1)}(\overline{r}'_\perp)-\frac{1}{2}k_z^2\xi^2\overline{a}^{(0)}(\overline{r}'_\perp)\right]$$
$$+[-\hat{v}(-k_z)\hat{h}(-k_z)+\hat{h}(-k_z)\hat{v}(-k_z)]$$
$$\cdot\left[\frac{1}{2}\overline{b}^{(2)}(\overline{r}'_\perp)+ik_z\xi(\overline{r}'_\perp)\overline{b}^{(1)}(\overline{r}'_\perp)-\frac{1}{2}k_z^2\xi^2\overline{b}^{(0)}(\overline{r}'_\perp)\right]\Bigg\} \quad (33a)$$

$$0 = \frac{1}{8\pi^2}\int d\overline{k}_\perp e^{i\overline{k}_\perp\cdot\overline{r}_\perp} e^{ik_{1z} z}\frac{k_1}{k_{1z}}\int d\overline{r}'_\perp e^{-i\overline{k}_\perp\cdot\overline{r}'_\perp}\exp\left[-ik_{1z} H\cos(\frac{2\pi}{p}x')\right]$$
$$\times\Bigg\{\frac{k}{k_1}[\hat{h}_1(k_{1z})\hat{h}_1(k_{1z})+\hat{v}_1(k_{1z})\hat{v}_1(k_{1z})]$$
$$\cdot\left[\frac{1}{2}\overline{a}^{(2)}(\overline{r}'_\perp)-ik_{1z}\xi(\overline{r}'_\perp)\overline{a}^{(1)}(\overline{r}'_\perp)-\frac{1}{2}k_{1z}^2\xi^2\overline{a}^{(0)}(\overline{r}'_\perp)\right]$$
$$+[-\hat{v}_1(k_{1z})\hat{h}_1(k_{1z})+\hat{h}_1(k_{1z})\hat{v}_1(k_{1z})]$$
$$\cdot\left[\frac{1}{2}\overline{b}^{(2)}(\overline{r}'_\perp)-ik_{1z}\xi(\overline{r}'_\perp)\overline{b}^{(1)}(\overline{r}'_\perp)-\frac{1}{2}k_{1z}^2\xi^2\overline{b}^{(0)}(\overline{r}'_\perp)\right]\Bigg\} \quad (33b)$$

Taking the Fourier transform of the surface currents and the roughness function ξ, we can solve the second-order equations in the spatial frequency domain.

$$A_x^{(2)}(\overline{k}_\perp) = \frac{1}{(2\pi)^2}\int d\overline{r}'_\perp a_x^{(2)}(\overline{r}'_\perp)e^{-i\overline{k}_{\perp i}\cdot\overline{r}'_\perp}e^{-i\overline{k}_\perp\cdot\overline{r}'_\perp} \quad (34a)$$

$$A_y^{(2)}(\overline{k}_\perp) = \frac{1}{(2\pi)^2}\int d\overline{r}'_\perp a_y^{(2)}(\overline{r}'_\perp)e^{-i\overline{k}_{\perp i}\cdot\overline{r}'_\perp}e^{-i\overline{k}_\perp\cdot\overline{r}'_\perp} \quad (34b)$$

$$B_x^{(2)}(\overline{k}_\perp) = \frac{1}{(2\pi)^2}\int d\overline{r}'_\perp b_x^{(2)}(\overline{r}'_\perp)e^{-i\overline{k}_{\perp i}\cdot\overline{r}'_\perp}e^{-i\overline{k}_\perp\cdot\overline{r}'_\perp} \quad (34c)$$

$$B_y^{(2)}(\overline{k}_\perp) = \frac{1}{(2\pi)^2}\int d\overline{r}'_\perp b_y^{(2)}(\overline{r}'_\perp)e^{-i\overline{k}_{\perp i}\cdot\overline{r}'_\perp}e^{-i\overline{k}_\perp\cdot\overline{r}'_\perp} \quad (34d)$$

Substituting (18), (21), (25), (26) and (34) into (33), we have the following matrix equation, which is given explicitly in Appendix C.

$$M^{(2)}\begin{bmatrix} A_x^{(2)} \\ A_y^{(2)} \\ B_x^{(2)} \\ B_y^{(2)} \end{bmatrix} = R_1^{(2)}\begin{bmatrix} F * A_x^{(1)}(\overline{k}_\perp - \overline{k}_{\perp mi}) \\ F * A_y^{(1)}(\overline{k}_\perp - \overline{k}_{\perp mi}) \\ F * B_x^{(1)}(\overline{k}_\perp - \overline{k}_{\perp mi}) \\ F * B_y^{(1)}(\overline{k}_\perp - \overline{k}_{\perp mi}) \end{bmatrix} + R_2^{(2)}\begin{bmatrix} F * F(\overline{k}_\perp - \overline{k}_{\perp mi}) \end{bmatrix} + R_3^{(2)}\begin{bmatrix} (k_x - k_{xmi})F * A_x^{(1)}(\overline{k}_\perp - \overline{k}_{\perp mi}) \\ (k_y - k_{ymi})F * A_y^{(1)}(\overline{k}_\perp - \overline{k}_{\perp mi}) \\ (k_x - k_{xmi})F * B_x^{(1)}(\overline{k}_\perp - \overline{k}_{\perp mi}) \\ (k_y - k_{ymi})F * B_y^{(1)}(\overline{k}_\perp - \overline{k}_{\perp mi}) \end{bmatrix} \tag{35}$$

where $A_x^{(2)}$, $A_y^{(2)}$, $B_x^{(2)}$, and $B_y^{(2)}$ are the column vectors formed by $A_x^{(2)}(\overline{k}_\perp - \overline{k}_{\perp mi})$, $A_y^{(2)}(\overline{k}_\perp - \overline{k}_{\perp mi})$, $B_x^{(2)}(\overline{k}_\perp - \overline{k}_{\perp mi})$ and $B_y^{(2)}(\overline{k}_\perp - \overline{k}_{\perp mi})$, respectively, and '$*$' denotes convolution.

The second-order equation is similar to the first order equation in that only the spatial frequency components with integer times $2\pi/p$ difference are coupled together. The second-order solution can be solved in terms of the zeroth-order solution, the first-order solution, and the Fourier components of the roughness function.

Keeping (14) to second order, we have the following expressions for the second-order scattered field:

$$\overline{E}_s^{(2)}(\overline{r}) = \int d\overline{k}_\perp e^{i\overline{k}_\perp\cdot\overline{r}_\perp} e^{ik_z z}\left\{\hat{h}(k_z)\sum_m E_{hsm}^{(2)} + \hat{v}(k_z)\sum_m E_{vsm}^{(2)}\right\} \tag{36}$$

$$\begin{bmatrix} E_{hs}^{(2)} \end{bmatrix} = H_1^{(2)}\begin{bmatrix} A_x^{(2)}(\overline{k}_\perp - \overline{k}_{\perp mi}) \\ A_y^{(2)}(\overline{k}_\perp - \overline{k}_{\perp mi}) \\ B_x^{(2)}(\overline{k}_\perp - \overline{k}_{\perp mi}) \\ B_y^{(2)}(\overline{k}_\perp - \overline{k}_{\perp mi}) \end{bmatrix} + H_2^{(2)}\begin{bmatrix} F * A_x^{(1)}(\overline{k}_\perp - \overline{k}_{\perp mi}) \\ F * A_y^{(1)}(\overline{k}_\perp - \overline{k}_{\perp mi}) \\ F * B_x^{(1)}(\overline{k}_\perp - \overline{k}_{\perp mi}) \\ F * B_y^{(1)}(\overline{k}_\perp - \overline{k}_{\perp mi}) \end{bmatrix} + H_3^{(2)}\begin{bmatrix} F * F(\overline{k}_\perp - \overline{k}_{\perp mi}) \end{bmatrix}$$

$$+ H_4^{(2)} \begin{bmatrix} (k_x - k_{xmi})F * A_x^{(1)}(\overline{k}_\perp - \overline{k}_{\perp mi}) \\ (k_y - k_{ymi})F * A_y^{(1)}(\overline{k}_\perp - \overline{k}_{\perp mi}) \\ (k_x - k_{xmi})F * B_x^{(1)}(\overline{k}_\perp - \overline{k}_{\perp mi}) \\ (k_y - k_{ymi})F * B_y^{(1)}(\overline{k}_\perp - \overline{k}_{\perp mi}) \end{bmatrix}$$

$$= \left(H_1^{(2)} M^{(2)-1} R_1^{(2)} + H_2^{(2)}\right) \begin{bmatrix} F * A_x^{(1)}(\overline{k}_\perp - \overline{k}_{\perp mi}) \\ F * A_y^{(1)}(\overline{k}_\perp - \overline{k}_{\perp mi}) \\ F * B_x^{(1)}(\overline{k}_\perp - \overline{k}_{\perp mi}) \\ F * B_y^{(1)}(\overline{k}_\perp - \overline{k}_{\perp mi}) \end{bmatrix}$$

$$+ \left(H_1^{(2)} M^{(2)-1} R_2^{(2)} + H_3^{(2)}\right) \begin{bmatrix} F * F(\overline{k}_\perp - \overline{k}_{\perp mi}) \end{bmatrix}$$

$$+ \left(H_1^{(2)} M^{(2)-1} R_3^{(2)} + H_4^{(2)}\right) \begin{bmatrix} (k_x - k_{xmi})F * A_x^{(1)}(\overline{k}_\perp - \overline{k}_{\perp mi}) \\ (k_y - k_{ymi})F * A_y^{(1)}(\overline{k}_\perp - \overline{k}_{\perp mi}) \\ (k_x - k_{xmi})F * B_x^{(1)}(\overline{k}_\perp - \overline{k}_{\perp mi}) \\ (k_y - k_{ymi})F * B_y^{(1)}(\overline{k}_\perp - \overline{k}_{\perp mi}) \end{bmatrix} \tag{37a}$$

$$\begin{bmatrix} E_{vs}^{(2)} \end{bmatrix} = V_1^{(2)} \begin{bmatrix} A_x^{(2)}(\overline{k}_\perp - \overline{k}_{\perp mi}) \\ A_y^{(2)}(\overline{k}_\perp - \overline{k}_{\perp mi}) \\ B_x^{(2)}(\overline{k}_\perp - \overline{k}_{\perp mi}) \\ B_y^{(2)}(\overline{k}_\perp - \overline{k}_{\perp mi}) \end{bmatrix} + V_2^{(2)} \begin{bmatrix} F * A_x^{(1)}(\overline{k}_\perp - \overline{k}_{\perp mi}) \\ F * A_y^{(1)}(\overline{k}_\perp - \overline{k}_{\perp mi}) \\ F * B_x^{(1)}(\overline{k}_\perp - \overline{k}_{\perp mi}) \\ F * B_y^{(1)}(\overline{k}_\perp - \overline{k}_{\perp mi}) \end{bmatrix}$$

$$+ V_3^{(2)} \begin{bmatrix} F * F(\overline{k}_\perp - \overline{k}_{\perp mi}) \end{bmatrix}$$

$$+ V_4^{(2)} \begin{bmatrix} (k_x - k_{xmi})F * A_x^{(1)}(\overline{k}_\perp - \overline{k}_{\perp mi}) \\ (k_y - k_{ymi})F * A_y^{(1)}(\overline{k}_\perp - \overline{k}_{\perp mi}) \\ (k_x - k_{xmi})F * B_x^{(1)}(\overline{k}_\perp - \overline{k}_{\perp mi}) \\ (k_y - k_{ymi})F * B_y^{(1)}(\overline{k}_\perp - \overline{k}_{\perp mi}) \end{bmatrix}$$

$$= \left(V_1^{(2)} M^{(2)-1} R_1^{(2)} + V_2^{(2)}\right) \begin{bmatrix} F * A_x^{(1)}(\overline{k}_\perp - \overline{k}_{\perp mi}) \\ F * A_y^{(1)}(\overline{k}_\perp - \overline{k}_{\perp mi}) \\ F * B_x^{(1)}(\overline{k}_\perp - \overline{k}_{\perp mi}) \\ F * B_y^{(1)}(\overline{k}_\perp - \overline{k}_{\perp mi}) \end{bmatrix}$$

$$+ \left(V_1^{(2)} M^{(2)-1} R_2^{(2)} + V_3^{(2)}\right) \begin{bmatrix} F * F(\overline{k}_\perp - \overline{k}_{\perp mi}) \end{bmatrix}$$

$$+\left(V_1^{(2)}M^{(2)-1}R_3^{(2)}+V_4^{(2)}\right)\begin{bmatrix}(k_x-k_{xmi})F*A_x^{(1)}(\overline{k}_\perp-\overline{k}_{\perp mi})\\(k_y-k_{ymi})F*A_y^{(1)}(\overline{k}_\perp-\overline{k}_{\perp mi})\\(k_x-k_{xmi})F*B_x^{(1)}(\overline{k}_\perp-\overline{k}_{\perp mi})\\(k_y-k_{ymi})F*B_y^{(1)}(\overline{k}_\perp-\overline{k}_{\perp mi})\end{bmatrix} \tag{37b}$$

where $E_{hs}^{(2)}$ and $E_{vs}^{(2)}$ are the column matrices formed by $E_{hsm}^{(2)}$ and $E_{vsm}^{(2)}$, respectively [Appendix C].

Thereafter (27) is solved and substituted into (37), we can express $E_{hs}^{(2)}$ and $E_{vs}^{(2)}$ in terms of a sum of this kind of components, $fF(\overline{k}_\perp-\overline{k}_{\perp mi})*gF(\overline{k}_\perp-\overline{k}_{\perp ni})$, where f and g are deterministic functions of $\overline{k}_\perp$. If the roughness is modeled as a Gaussian random process, then it can be shown that

$$\begin{aligned}< fF(\overline{k}_\perp-\overline{k}_{\perp mi})*gF(\overline{k}_\perp-\overline{k}_{\perp ni}) > &= 4\pi^2\delta(\overline{k}_\perp-\overline{k}_{\perp mi}+\overline{k}_{\perp ni})\\ &\times\int d\overline{k}'_\perp f(\overline{k}'_\perp)g(\overline{k}_\perp-\overline{k}'_\perp)W(\overline{k}'_\perp-\overline{k}_{\perp mi})\end{aligned} \tag{38}$$

where $W(|\overline{k}_\perp-\overline{k}_{\perp mi}|)$ is the spectral density of the random roughness and is the Fourier transform of the correlation function. The spectral density is given by

$$W(\overline{k}_\perp)=\frac{\sigma^2}{(2\pi)^2}\int d\overline{r}_\perp e^{i\overline{k}_\perp\cdot\overline{r}_\perp}C(\overline{r}_\perp) \tag{39}$$

and satisfies the relation

$$\left\langle F(\overline{k}'_\perp)F^*(\overline{k}_\perp)\right\rangle=\delta(\overline{k}'_\perp-\overline{k}_\perp)W(|\overline{k}'_\perp|) \tag{40}$$

Then from (36) and (37) the ensemble average of the second-order scattered field reduces to a set of plane waves which propagate in the directions of zeroth-order scattered fields. This implies that the coherent second-order scattered fields are the correction to the zeroth-order solution in the Bragg diffraction directions of the periodic surface.

b. Bistatic and Backscattering Coefficients

The lowest order incoherent scattering coefficients can be derived from the first-order scattered fields by considering the horizontally and vertically polarized incidence and calculating the horizontally and vertically polarized scattered fields. The lowest order incoherent Poynting power flow is averaged over one period and given by

$$\overline{S}^{(1)} = \frac{1}{p}\int_0^p Re\{<\overline{E}_s^{(1)} \times \overline{H}_s^{(1)*}>\}dx$$
$$= \int d\overline{k}_\perp \frac{\overline{k}}{\omega\mu} \sum_{m=-\infty}^{\infty} W(|\overline{k}_\perp - \overline{k}_{\perp mi}|)(|A_{hsm}^{(1)}|^2 + |A_{vsm}^{(1)}|^2) \quad (41)$$

For a Gaussian correlation function $C(\overline{r}_\perp)$, the spectral density is given by

$$W(|\overline{k}_\perp - \overline{k}_{\perp mi}|) = \frac{1}{4\pi}\sigma^2 l^2 \exp[-\frac{1}{4}|\overline{k}_\perp - \overline{k}_{\perp mi}|^2 l^2] \quad (42)$$

where σ is the standard deviation of the random roughness, and l is the correlation length for $\xi(\overline{r}_\perp)$ in the transverse plane.

The bistatic scattering coefficients $\gamma_{ba}(\hat{k}_s, \hat{k}_i)$ are defined as the ratio of scattered power of polarization b_s per unit solid angle in direction $\hat{k}_s$ and the intercepted power of polarization a_i with amplitude E_{ao} in direction $\hat{k}_i$ averaged over 4π radians. Thus, we have

$$\gamma_{ba}(\hat{k}_s, \hat{k}_i) = 4\pi \frac{k^2 \cos^2\theta_s \sum_{m=-\infty}^{\infty} |A_{bsm}^{(1)}|^2 W(|\overline{k}_{\perp s} - \overline{k}_{\perp mi}|)}{\cos\theta_i |E_{ao}|^2}$$
$$a, b = h, v \quad (43)$$

In the backscattering direction, $\hat{k}_s = -\hat{k}_i$, the backscattering cross section per unit area is defined to be

$$\sigma_{ba}(\hat{k}_i) = \cos\theta_i\, \gamma_{ba}(-\hat{k}_i, \hat{k}_i) \quad (44)$$

where

$$\hat{k}_i = \sin\theta_i \cos\phi_i \hat{x} + \sin\theta_i \sin\phi_i \hat{y} - \cos\theta_i \hat{z} \quad (45a)$$
$$\hat{k}_s = \sin\theta_s \cos\phi_s \hat{x} + \sin\theta_s \sin\phi_s \hat{y} + \cos\theta_s \hat{z} \quad (45b)$$

In the next section, the numerical results of bistatic scattering coefficients and backscattering cross sections per unit area are illustrated as functions of incident and scattered directions.

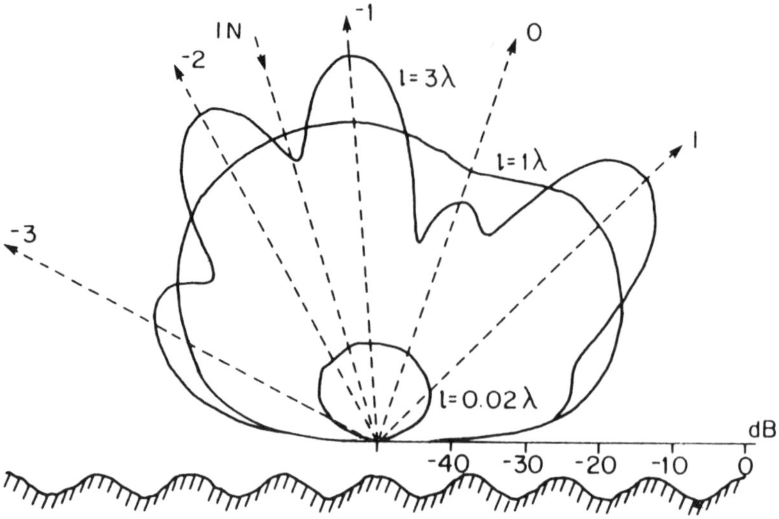

Figure 4.2.2 Scattering pattern γ_{hh} versus θ_s for various correlation lengths $l = 0.02$, 1.0, and 3.0λ with $\theta_i = 18°$, $\phi_i = 0°$, $\sigma = 0.01\lambda$, $H = 100$ nm, $P = 1205$ nm, $\lambda = 482$ nm, and $\epsilon_1 = (-7.29139 + 0.2943871)\epsilon_0$

c. Numerical Results and Discussions

The theoretical results obtained using the EBC/SPM method are illustrated in this section by calculating the bistatic scattering coefficients and the backscattering cross sections per unit area. In Figs. 4.2.2 and 4.2.3 the bistatic scattering coefficients $\gamma_{ba}(\theta_s, \phi_s)$ are plotted in a polar plot versus the scattering angle θ_s for several correlation lengths l. The plane wave is incident perpendicular to the row direction of a sinusoidal silver grating with random roughness. The scattering patterns for horizontally and vertically polarized incident fields are given in Figs. 4.2.2 and 4.2.3, respectively. The electromagnetic wave scattering from this silver grating without roughness has been studied by Chuang and Kong [1981].

In Fig. 4.2.2, the scattering pattern for horizontally polarized incidence is plotted on the incident plane with θ_s varied. When the correlation length is large, there is one peak associated with each Bragg diffraction direction. As the correlation length decreases from $l = 3\lambda$ to $l = 1\lambda$, the beams become broader. This beam shape is determined by (43). The beam width depends on the correlation length and the function $A_{bsm}^{(1)}$. The effect of correlation length on the beam width

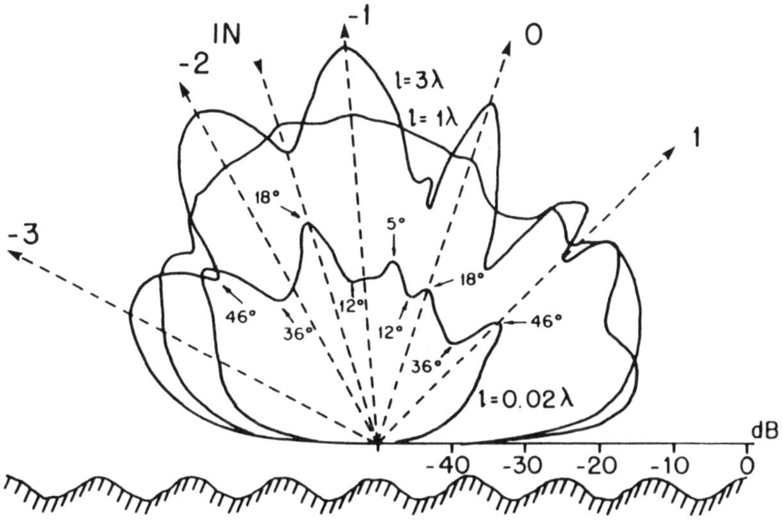

Figure 4.2.3 Scattering pattern γ_{vv} versus θ_s for various correlation lengths $l = 0.02$, 1.0, and 3.0λ with $\theta_i = 18°$, $\phi_i = 0°$, $\sigma = 0.01\lambda$, $H = 100$ nm, $P = 1205$ nm, $\lambda = 482$ nm, and $\epsilon_1 = (-7.29139 + 0.2943871)\epsilon_0$

can be explained as follows. Smaller correlation lengths means that the standard deviation of the slope of the roughness is larger, and the power can be scattered into wider directions. As shown in Fig. 4.2.2, when the correlation length becomes small, say $l = 0.02\lambda$, the power seems to be scattered isotropically into all the directions.

The corresponding scattering pattern for vertically polarized incidence is shown in Fig. 4.2.3. The result is similar to horizontal polarization incidence for large correlation lengths. Again several beams show up for $l = 3\lambda$. However, when the correlation length is $l = 0.02\lambda$, several peaks at $(\theta_s, \phi_s) = (46°, 180°), (18°, 180°), (5°, 0°), (18°, 0°)$, and $(46°, 0°)$ show up in the scattering pattern. As shown in [Chuang and Kong, 1981], the scattered wave is strongly affected by the surface plasmons at $\theta_s = 18°$ and $46°$ which are related to the propagation constants of the guided waves supportable by the periodic surface, and the diffraction efficiencies of mode -1 and -2 show maximum and minimum values, respectively, at $\theta_i = 5°$. If we take a close look at Fig. 4.2.3, we notice that the scattering pattern shows somewhat unsmooth behavior at $(\theta_s, \phi_s) = (12°, 180°), (36°, 180°), (12°, 0°)$ and $(36°, 0°)$. These are related to the Rayleigh wavelength anomalies [Hessel and

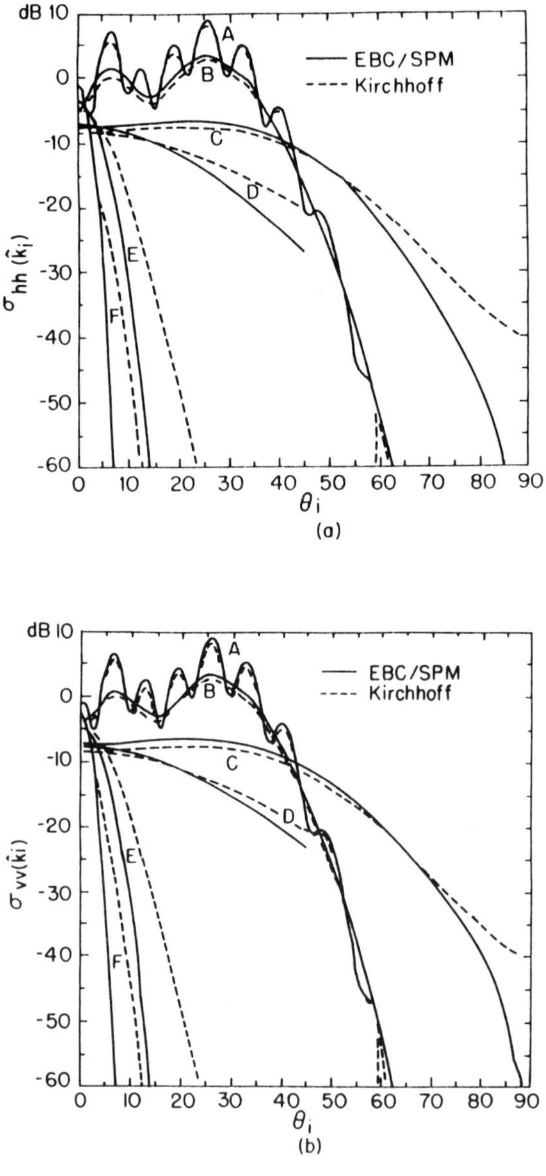

Figure 4.2.4 Backscattering cross section per unit area for various correlation lengths $l = 10$, 50, 100 cm with $\sigma = 1$ cm, $H = 10$ cm, $P = 100$ cm, $f = 1.4$ GHz and $\epsilon_1 = (6.0+i0.6)\epsilon_o$. (a) σ_{hh} versus θ_i with $\phi_i = 0°$ and 90°. (b) σ_{vv} versus θ_i with $\phi_i = 0°$and 90°. (Curves A: $l = 100$ cm, $\phi_i = 0°$; B: $l = 50$ cm, $\phi_i = 0°$; C: $l = 10$ cm, $\phi_i = 0°$; D: $l = 100$ cm, $\phi_i = 90°$; E: $l = 50$ cm, $\phi_i = 90°$; F: $l = 10$ cm, $\phi_i = 90°$); (——EBC/SPM, – – –Kirchhoff).

Oliner, 1965]: Floquet mode 2 becomes evanescent, and Floquet mode -3 becomes radiating at $11.5°$ and, similarly, for Floquet mode 1 and Floquet mode -4 at $36.9°$ in Ref. 2. Because of the roughness on the top of the periodic surface, even though the surface is illuminated by a plane wave with single spatial frequency, the surface currents and scattered fields of all spatial frequencies can be generated. The scattered fields with the same spatial frequencies as those of Wood's anomalies [Hessel and Oliner, 1965] will interact strongly with the periodic surface. This shows that the first-order scattered field pattern preserves every scattering characteristic of the periodic surface which is slightly different from those of the horizontally polarized incidence case as shown in Fig. 4.2.2.

The theoretical results for backscattering cross sections per unit area are illustrated in Figs. 4.2.4 and 4.2.5. This case has also been studied using the Kirchhoff approximation in [Shin and Kong, 1984], and the parameters used correspond to a typical terrain plow field.

In Fig. 4.2.4, the backscattering cross sections per unit area are plotted as a function of incident angle θ_i for azimuth angles $\phi_i = 0°$ and $90°$ and for correlation lengths $l = 10$, 50, and 100 cm. The results obtained from Kirchhoff approximation [Shin and Kong, 1984] show very good agreement with the results obtained using the EBC/SPM method at small angles of incidence. When the results in Fig. 4.2.4(a) are compared with those in Fig. 4.2.4(b), the EBC/SPM method gives different returns for σ_{hh} and σ_{vv} while the Kirchhoff approximation gives the identical results for both like-polarized returns. However, the difference between σ_{hh} and σ_{vv} is seen to be small. In Fig. 4.2.4 we note that several peaks show up for the large correlation length case where the incident angles satisfy the Bragg diffraction condition [Shin and Kong, 1984]

$$2k_{xi} = n2\pi/P \tag{46}$$

When the above condition is satisfied, the backscattering direction is coincident with the zeroth-order diffracted mode.

In Fig. 4.2.5, the backscattering cross sections per unit area are plotted as a function of azimuth incidence angle ϕ_i for $\theta_i = 25°$. The results for the horizontally and vertically polarized incidence are shown in Figs. 4.2.5(a) and 4.2.5(b), respectively. Again, we note a good agreement for σ_{hh} and σ_{vv} between Kirchhoff approximation and EBC/SPM method. However, the Kirchhoff approximation does

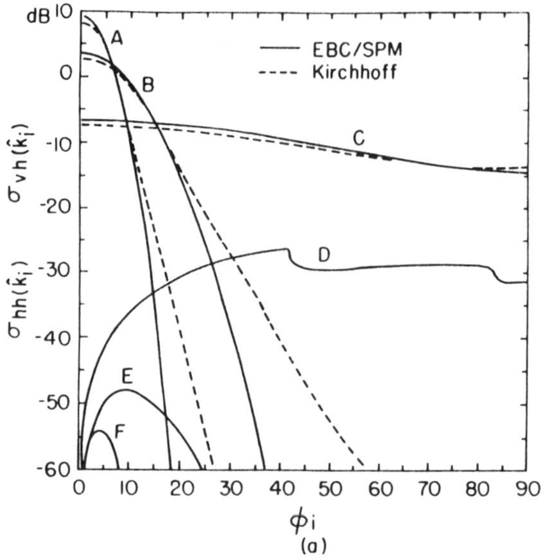

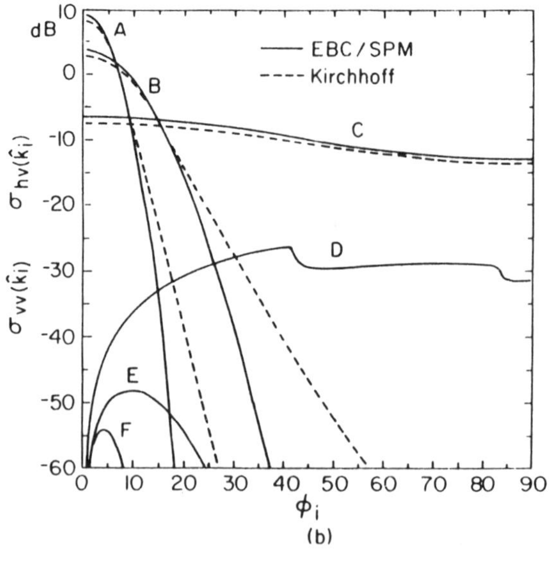

Figure 4.2.5 Backscattering cross section per unit area for various correlation lengths $l = 10$, 50, 100 cm with $\sigma = 1$ cm, $H = 10$ cm, $P = 100$ cm, $f = 1.4$ GHz and $\epsilon_1 = (6.0 + i0.6)\epsilon_o$. (a) σ_{hh} and σ_{vh} versus ϕ_i with $\theta_i = 25°$. (b) σ_{vv} and σ_{hv} versus ϕ_i with $\theta_i = 25°$. (Curves A: $l = 100$ cm, σ_{hh} or σ_{vv}; B: $l = 50$ cm, σ_{hh} or σ_{vv}; C: $l = 10$ cm, σ_{hh} or σ_{vv}; D: $l = 100$ cm, σ_{vh} or σ_{hv}; E: $l = 50$ cm, σ_{vh} or σ_{hv}; F: $l = 10$ cm, σ_{vh} or σ_{hv}); (——EBC/SPM, – – –Kirchhoff).

not give cross-polarized returns σ_{vh} in Fig. 4.2.5(a) and σ_{hv} in Fig. 4.2.5(b). The reason is that Kirchhoff approximation [Shin and Kong, 1984] makes use of the tangent plane approximation where the original rough surface can be seen as being approximated by many connected and locally perturbed tangent planes. Since it is well known that there is no cross-polarization in the first-order backscattering return from a randomly perturbed planar surface [Tsang et al., 1985], therefore Kirchhoff approximation predicts no depolarization in the backscattering direction. In this paper, the EBC/SPM method takes into account the curvature and tilting effect of the large-scale periodic surface exactly, and does give cross-polarized returns which can be significant when the incident wave vector is not perpendicular to the row directions. At $\phi_i \simeq 41°$ and $\phi_i \simeq 83°$ in the cross-polarized return (curve D: $l = 10$ cm), we can see two drops which correspond to the Rayleigh wavelength anomalies [Hessel and Oliner, 1965]: Floquet modes 3 and 4 become radiating modes at $\phi_i = 41.32°$ and $\phi_i = 83.13°$, respectively. Finally we note that the cross-polarized returns σ_{vh} and σ_{hv} presented in Figs. 4.2.5(a) and 4.2.5(b) are equal, which is consistent with the principle of reciprocity for the backscattering of electromagnetic wave from a reciprocal medium.

4.3 Scattering by a Randomly Perturbed Quasiperiodic Surface

a. Formulation

Consider a plane wave incident on a randomly perturbed quasiperiodic surface (Fig. 4.3.1). The electric field of the incident wave is given by

$$\overline{E}_i(\overline{r}) = \hat{e}_i E_o \exp(i\overline{k}_i \cdot \overline{r}) \tag{1}$$

where $\overline{k}_i$ denotes the incident wave vector and $\hat{e}_i$ the polarization of the incident electric field vector. The rough surface is characterized by a height distribution $Z = f(x, y)$, which is given by

$$f(x,y) = \xi(x,y) + A(x)\cos\left(\frac{2\pi}{P}x + \psi(x)\right) + B\cos\left(\frac{2\pi}{P}x + \phi\right) \tag{2}$$

where $\xi(x, y)$ is a Gaussian random variable with zero mean

$$\langle \xi(x,y) \rangle = 0 \tag{3}$$

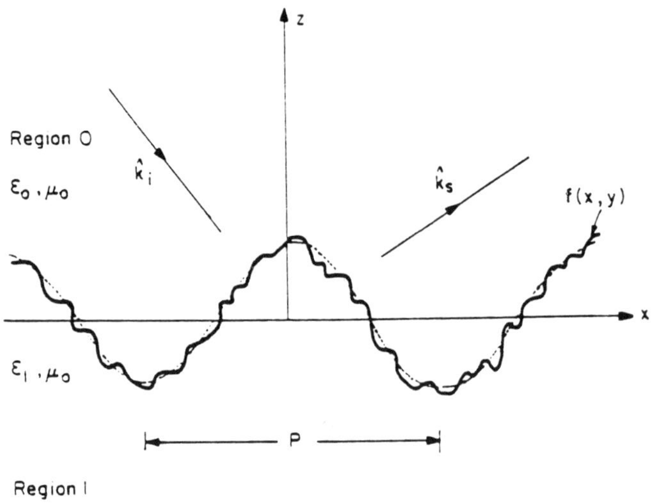

Figure 4.3.1 Geometrical configuration of the problem.

$A(x)\cos[(2\pi/P)x+\psi(x)]$ is described by a narrow-band Gaussian random process [Davenport and Root, 1958; Appendix D] centered around the spatial frequency of $2\pi/P$ where the variations of the envelope $A(x)$ and the phase $\psi(x)$ are slow compared to those of $\cos(2\pi x/P)$, and B and ϕ are assumed to be constants. Using the vector Kirchhoff approach, the scattered electric field $\overline{E}_s(\overline{r})$ can be expressed in the following form [Stogryn, 1967; Sancer, 1969; Leader, 1971; Tsang and Kong, 1980a]:

$$\overline{E}_s(\overline{r}) = \frac{ik\exp(ikr)}{4\pi r}E_o(\overline{I} - \hat{k}_s\hat{k}_s)\int_{A_o}\overline{F}(\alpha,\beta)\exp(i\overline{k}_d\cdot\overline{r}')d\overline{r}'_\perp \tag{4}$$

where A_o is the area of the rough surface projected onto the $x-y$ plane, $\hat{k}_s$ is the unit vector in the observation direction, and

$$\overline{k}_d = \overline{k}_i - \overline{k}_s \tag{5}$$

$$\begin{aligned}\overline{F}(\alpha,\beta) = (1+\alpha^2+\beta^2)^{1/2}\Big\{&-(1-R_h)(\hat{n}\cdot\hat{k}_i)(\hat{e}_i\cdot\hat{q}_i)\hat{q}_i \\ &+(\hat{e}_i\cdot\hat{p}_i)(\hat{n}\times\hat{q}_i)(1+R_v)+(\hat{e}_i\cdot\hat{q}_i)[\hat{k}_s\times(\hat{n}\times\hat{q}_i)](1+R_h) \\ &+(\hat{e}_i\cdot\hat{p}_i)(\hat{n}_i\cdot\hat{k}_i)(\hat{k}_s\times\hat{q}_i)(1-R_v)\Big\}\end{aligned} \tag{6}$$

In (6), α and β are the local slopes in the x and y directions,

$$\alpha = \frac{\partial f(x', y')}{\partial x'} \tag{7a}$$

$$\beta = \frac{\partial f(x', y')}{\partial y'} \tag{7b}$$

$\hat{n}$ is the local normal to the rough surface,

$$\hat{n} = \frac{-\hat{x}\alpha - \hat{y}\beta + \hat{z}}{(1 + \alpha^2 + \beta^2)^{1/2}} \tag{8}$$

$\hat{p}_i$ and $\hat{q}_i$ are, respectively, the unit vectors in the directions of the local parallel and perpendicular polarizations of the incident wave,

$$\hat{q}_i = \frac{\hat{k}_i \times \hat{n}}{|\hat{k}_i \times \hat{n}|} \tag{9a}$$

$$\hat{p}_i = \hat{q}_i \times \hat{k}_i \tag{9b}$$

R_v and R_h are the Fresnel reflection coefficients for TM and TE waves with local incidence angle:

$$R_h = \frac{-(\hat{n} \cdot \hat{k}_i) - [n_1^2 - 1 + (\hat{n} \cdot \hat{k}_i)^2]^{1/2}}{-(\hat{n} \cdot \hat{k}_i) + [n_1^2 - 1 + (\hat{n} \cdot \hat{k}_i)^2]^{1/2}} \tag{10a}$$

$$R_v = \frac{-n_1^2(\hat{n} \cdot \hat{k}_i) - [n_1^2 - 1 + (\hat{n} \cdot \hat{k}_i)^2]^{1/2}}{-n_1^2(\hat{n} \cdot \hat{k}_i) + [n_1^2 - 1 + (\hat{n} \cdot \hat{k}_i)^2]^{1/2}} \tag{10b}$$

with

$$\begin{aligned} n_1 &= k_1/k \\ k_1 &= \omega\sqrt{\mu_o \epsilon_1} \\ k &= \omega\sqrt{\mu_o \epsilon_o} \end{aligned} \tag{11}$$

and the orthonormal system for the incident and scattered fields are given by

$$\hat{k}_i = \hat{x} \sin\theta_i \cos\phi_i + \hat{y} \sin\theta_i \sin\phi_i - \hat{z} \cos\theta_i \tag{12a}$$

$$\hat{h}_i = -\hat{x} \sin\phi_i + \hat{y} \cos\phi_i \tag{12b}$$

$$\hat{v}_i = -\hat{x} \cos\theta_i \cos\phi_i - \hat{y} \cos\theta_i \sin\phi_i - \hat{z} \sin\theta_i \tag{12c}$$

$$\hat{k}_s = \hat{x} \sin\theta_s \cos\phi_s + \hat{y} \sin\theta_s \sin\phi_s + \hat{z} \cos\theta_s \tag{13a}$$

$$\hat{h}_s = -\hat{x} \sin\phi_s + \hat{y} \cos\phi_s \tag{13b}$$

$$\hat{v}_s = \hat{x} \cos\theta_s \cos\phi_s + \hat{y} \cos\theta_s \sin\phi_s - \hat{z} \sin\theta_s \tag{13c}$$

The Kirchhoff approximated diffraction integral in its present form is still difficult to evaluate, and further approximation is necessary. One commonly used approximation is to expand $\overline{F}(\alpha,\beta)$ in the power series of slope terms about the zero slope and to keep only the first few terms [Leader, 1971; Tsang and Newton, 1982]. However, in this section we shall expand $\overline{F}(\alpha,\beta)$ about the slopes at the stationary phase point α_o and β_o,

$$\alpha_o = -\frac{k_{dx}}{k_{dz}} \tag{14a}$$

$$\beta_o = -\frac{k_{dy}}{k_{dz}} \tag{14b}$$

where

$$k_{dx} = k(\sin\theta_i \cos\phi_i - \sin\theta_s \cos\phi_s) \tag{15a}$$

$$k_{dy} = k(\sin\theta_i \sin\phi_i - \sin\theta_s \sin\phi_s) \tag{15b}$$

$$k_{dz} = -k(\cos\theta_i + \cos\theta_s) \tag{15c}$$

Therefore, we expand $\overline{F}(\alpha,\beta)$ as follows:

$$\overline{F}(\alpha,\beta) = \overline{F}(\alpha_o,\beta_o) + \left.\frac{\partial \overline{F}}{\partial \alpha}\right|_{\alpha_o\beta_o} (\alpha - \alpha_o) + \left.\frac{\partial \overline{F}}{\partial \beta}\right|_{\alpha_o\beta_o} (\beta - \beta_o) + \cdots \tag{16}$$

Keeping only the first term in the above equation, we obtain from (4),

$$\overline{E}_s(\overline{r}) = \frac{ik\exp(ikr)}{4\pi r} E_o(\overline{I} - \hat{k}_s\hat{k}_s)\overline{F}(\alpha_o,\beta_o)I \tag{17}$$

where the integral I is given by

$$I = \int_{A_o} \exp(i\overline{k}_d \cdot \overline{r}')d\overline{r}'_\perp \tag{18}$$

Then the scattered field $\overline{E}_s(\overline{r})$ is separated into a mean field $\overline{E}_m(\overline{r})$ and a fluctuating part of the field $\overline{\mathcal{E}}(\overline{r})$

$$\overline{E}_s(\overline{r}) = \overline{E}_m(\overline{r}) + \overline{\mathcal{E}}(\overline{r}) \tag{19}$$

with

$$\langle \overline{\mathcal{E}}(\overline{r}) \rangle = 0 \tag{20}$$

and

$$\langle \overline{E}_s(\overline{r}) \rangle = \overline{E}_m(\overline{r}) \tag{21}$$

so that the total scattered intensity is a sum of coherent and incoherent scattered intensities.

$$\langle |\overline{E}_s|^2 \rangle = |\overline{E}_m(\overline{r})|^2 + \langle |\overline{\mathcal{E}}(\overline{r})|^2 \rangle \tag{22}$$

From (17) and (18), we have

$$\langle |\overline{E}_m(\overline{r})|^2 \rangle = \frac{k^2 |E_o|^2}{16\pi^2 r^2} \left[|\hat{v}_s \cdot \overline{F}(\alpha_o, \beta_o)|^2 + \left|\hat{h}_s \cdot \overline{F}(\alpha_o, \beta_o)\right|^2 \right] |\langle I \rangle|^2 \tag{23}$$

and

$$\langle |\overline{\mathcal{E}}(\overline{r})|^2 \rangle = \frac{k^2 |E_o|^2}{16\pi^2 r^2} \left[|\hat{v}_s \cdot \overline{F}(\alpha_o, \beta_o)|^2 + \left|\hat{h}_s \cdot \overline{F}(\alpha_o, \beta_o)\right|^2 \right] D_I \tag{24}$$

where

$$D_I = \langle |I|^2 \rangle - |\langle I \rangle|^2 \tag{25}$$

The explicit expressions for $|\langle I \rangle|^2$ and D_I for the randomly perturbed quasiperiodic surface are derived and expressed in terms of the statistical moments of the height distribution [Appendix E]. The advantage of expanding $\overline{F}(\alpha, \beta)$ around the stationary phase point (α_o, β_o) is that the bistatic scattering coefficients derived from (24) and (25) satisfy the principle of reciprocity, and at high frequency limit the geometrical optics solutions can be obtained from (25) without making any modifications [Stogryn, 1967]. Also, since $\overline{F}(\alpha, \beta)$ is evaluated at the stationary phase point, the same solution is obtained using total or reflected field on the surface [Holzer and Sung, 1978].

b. Coherent and Incoherent Scattering Coefficients

The bistatic scattering coefficients are defined as [Peake, 1959]

$$\gamma_{ab}(\hat{k}_s, \hat{k}_i) = \frac{4\pi r^2 I_{sa}}{A_o \cos\theta_i I_{ib}} \qquad (a, b = v, h) \tag{26}$$

where I_{sa} and I_{ib} are, respectively, the intensity of the scattered wave in polarization a and the intensity of the incident wave in polarization b. From (23) and (24) we first calculate the vertically and horizontally polarized coherent and incoherent scattered intensities for the cases of vertically and horizontally polarized incident fields. For an incident field with polarization b_i, the scattered intensities with polarization a_s are given by [Stogryn, 1967]

$$|E_m(\overline{r})|^2 = \frac{k^2 |E_o|^2}{16\pi^2 r^2} \left|\hat{a}_s \cdot \overline{F}_b(\alpha_o, \beta_o)\right|^2 |\langle I \rangle|^2 \tag{27}$$

$$|\mathcal{E}(\overline{r})|^2 = \frac{k^2 |E_o|^2}{16\pi^2 r^2} \left|\hat{a}_s \cdot \overline{F}_b(\alpha_o, \beta_o)\right|^2 D_I \tag{28}$$

where

$$\overline{F}_b(\alpha_o, \beta_o) = \overline{F}(\alpha_o, \beta_o)\Big|_{\hat{e}_i = \hat{b}_i} \tag{29}$$

and

$$\left|\hat{a}_s \cdot \overline{F}_b(\alpha_o, \beta_o)\right|^2 = \frac{\left|\overline{k}_d\right|^4}{k^2 \left|\hat{k}_i \times \hat{k}_s\right|^4 k_{dz}^2} f_{ba} \tag{30}$$

with

$$f_{vv} = \left|(\hat{h}_s \cdot \hat{k}_i)(\hat{h}_i \cdot \hat{k}_s) R_h + (\hat{v}_s \cdot \hat{k}_i)(\hat{v}_i \cdot \hat{k}_s) R_v\right|^2 \tag{31a}$$

$$f_{vh} = \left|(\hat{h}_s \cdot \hat{k}_i)(\hat{v}_i \cdot \hat{k}_s) R_h - (\hat{v}_s \cdot \hat{k}_i)(\hat{h}_i \cdot \hat{k}_s) R_v\right|^2 \tag{31b}$$

$$f_{hv} = \left|(\hat{v}_s \cdot \hat{k}_i)(\hat{h}_i \cdot \hat{k}_s) R_h - (\hat{h}_s \cdot \hat{k}_i)(\hat{v}_i \cdot \hat{k}_s) R_v\right|^2 \tag{31c}$$

$$f_{hh} = \left|(\hat{v}_s \cdot \hat{k}_i)(\hat{v}_i \cdot \hat{k}_s) R_h + (\hat{h}_s \cdot \hat{k}_i)(\hat{h}_i \cdot \hat{k}_s) R_v\right|^2 \tag{31d}$$

and R_v and R_h, which are given by (10), are evaluated at

$$\hat{n} = \frac{\hat{x} k_{dx}/k_{dz} + \hat{y} k_{dy}/k_{dz} + \hat{z}}{(k_{dx}^2/k_{dz}^2 + k_{dy}^2/k_{dz}^2 + 1)^{1/2}} \tag{32}$$

In view of (27) and (28), the bistatic scattering coefficients can be decomposed into a coherent part γ_{ab}^c and an incoherent part γ_{ab}^i. Substituting in the expressions for $|\langle I \rangle|^2$ and D_I, (E10) and (E18), into (27) and (28), we have

$$\gamma_{ab}(\hat{k}_s, \hat{k}_i) = \gamma_{ab}^c(\hat{k}_s, \hat{k}_i) + \gamma_{ab}^i(\hat{k}_s, \hat{k}_i) \tag{33}$$

where

$$\gamma_{ab}^{c}(\hat{k}_s, \hat{k}_i) = \frac{\pi|\overline{k}_d|^4}{\cos\theta_i \left|\hat{k}_i \times \hat{k}_s\right|^4 k_{dz}^2} f_{ba} \exp[-k_{dz}^2(\sigma^2 + \sigma_x^2)]$$
$$\times \sum_{n=-\infty}^{\infty} |J_n(k_{dz}B)|^2 \delta\left(k_{dx} + n\frac{2\pi}{P}\right) \delta(k_{dy}) \qquad (34)$$

$$\gamma_{ab}^{i}(\hat{k}_s, \hat{k}_i) = \frac{|\overline{k}_d|^4}{\cos\theta_i \, k^2 \left|\hat{k}_i \times \hat{k}_s\right|^4 k_{dz}^2} f_{ba} \left(D_{I_1} + D_{I_2}\right) \qquad (35)$$

In (34), J_n is the nth-order Bessel function, and δ is the Dirac delta function. It can easily be shown that the bistatic scattering coefficients satisfy the principle of reciprocity,

$$\cos\theta_i \, \gamma_{ab}(\hat{k}_s, \hat{k}_i) = \cos\theta_s \, \gamma_{ba}(\hat{k}_i, \hat{k}_s) \qquad (36)$$

When the incident wave vector is not perpendicular to the row direction of the periodic surface ($k_{yi} \neq 0$), the coherent scattering coefficient, (34), gives rise to scattered intensities along the directions of Floquet modes, forming a cone. This conical diffraction is a characteristic of scattering from a periodic surface [Chuang and Kong, 1982]. A part of incoherent scattering coefficient will also give rise to conical diffraction. The second term on the right-hand side of (35) has $\delta(k_{dy})$ dependence, and this will give rise to scattered intensities only in the direction $k_{ys} = k_{yi}$ forming a cone shape. However, unlike the coherent term, which only scatters into a set of discrete directions, this term will scatter intensities in all k_{xs} directions.

In the backscattering direction $\theta_s = \theta_i$ and $\phi_s = \phi_i + \pi$. The backscattering cross sections per unit area are defined to be

$$\sigma_{ab}(\hat{k}_i) = \cos\theta_i \, \gamma_{ab}(-\hat{k}_i, \hat{k}_i) \qquad (37)$$

From (33)–(35) we obtain

$$\sigma_{vv}(\hat{k}_i) = \sigma_{hh}(\hat{k}_i) = \sigma^c(\hat{k}_i) + \sigma_1^i(\hat{k}_i) + \sigma_2^i(\hat{k}_i) \qquad (38)$$

$$\sigma_{hv}(\hat{k}_i) = \sigma_{vh}(\hat{k}_i) = 0 \qquad (39)$$

where

$$\sigma^c(\hat{k}_i) = \frac{4\pi k^2 |R|^2}{\cos^2\theta_i} \exp\left[-4k_{zi}^2(\sigma^2 + \sigma_x^2)\right] \sum_{n=-\infty}^{\infty} |J_n(2k_{zi}B)|^2 \times \delta\left(2k_{xi} + n\frac{2\pi}{P}\right)\delta(2k_{yi}) \tag{40}$$

$$\begin{aligned}\sigma_1^i(\hat{k}_i) = &\frac{k^2 |R|^2}{\cos^2\theta_i} \sum_{\mu=-\infty}^{\infty} |J_n(2k_{zi}B)|^2 \exp\left[-4k_{zi}^2(\sigma^2 + \sigma_x^2)\right] \\ &\times \sum_{m=1}^{\infty} \frac{(4k_{zi}^2\sigma^2)^m}{m!}\frac{l^2}{m}\left[\exp\left(-\left[\left(\frac{k_{xi}+\mu\pi}{P}\right)^2 + k_{yi}^2\right]\frac{l^2}{m}\right)\right. \\ &+ \sum_{m'=1}^{\infty}\sum_{n=0}^{m'} \frac{1}{2^{m'}}\binom{m'}{n}\frac{(4k_{zi}^2\sigma_x^2)^{m'}}{m'!}\frac{1}{q} \\ &\left.\times \exp\left(-\left[\frac{k_{xi} + (\mu + m' - 2n)\pi}{P}\right]^2\frac{l^2}{mq} - k_{yi}^2\frac{l^2}{m}\right)\right]\end{aligned} \tag{41}$$

with

$$q = \sqrt{1 + \frac{l^2 m'}{l_x^2 m}}$$

$$\begin{aligned}\sigma_2^i(\hat{k}_i) = &\frac{2k^2 |R|^2}{\cos^2\theta_i}\delta(2k_{yi}) \sum_{\mu=-\infty}^{\infty} |J_n(2k_{zi}B)|^2 \exp\left[-4k_{zi}^2(\sigma^2 + \sigma_x^2)\right] \\ &\times \sum_{m=1}^{\infty}\sum_{n=0}^{m} \frac{1}{2^m}\binom{m}{n}\sqrt{\frac{\pi}{m}}\, l_x \frac{(4k_{zi}^2\sigma_x^2)^m}{m!} \\ &\times \exp\left(-\left[\frac{k_{xi} + (\mu + m - 2n)\pi}{P}\right]^2\frac{l_x^2}{m}\right)\end{aligned} \tag{42}$$

and R is the Fresnel reflection coefficient at normal incidence. We note that there is no depolarization in the backscattering direction, and because of the $\delta(2k_{yi})$ dependence, $\sigma^c(\hat{k}_i)$ terms contribute only when the incident wave vector is perpendicular to the row direction of the periodic surface.

c. Geometrical Optics Solution

Under the geometrical optics limit as $k \to \infty$, further simplifications can be made for the expressions of $|\langle I\rangle|^2$ and D_I. Since $k_{dz}\sigma, k_{dz}\sigma_x \gg 1$, the coherent component of the scattered fields are negligible, and only the incoherent scattering coefficients will remain. Under the stationary phase approximation the bistatic scattering coefficients simplify to

$$\gamma_{ab}(\hat{k}_s, \hat{k}_i) = \frac{|\overline{k}_d|^4}{\cos\theta_i \left|\hat{k}_i \times \hat{k}_s\right|^4 k_{dz}^4} f_{ba} \frac{1}{2\pi} \int_0^{2\pi} d\phi' \frac{1}{2s_x s_y} \exp\left(-\frac{\left(\frac{k_{dx}}{k_{dz}} - 2\pi\frac{B}{P}\sin\phi'\right)^2}{2s_x^2} - \frac{\left(\frac{k_{dy}}{k_{dz}}\right)^2}{2s_y^2}\right) \tag{43}$$

where s_x^2 and s_y^2 are, respectively, the mean square surface slopes in the x and y directions

$$s_x^2 = \sigma^2 |C''(0)| + \sigma_x^2[(2\pi/P)^2 + |C_x''(0)|] \tag{44a}$$

$$s_y^2 = \sigma^2 |C''(0)| \tag{44b}$$

In the above equations, C'' and C_x'' are the second derivatives of the correlation functions and for Gaussian correlation functions assumed

$$|C''(0)| = 2/l^2 \tag{45a}$$

$$|C_x''(0)| = 2/l_x^2 \tag{45b}$$

The probability of finding slopes (α, β) at point $\overline{r}_\perp$ on the surface can be calculated to be

$$P[\alpha(\overline{r}_\perp), \beta(\overline{r}_\perp)] = \frac{1}{2\pi s_x s_y} \exp\left(-\frac{\left[\alpha + 2\pi\frac{B}{P}\sin\left(\frac{2\pi}{P}x + \phi\right)\right]^2}{2s_x^2} - \frac{\beta^2}{2s_y^2}\right) \tag{46}$$

Averaging the above expression over one period we obtain the averaged PDF for α, β

$$P(\alpha, \beta) = \frac{1}{P}\int_0^P dx \frac{1}{2\pi s_x s_y} \exp\left(-\frac{\left[\alpha + 2\pi\frac{B}{P}\sin\left(\frac{2\pi}{P}x + \phi\right)\right]^2}{2s_x^2} - \frac{\beta^2}{2s_y^2}\right) \tag{47}$$

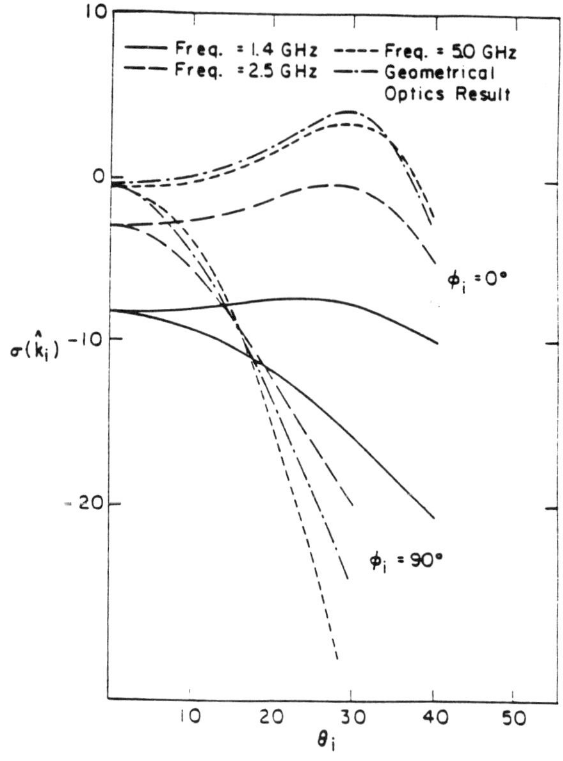

Figure 4.3.2 $\sigma(\hat{k}_i)$ **as a function of** θ_i **for different frequencies with** $\sigma =$ **1 cm,** $l = 10$ **cm,** $B = 10$ **cm,** $P = 100$ **cm,** $\sigma_x = 0$**, and** $\epsilon_1 = (6.0 + i0.6)\epsilon_o$.

which is proportional to the geometrical optics solution. Therefore, the geometrical optics result states that the scattered intensity is proportional to the probability of the occurrence of the slopes which will specularly reflect the incident wave into the direction of the scattered wave [Barrick, 1968]. We also note that in the geometrical optics limit there is no difference between the above solution and the solution obtained using the incoherent model [Ulaby *et al.*, 1982] except for the factors due to quasiperiodicity.

d. Results and Discussion

The backscattering cross sections per unit area $\sigma_1^i(\hat{k}_i)$ are calculated and illustrated for various cases. The conically diffracted coherent

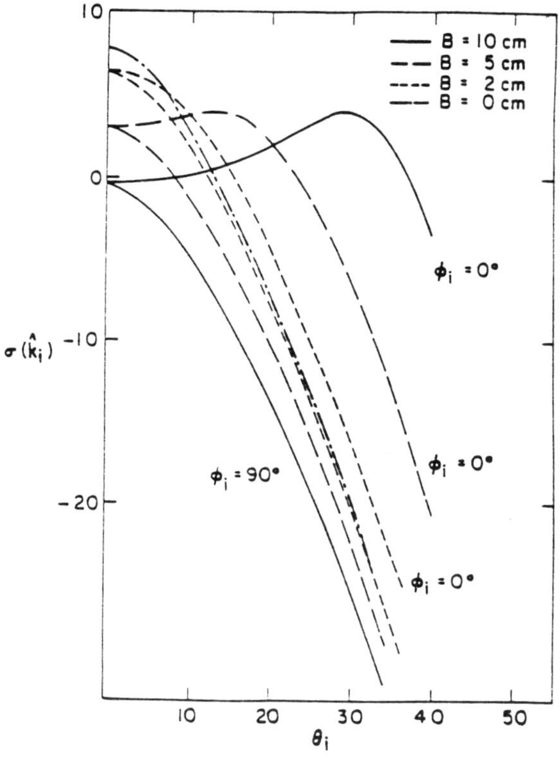

Figure 4.3.3 $\sigma(\hat{k}_i)$ **as a function of** θ_i **for different values of** B **at 5.0 GHz with** $P = 100$ **cm,** $\sigma = 1$ **cm,** $l = 10$ **cm,** $\sigma_x = 0$**, and** $\epsilon_1 = (6.0 + i0.6)\epsilon_o$**.**

and incoherent components, $\sigma^c(\hat{k}_i)$ and $\sigma_2^i(\hat{k}_i)$, which only contribute when the incident wave is perpendicular to the row direction of the periodic surface, are not included in the calculations. In order to correctly incorporate the contributions from these components, the characteristics of the antenna used to make the measurements must be taken into account.

The results of randomly perturbed sinusoidal surface cases, $\sigma_x = 0$, are first illustrated in Figs. 4.3.2–4.3.12. In Fig. 4.3.2, the backscattering cross sections per unit area $\sigma(\hat{k}_i)$ are plotted as a function of incidence angle for different frequencies. As the frequency is increased, the solution approaches the geometrical optics result as expected. The difference between the cases where the incident wave vector is parallel, $\phi_i = 90°$, or perpendicular, $\phi_i = 0°$, to the row direction is seen to

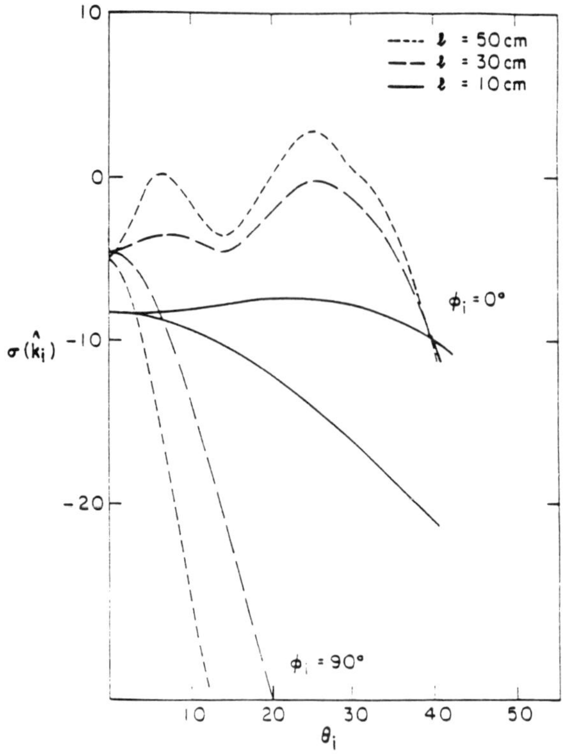

Figure 4.3.4 $\sigma(\hat{k}_i)$ **as a function of** θ_i **for different values of** l **at 1.4 GHz with** $\sigma = 1$ **cm,** $B = 10$ **cm,** $P = 100$ **cm,** $\sigma_x = 0$**, and** $\epsilon_1 = (6.0 + i0.6)\epsilon_o$**.**

be large. For the $\phi_i = 0°$ case the maximum value of $\sigma(\hat{k}_i)$ is shown to be not at normal incidence. In Fig. 4.3.3 the effect of change in the amplitude of sinusoidal variation B is illustrated for 5.0 GHz. As B is decreased the results of $\phi_i = 0°$ and $\phi_i = 90°$ cases approach each other, and when $B = 0$ we reproduce the random rough surface result which is independent of azimuthal incident angle ϕ_i.

In Fig. 4.3.4 the effect of the correlation length l at 1.4 GHz is illustrated. As l is increased, the $\sigma(\hat{k}_i)$ falls off faster as a function of θ_i for $\phi_i = 90°$, and there is an appearance of peaks for $\phi_i = 0°$. The change in $\sigma(\hat{k}_x)$ as ϕ_i is varied is shown in Fig. 4.3.5. The appearance of the peaks for $\phi_i = 0°$ can be explained as follows.

The result for a randomly perturbed sinusoidal surface is related to the convolution of the results for the sinusoidal surface with those of the

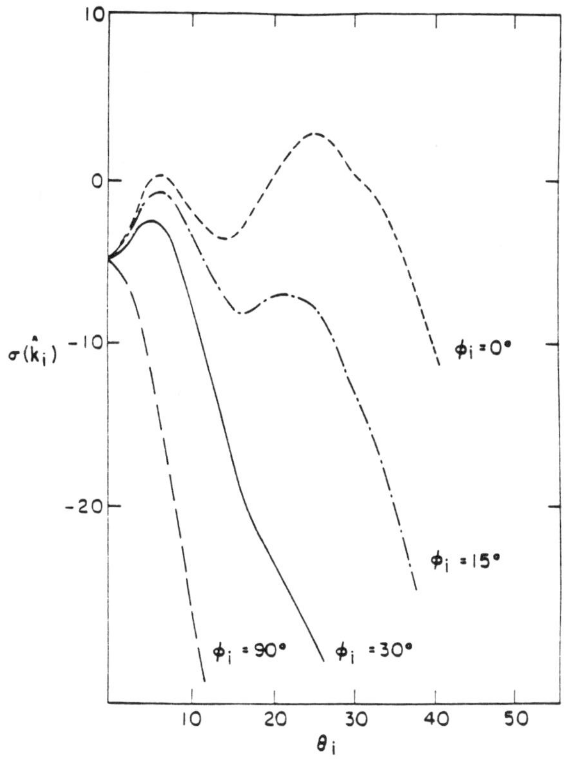

Figure 4.3.5 $\sigma(\hat{k}_i)$ **as a function of** θ_i **for different azimuthal angles of incidence** ϕ_i **at 1.4 GHz with** $\sigma = 1$ **cm,** $l = 50$ **cm,** $B = 10$ **cm,** $P = 100$ **cm,** $\sigma_x = 0$**, and** $\epsilon_1 = (6.0 + i0.6)\epsilon_o$**.**

random rough surface. For a sinusoidal surface, we have contribution in the backscattering direction only when it coincides with one of the Floquet modes direction.

$$2k_{xi} = n\frac{2\pi}{P} \tag{48}$$

As l is increased the scattering pattern from a random rough surface is sharply peaked around the specular direction. Therefore, by making l sufficiently large, we obtain the result which is sharply peaked at the mode directions given by (48). This is illustrated in Fig. 4.3.6(a). The locations and amplitudes of the Floquet modes are plotted in Fig. 4.3.6(a). Notice that for the cases of $l = 100\,\text{cm}$, we see from Fig.

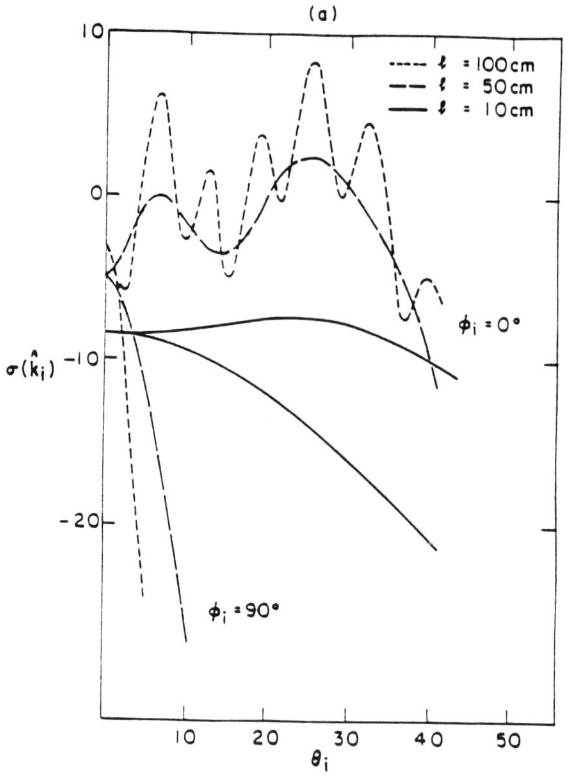

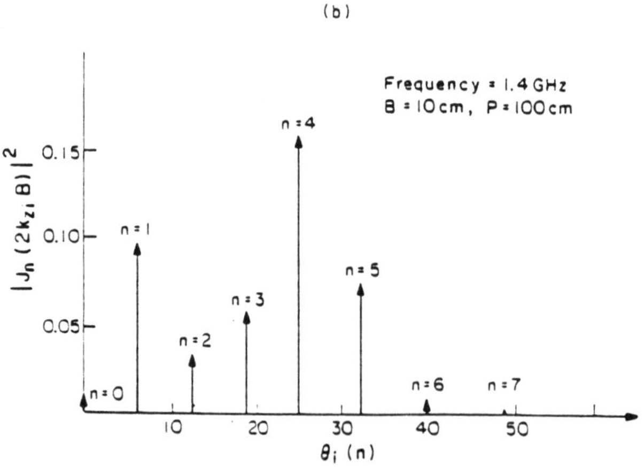

Figure 4.3.6 **(a)** $\sigma(\hat{k}_i)$ **as a function of** θ_i **for different values of** l **at 1.4 GHz with** $\sigma = 1$ **cm,** $B = 10$ **cm,** $P = 100$ **cm,** $\sigma_x = 0$**, and** $\epsilon_1 = (6.0 + i0.6)\epsilon_o$**. (b) Locations and amplitudes of the modes for** $B = 10$ **cm.**

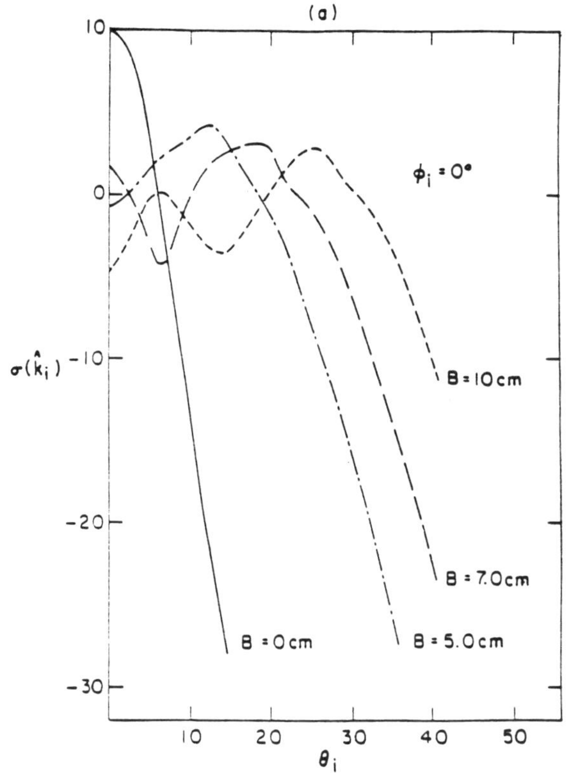

Figure 4.3.7 (a) $\sigma(\hat{k}_i)$ **as a function of** θ_i **for different values of** B **at 1.4 GHz with** $\sigma = 1$ **cm,** $l = 50$ **cm,** $P = 100$ **cm,** $\sigma_x = 0$**, and** $\epsilon_1 = (6.0+i0.6)\epsilon_o$**.**

4.3.6(a) that the peaks are visibly illustrated. When l is smaller the scattering pattern of the random rough surface becomes broader, and we do not reproduce all the peaks. However, the peaks around the two dominant modes, $n = 1$ and $n = 4$, are still reproduced for $l = 50$ cm. When l is further decreased none of the peaks are reproduced, and we have a fairly flat behavior.

In Fig. 4.3.7(a), the effect of change in B at 1.4 GHz is illustrated. Note that as B is decreased, there seems to be a shifting of the peaks. Since the period P is not changed, the locations of the modes do not change. However, as we can see from Fig. 4.3.7(b), the amplitude of each mode is changed as B is changed. The location of the mode with the maximum amplitude is shifted as B is varied, and the results in Fig. 4.3.7(a) reflect this effect. When $B = 0$ only the amplitude of

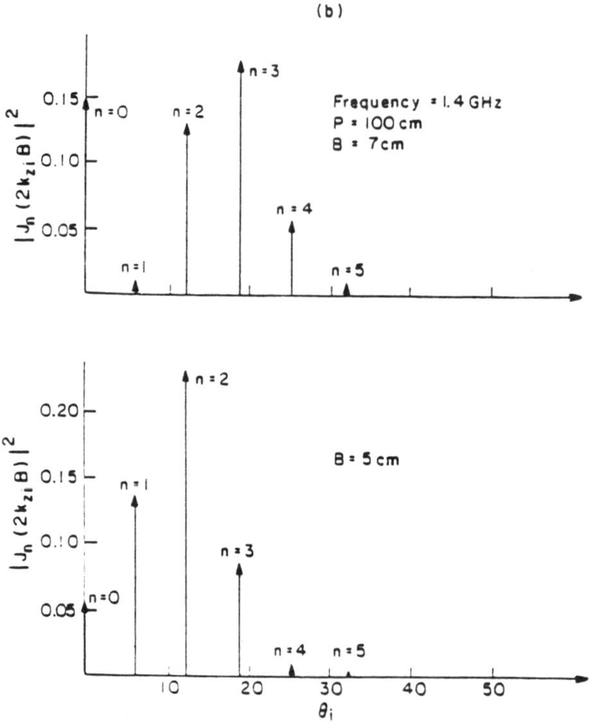

Figure 4.3.7 (b) Locations and amplitudes of the modes for $B = 7$ cm and $B = 5$ cm.

the $n = 0$ mode is nonzero, and the random rough surface result is reproduced.

The effect of change in the period P is illustrated in Fig. 4.3.8. The locations of the modes will change as P changes, while the amplitude of each mode will not change since B is the same. As can be seen from Fig. 4.3.8, when P is increased the modes are spaced closer together, and when P is decreased the modes become further apart.

The effect of change in σ is shown in Fig. 4.3.9(a). Initially, as σ is increased, the backscattering cross section $\sigma(\hat{k}_i)$ is increased. Then, as σ is further increased, there is a decrease near normal incidence and a disappearance of one of the peaks. This is due to the change in the scattering characteristics of the random rough surface in the absence of sinusoidal variation. In Fig. 4.3.9(b) the backscattering cross section

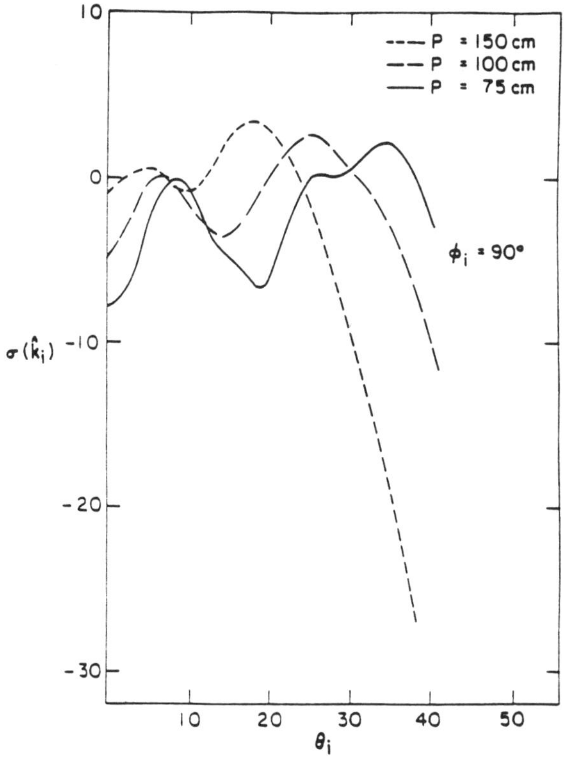

Figure 4.3.8 $\sigma(\hat{k}_i)$ **as a function of** θ_i **for different values of** P **at 1.4 GHz with** $\sigma = 1$ **cm,** $l = 50$ **cm,** $B = 10$ **cm,** $P = 100$ **cm,** $\sigma_x = 0$**, and** $\epsilon_1 = (6.0 + i0.6)\epsilon_o$**.**

for the random rough surface is plotted. Note that for $\sigma = 5\,\text{cm}$ there is a decrease near normal incidence and a broadening of the scattering pattern which explains the trends in Fig. 4.3.9(a).

In Fig. 4.3.10(a) and 4.3.10(b) we illustrate the results for randomly perturbed quasi-periodic surfaces, $B = 0$ and $\sigma_x \neq 0$. In Fig. 4.3.10(a) the backscattering cross sections are plotted for different correlation lengths l. Again, as l is increased, there is an appearance of peaks. But unlike the sinusoidal case, the values of the peaks are monotonically decreasing with increasing angle of incidence. It is interesting to look at the solution in the limit $l_x \to \infty$ since we obtain

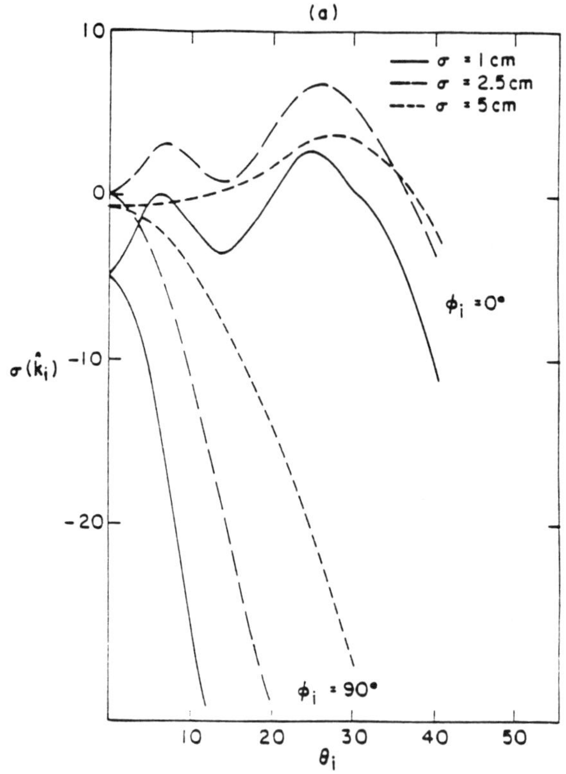

Figure 4.3.9 **(a)** $\sigma(\hat{k}_i)$ **as a function of** θ_i **for different values of** σ **at 1.4 GHz with** $l = 50$ **cm,** $B = 10$ **cm,** $P = 100$ **cm,** $\sigma_x = 0$**, and** $\epsilon_1 = (6.0 + i0.6)\epsilon_o$**.**

a much simpler analytical solution. From (E14) we obtain, for $B = 0$,

$$\langle II^* \rangle = \int_{-2L_x}^{2L_x} dx \int_{-2L_y}^{2L_y} dy\, e^{i\bar{k}_{d\perp}\cdot\bar{r}_\perp}[2L_x - |x|][2L_y - |y|]$$
$$\exp\left[-k_{dz}^2\sigma^2 + k_{dz}^2\sigma^2 C(\bar{r}_\perp)\right]$$
$$\times \exp\left[-k_{dz}^2\sigma_x^2 + k_{dz}^2\sigma_x^2 C_x(x)\right] \tag{49}$$

In the limit $l_x \to \infty$, we obtain

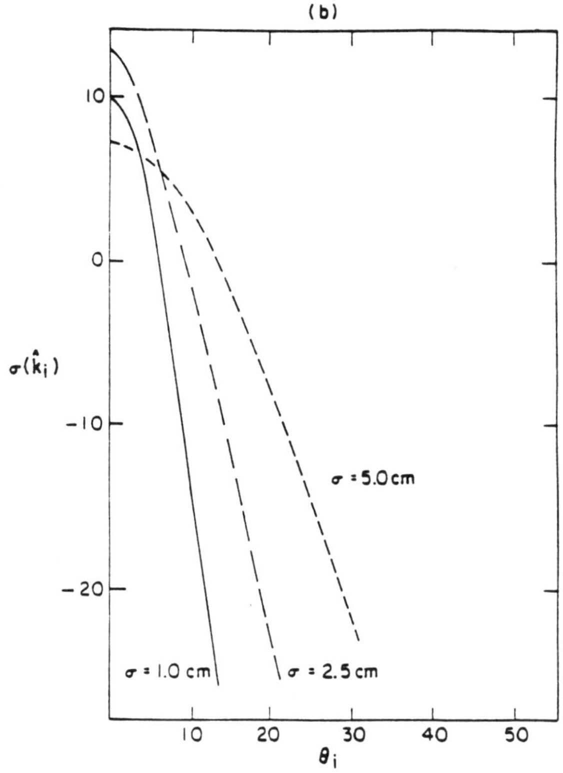

Figure 4.3.9 (b) $\sigma(\hat{k}_i)$ as a function of θ_i for different values of σ at 1.4 GHz with $l = 50$ cm, $B = 0$, $\sigma_x = 0$, and $\epsilon_1 = (6.0 + i0.6)\epsilon_o$.

$$\langle II^* \rangle = \int_{-2L_x}^{2L_x} dx \int_{-2L_y}^{2L_y} dy \sum_{n=-\infty}^{\infty} \exp\left[-k_{dz}^2\sigma_x^2\right] I_n(k_{dz}^2\sigma_x^2) \exp\left[i\left(\overline{k}_{d\perp} + \hat{x}n\frac{2\pi}{P}\right)\cdot\overline{r}_\perp\right] \times [2L_x - |x|][2L_y - |y|] \exp\left[-k_{dz}^2\sigma^2 + k_{dz}^2\sigma^2 C(\overline{r}_\perp)\right] \quad (50)$$

where I_n is the nth order modified Bessel function. This is similar to the randomly perturbed sinusoidal surface result. In this case the amplitudes of the modes are given by the modified Bessel functions, whereas before they were given in terms of the Bessel functions. The

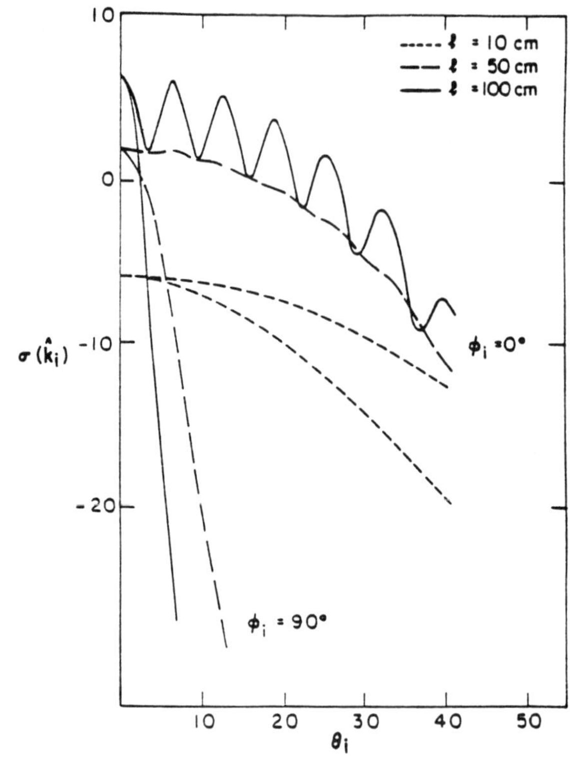

Figure 4.3.10 (a) $\sigma(\hat{k}_i)$ **as a function of** θ_i **for different values of** l **at 1.4 GHz with** $\sigma = 1$ **cm,** $B = 0$**,** $P = 100$ **cm,** $\sigma_x = 5$ **cm,** $l_x = 300$ **cm, and** $\epsilon_1 = (6.0 + i0.6)\epsilon_o$**.**

amplitudes of the modes are plotted in Fig. 4.3.10(b), and we can see that they are monotonically decreasing as n is increased, which explains the results in Fig. 4.3.10(a). Also note that as σ_x is decreased, only the first few modes have larger amplitudes; and as n is increased, they decay much faster. When $\sigma_x = 0$, only the $n = 0$ mode remains, and we reproduce the random rough surface results.

The above result in the limit $\sigma_x \to \infty$, (50), can also be related to the randomly perturbed sinusoidal surface case. When $\sigma_x = 0$, we obtain

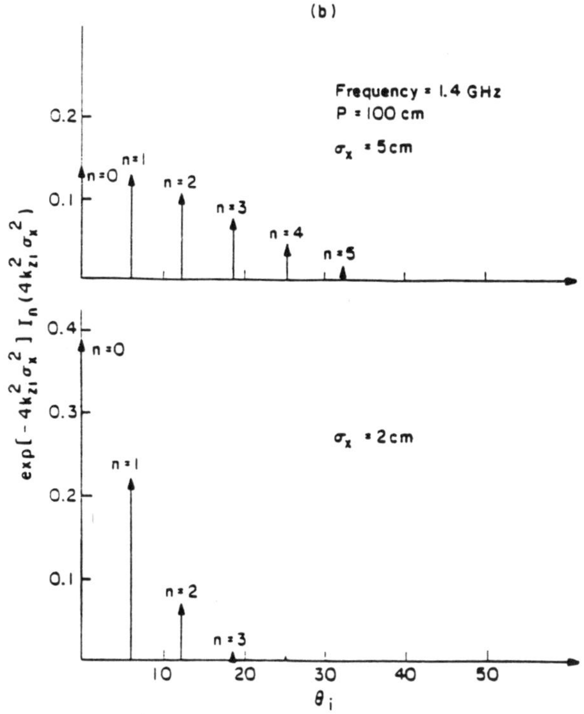

Figure 4.3.10 (b) Locations and amplitudes of the modes for $\sigma_x = 5$ cm and $\sigma_x = 2$ cm.

$$\langle II^* \rangle = \int_{-2L_x}^{2L_x} dx \int_{-2L_y}^{2L_y} dy \sum_n J_n^2(k_{dz}B) \exp\left[i\left(\overline{k}_{d\perp} + \hat{x}n\frac{2\pi}{P}\right)\cdot \overline{r}_\perp\right]$$
$$\times [2L_x - |x|][2L_y - |y|] \exp\left[-k_{dz}^2\sigma^2 + k_{dz}^2\sigma^2 C(\overline{r}_\perp)\right] \tag{51}$$

For a narrow-band Gaussian random process, $A(x)\cos[(2\pi/P)x + \psi(x)]$, the PDF for $A(x)$ and $\psi(x)$ is given by

$$P(A,\psi) = \begin{cases} \frac{A}{2\pi\sigma_x^2}\exp\left(-\frac{A^2}{2\sigma_x^2}\right) & \text{for } A > 0, \quad 0 \le \psi \le 2\pi \\ 0 & \text{otherwise} \end{cases} \tag{52}$$

Therefore, if we treat the amplitude B and the phase ψ of sinusoidal variation as random variables with PDF given by (52), and take the

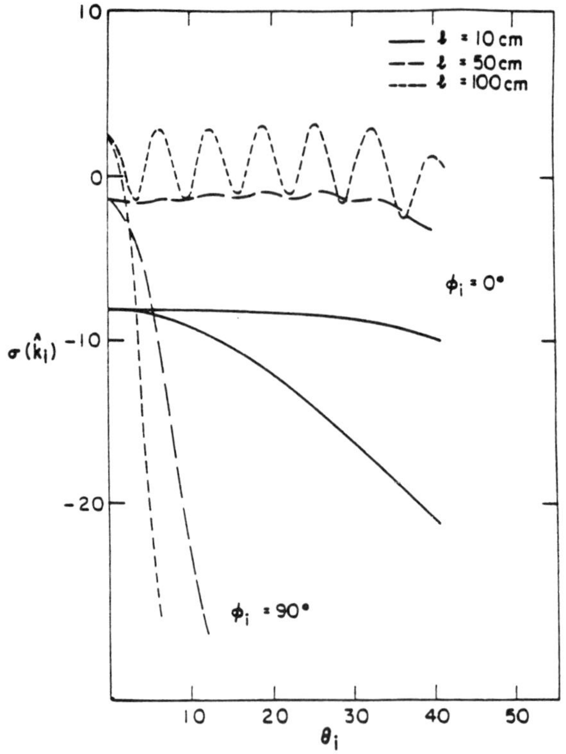

Figure 4.3.11 $\sigma(\hat{k}_i)$ **as a function of** θ_i **for different values of** l **at 1.4 GHz with** $\sigma = 1$ **cm,** $B = 10$ **cm,** $P = 100$ **cm,** $\sigma_x = 5$ **cm,** $l_x = 300$ **cm, and** $\epsilon_1 = (6.0 + i0.6)\epsilon_o$.

average of (51) with respect to B and ψ, we obtain the randomly perturbed quasiperiodic surface result, given by (50), by making use of

$$\int_0^\infty dB\, J_n^2(k_{dz}B)\frac{B}{\sigma_x^2}\exp\left(-\frac{B^2}{2\sigma_x^2}\right) = \exp(-k_{dz}^2\sigma_x^2)I_n(-k_{dz}^2\sigma_x^2) \quad (53)$$

In Figs. 4.3.11 and 4.3.12 we illustrate the combined effect of the previous cases. In Fig. 4.3.11 the backscattering cross sections are plotted for different l for the case $B = 10\,\text{cm}$ and $\sigma_x = 5\,\text{cm}$. The $\phi_i = 0°$ results are seen to be much flatter as a function of incident angle than the corresponding cases in Fig. 4.3.10(a). The effect of varying σ_x is illustrated in Fig. 4.3.12. Therefore, by varying σ_x and B, we can

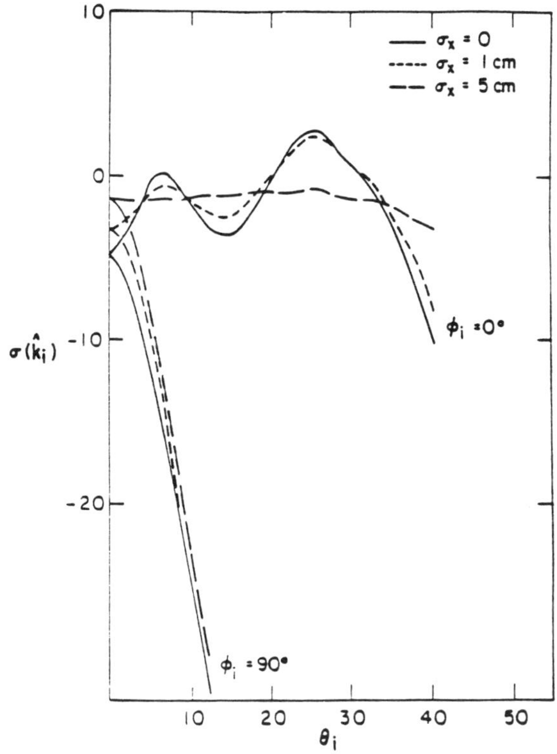

Figure 4.3.12 $\sigma(\hat{k}_i)$ **as a function of** θ_i **for different values of** σ_x **at 1.4 GHz with** $\sigma = 1$ **cm,** $l = 50$ **cm,** $B = 10$ **cm,** $P = 100$ **cm,** $l_x = 300$ **cm, and** $\epsilon_1 = (6.0 + i0.6)\epsilon_o$.

obtain different combinations of the previous two cases when $\sigma_x = 0$ or $B = 0$.

Appendix A. Zeroth-Order Equation and Solution

The zeroth-order equations are given as follows.

$$\hat{h}(-k_{zi}) \cdot \hat{e}_i \delta_m = \sum_{n=-\infty}^{\infty} \frac{k}{2k_{zmi}} Q^-_{mn}(k_{zmi}) \left\{ \frac{k_{ymi}}{k_{\rho mi}} \alpha^{(0)}_{xn} \right.$$

$$-\frac{k_{xmi}}{k_{\rho mi}}\alpha_{yn}^{(0)} + \left[\frac{k_{zmi}}{k}\frac{k_{xmi}}{k_{\rho mi}} - \frac{k_{\rho mi}}{k}\frac{(n-m)\frac{2\pi}{p}}{k_{zmi}}\right]\beta_{xn}^{(0)}$$

$$\left. + \frac{k_{zmi}}{k}\frac{k_{ymi}}{k_{\rho mi}}\beta_{yn}^{(0)}\right\} \qquad (A1a)$$

$$\hat{v}(-k_{zi})\cdot\hat{e}_i\delta_m = \sum_{n=-\infty}^{\infty}\frac{k}{2k_{zmi}}Q_{mn}^{-}(k_{zmi})$$

$$\left\{\left[\frac{k_{zmi}}{k}\frac{k_{xmi}}{k_{\rho mi}} - \frac{k_{\rho mi}}{k}\frac{(n-m)\frac{2\pi}{p}}{k_{zmi}}\right]\alpha_{xn}^{(0)}\right.$$

$$\left. + \frac{k_{zmi}}{k}\frac{k_{ymi}}{k_{\rho mi}}\alpha_{yn}^{(0)} - \frac{k_{ymi}}{k_{\rho mi}}\beta_{xn}^{(0)} + \frac{k_{xmi}}{k_{\rho mi}}\beta_{yn}^{(0)}\right\} \qquad (A1b)$$

$$0 = \sum_{n=-\infty}^{\infty}Q_{mn}^{+}(k_{1zmi})$$

$$\left\{\frac{k}{k_1}\frac{k_{ymi}}{k_{\rho mi}}\alpha_{xn}^{(0)} - \frac{k}{k_1}\frac{k_{xmi}}{k_{\rho mi}}\alpha_{yn}^{(0)} + \left[-\frac{k_{1zmi}}{k_1}\frac{k_{xmi}}{k_{\rho mi}}\right.\right.$$

$$\left.\left. + \frac{k_{\rho mi}}{k_1}\frac{(n-m)\frac{2\pi}{p}}{k_{1zmi}}\right]\beta_{xn}^{(0)} - \frac{k_{1zmi}}{k_1}\frac{k_{ymi}}{k_{\rho mi}}\beta_{yn}^{(0)}\right\} \qquad (A1c)$$

$$0 = \sum_{n=-\infty}^{\infty}Q_{mn}^{+}(k_{1zmi})$$

$$\left\{\frac{k}{k_1}\left[-\frac{k_{1zmi}}{k_1}\frac{k_{xmi}}{k_{\rho mi}} + \frac{k_{\rho mi}}{k_1}\frac{(n-m)\frac{2\pi}{p}}{k_{1zmi}}\right]\alpha_{xn}^{(0)}\right.$$

$$\left. - \frac{k}{k_1}\frac{k_{1zmi}}{k_1}\frac{k_{ymi}}{k_{\rho mi}}\alpha_{yn}^{(0)} - \frac{k_{ymi}}{k_{\rho mi}}\beta_{xn}^{(0)} + \frac{k_{xmi}}{k_{\rho mi}}\beta_{yn}^{(0)}\right\}$$

$$m = -\infty, \ldots, \infty \qquad (A1d)$$

where k_{xmi}, k_{ymi}, $k_{\rho mi}$, k_{zmi}, k_{1zmi}, and $Q_{mn}^{\pm}(k_z)$ are given by

$$k_{xmi} = k_{xi} + \frac{2m\pi}{p}, \quad k_{ymi} = k_{yi}, \quad k_{\rho mi} = \sqrt{k_{xmi}^2 + k_{ymi}^2} \qquad (A2a)$$

$$k_{zmi} = \sqrt{k^2 - k_{\rho mi}^2}, \quad k_{1zmi} = \sqrt{k_1^2 - k_{\rho mi}^2} \qquad (A2b)$$

$$
\begin{aligned}
Q_{mn}^{\pm}(k_z) &= \frac{1}{p}\int_0^p dx \exp\left[-i(m-n)\frac{2\pi}{p}x \mp ik_zH\cos\left(\frac{2\pi}{p}x\right)\right] \\
&= (\mp i)^{|m-n|}J_{|m-n|}(k_zH) \qquad (A3)
\end{aligned}
$$

The components of the zeroth-order solution are given explicitly as follows.

$$
\begin{aligned}
E_{hsm}^{(0)} =& \frac{k}{2k_{zmi}}\sum_{n=-\infty}^{\infty} Q_{mn}^{+}(k_{zmi})\left\{-\frac{k_{ymi}}{k_{\rho mi}}\alpha_{xn}^{(0)} + \frac{k_{xmi}}{k_{\rho mi}}\alpha_{yn}^{(0)}\right. \\
&\left. + \left(\frac{k_{zmi}}{k}\frac{k_{xmi}}{k_{\rho mi}} - \frac{k_{\rho mi}}{k}\frac{(n-m)\frac{2\pi}{p}}{k_{zmi}}\right)\beta_{xn}^{(0)} + \frac{k_{zmi}}{k}\frac{k_{ymi}}{k_{\rho mi}}\beta_{yn}^{(0)}\right\} \qquad (A4a)
\end{aligned}
$$

$$
\begin{aligned}
E_{vsm}^{(0)} =& \frac{k}{2k_{zmi}}\sum_{n=-\infty}^{\infty} Q_{mn}^{+}(k_{zmi})\left\{\left(\frac{k_{zmi}}{k}\frac{k_{xmi}}{k_{\rho mi}} - \frac{k_{\rho mi}}{k}\frac{(n-m)\frac{2\pi}{p}}{k_{zmi}}\right)\alpha_{xn}^{(0)}\right. \\
&\left. + \frac{k_{zmi}}{k}\frac{k_{ymi}}{k_{\rho mi}}\alpha_{yn}^{(0)} + \frac{k_{ymi}}{k_{\rho mi}}\beta_{xn}^{(0)} - \frac{k_{xmi}}{k_{\rho mi}}\beta_{yn}^{(0)}\right\} \qquad (A4b)
\end{aligned}
$$

Appendix B. First-Order Equation and Solution

The first-order equations for the surface fields are given explicitly as follows.

$$
\begin{aligned}
&\sum_{m=-\infty}^{\infty} Q_{m0}^{-}(k_{zl})\left\{\frac{k_{yl}}{k_{\rho l}}A_x^{(1)}(\overline{k}_{\perp l} - \overline{k}_{\perp mi})\right. \\
&\quad - \frac{k_{xl}}{k_{\rho l}}A_y^{(1)}(\overline{k}_{\perp l} - \overline{k}_{\perp mi}) + \left(\frac{k_{zl}}{k}\frac{k_{xl}}{k_{\rho l}} + \frac{k_{\rho l}}{k}\frac{m\frac{2\pi}{p}}{k_{zl}}\right)B_x^{(1)}(\overline{k}_{\perp l} - \overline{k}_{\perp mi}) \\
&\quad \left. + \frac{k_{zl}}{k}\frac{k_{yl}}{k_{\rho l}}B_y^{(1)}(\overline{k}_{\perp l} - \overline{k}_{\perp mi})\right\} \\
&= -\sum_{m=-\infty}^{\infty} ik_{zl}F(\overline{k}_{\perp l} - \overline{k}_{\perp mi})\sum_{n=-\infty}^{\infty} Q_{mn}^{-}(k_{zl}) \\
&\quad \left\{\frac{k_{yl}}{k_{\rho l}}\alpha_{xn}^{(0)} - \frac{k_{xl}}{k_{\rho l}}\alpha_{yn}^{(0)} + \left[\frac{k_{zl}}{k}\frac{k_{xl}}{k_{\rho l}} - \frac{k_{\rho l}}{k}\frac{(n-m)\frac{2\pi}{p}}{k_{zl}}\right.\right.
\end{aligned}
$$

$$+ \frac{k_{\rho l}}{k} \frac{(k_{xl} - k_{xmi})}{k_{zl}} \Big] \beta_{xn}^{(0)} + \left[\frac{k_{zl}}{k} \frac{k_{yl}}{k_{\rho l}} + \frac{k_{\rho l}}{k} \frac{(k_{yl} - k_{yi})}{k_{zl}} \right] \beta_{yn}^{(0)} \Bigg\} \tag{B1a}$$

$$\sum_{m=-\infty}^{\infty} Q_{m0}^{-}(k_{zl}) \Bigg\{ \left(\frac{k_{zl}}{k} \frac{k_{xl}}{k_{\rho l}} + \frac{k_{\rho l}}{k} \frac{m \frac{2\pi}{p}}{k_{zl}} \right) A_x^{(1)}(\overline{k}_{\perp l} - \overline{k}_{\perp mi})$$

$$+ \frac{k_{zl}}{k} \frac{k_{yl}}{k_{\rho l}} A_y^{(1)}(\overline{k}_{\perp l} - \overline{k}_{\perp mi})$$

$$- \frac{k_{yl}}{k_{\rho l}} B_x^{(1)}(\overline{k}_{\perp l} - \overline{k}_{\perp mi}) + \frac{k_{xl}}{k_{\rho l}} B_y^{(1)}(\overline{k}_{\perp l} - \overline{k}_{\perp mi}) \Bigg\}$$

$$= - \sum_{m=-\infty}^{\infty} ik_{zl} F(\overline{k}_{\perp l} - \overline{k}_{\perp mi}) \sum_{n=-\infty}^{\infty} Q_{mn}^{-}(k_{zl})$$

$$\Bigg\{ \left[\frac{k_{zl}}{k} \frac{k_{xl}}{k_{\rho l}} - \frac{k_{\rho l}}{k} \frac{(n-m)\frac{2\pi}{p}}{k_{zl}} + \frac{k_{\rho l}}{k} \frac{(k_{xl} - k_{xmi})}{k_{zl}} \right] \alpha_{xn}^{(0)}$$

$$+ \left[\frac{k_{zl}}{k} \frac{k_{yl}}{k_{\rho l}} + \frac{k_{\rho l}}{k} \frac{(k_{yl} - k_{yi})}{k_{zl}} \right] \alpha_{yn}^{(0)} - \frac{k_{yl}}{k_{\rho l}} \beta_{xn}^{(0)} + \frac{k_{xl}}{k_{\rho l}} \beta_{yn}^{(0)} \Bigg\} \tag{B1b}$$

$$\sum_{m=-\infty}^{\infty} Q_{m0}^{+}(k_{1zl}) \Bigg\{ \frac{k}{k_1} \frac{k_{yl}}{k_{\rho l}} A_x^{(1)}(\overline{k}_{\perp l} - \overline{k}_{\perp mi})$$

$$- \frac{k}{k_1} \frac{k_{xl}}{k_{\rho l}} A_y^{(1)}(\overline{k}_{\perp l} - \overline{k}_{\perp mi}) - \left(\frac{k_{1zl}}{k_1} \frac{k_{xl}}{k_{\rho l}} + \frac{k_{\rho l}}{k_1} \frac{m \frac{2\pi}{p}}{k_{1zl}} \right)$$

$$\times B_x^{(1)}(\overline{k}_{\perp l} - \overline{k}_{\perp mi}) - \frac{k_{1zl}}{k_1} \frac{k_{yl}}{k_{\rho l}} B_y^{(1)}(\overline{k}_{\perp l} - \overline{k}_{\perp mi}) \Bigg\}$$

$$= - \sum_{m=-\infty}^{\infty} ik_{1zl} F(\overline{k}_{\perp l} - \overline{k}_{\perp mi}) \sum_{n=-\infty}^{\infty} Q_{mn}^{+}(k_{1zl})$$

$$\Bigg\{ - \frac{k}{k_1} \frac{k_{yl}}{k_{\rho l}} \alpha_{xn}^{(0)} + \frac{k}{k_1} \frac{k_{xl}}{k_{\rho l}} \alpha_{yn}^{(0)} + \Bigg[\frac{k_{1zl}}{k_1} \frac{k_{xl}}{k_{\rho l}} - \frac{k_{\rho l}}{k_1} \frac{(n-m)\frac{2\pi}{p}}{k_{1zl}}$$

$$+ \frac{k_{\rho l}}{k_1} (\frac{k_{xl} - k_{xmi})}{k_{1zl}} \Bigg] \beta_{xn}^{(0)} + \left[\frac{k_{1zl}}{k_1} \frac{k_{yl}}{k_{\rho l}} + \frac{k_{\rho l}}{k_1} \frac{(k_{yl} - k_{yi})}{k_{1zl}} \right] \beta_{yn}^{(0)} \Bigg\} \tag{B1c}$$

$$\sum_{m=-\infty}^{\infty} Q_{m0}^{+}(k_{1zl}) \Bigg\{ - \frac{k}{k_1} (\frac{k_{1zl}}{k_1} \frac{k_{xl}}{k_{\rho l}} + \frac{k_{\rho l}}{k_1} \frac{m \frac{2\pi}{p}}{k_{1zl}}) A_x^{(1)}(\overline{k}_{\perp l} - \overline{k}_{\perp mi})$$

$$-\frac{k}{k_1}\frac{k_{1zl}}{k_1}\frac{k_{yl}}{k_{\rho l}}A_y^{(1)}(\overline{k}_{\perp l}-\overline{k}_{\perp mi})-\frac{k_{yl}}{k_{\rho l}}B_x^{(1)}(\overline{k}_{\perp l}-\overline{k}_{\perp mi})$$

$$+\frac{k_{xl}}{k_{\rho l}}B_y^{(1)}(\overline{k}_{\perp l}-\overline{k}_{\perp mi})\Bigg\}$$

$$=-\sum_{m=-\infty}^{\infty} ik_{1zl}F(\overline{k}_{\perp l}-\overline{k}_{\perp mi})\sum_{n=-\infty}^{\infty} Q_{mn}^{+}(k_{1zl})$$

$$\Bigg\{\frac{k}{k_1}\left[\frac{k_{1zl}}{k_1}\frac{k_{xl}}{k_{\rho l}}-\frac{k_{\rho l}}{k_1}\frac{(n-m)\frac{2\pi}{p}}{k_{1zl}}+\frac{k_{\rho l}}{k_1}\frac{(k_{xl}-k_{xmi})}{k_{1zl}}\right]\alpha_{xn}^{(0)}$$

$$+\frac{k}{k_1}\left[\frac{k_{1zl}}{k_1}\frac{k_{yl}}{k_{\rho l}}+\frac{k_{\rho l}}{k_1}\frac{(k_{yl}-k_{yi})}{k_{1zl}}\right]\alpha_{yn}^{(0)}+\frac{k_{yl}}{k_{\rho l}}\beta_{xn}^{(0)}-\frac{k_{xl}}{k_{\rho l}}\beta_{yn}^{(0)}\Bigg\}$$

$$l=-\infty,\ldots,\infty \qquad (B1d)$$

where $Q_{mn}^{\pm}$ are given by (A3), and

$$\overline{k}_{\perp l}=k_{xl}\hat{x}+k_{yl}\hat{y}=\left(k_x+\frac{2l\pi}{p}\right)\hat{x}+k_y\hat{y} \qquad (B2a)$$

$$k_{\rho l}=\sqrt{k_{xl}^2+k_{yl}^2} \qquad (B2b)$$

$$k_{zl}=\sqrt{k^2-k_{\rho l}^2},\quad k_{1zl}=\sqrt{k_1^2-k_{\rho l}^2} \qquad (B2c)$$

The components of the first-order scattered field are given as follows.

$$E_{hsm}^{(1)}=\frac{k}{2k_z}\Bigg\{Q_{m0}^{+}(k_z)\Bigg[-\frac{k_y}{k_\rho}A_x^{(1)}(\overline{k}_\perp-\overline{k}_{\perp mi})+\frac{k_x}{k_\rho}A_y^{(1)}(\overline{k}_\perp-\overline{k}_{\perp mi})$$

$$+\left(\frac{k_z}{k}\frac{k_x}{k_\rho}+\frac{k_\rho}{k}\frac{m\frac{2\pi}{p}}{k_z}\right)B_x^{(1)}(\overline{k}_\perp-\overline{k}_{\perp mi})+\frac{k_z}{k}\frac{k_y}{k_\rho}B_y^{(1)}(\overline{k}_\perp-\overline{k}_{\perp mi})\Bigg]$$

$$+ik_zF(\overline{k}_\perp-\overline{k}_{\perp mi})\sum_{n=-\infty}^{\infty} Q_{mn}^{+}(k_z)\Bigg\{\frac{k_y}{k_\rho}\alpha_{xn}^{(0)}-\frac{k_x}{k_\rho}\alpha_{yn}^{(0)}$$

$$-\left[\frac{k_z}{k}\frac{k_x}{k_\rho}-\frac{k_\rho}{k}\frac{(n-m)\frac{2\pi}{p}}{k_z}+\frac{k_\rho}{k}\frac{(k_x-k_{xmi})}{k_z}\right]\beta_{xn}^{(0)}$$

$$-\left[\frac{k_z}{k}\frac{k_y}{k_\rho}+\frac{k_\rho}{k}\frac{(k_y-k_{yi})}{k_z}\right]\beta_{yn}^{(0)}\Bigg\}\Bigg\} \qquad (B3a)$$

$$E_{vsm}^{(1)} = \frac{k}{2k_z}\Bigg\{Q_{m0}^{+}(k_z)\Bigg[\left(\frac{k_z}{k}\frac{k_x}{k_\rho} + \frac{k_\rho}{k}\frac{m\frac{2\pi}{p}}{k_z}\right) A_x^{(1)}(\bar{k}_\perp - \bar{k}_{\perp mi})$$
$$+ \frac{k_z}{k}\frac{k_y}{k_\rho} A_y^{(1)}(\bar{k}_\perp - \bar{k}_{\perp mi})$$
$$+ \frac{k_y}{k_\rho} B_x^{(1)}(\bar{k}_\perp - \bar{k}_{\perp mi}) - \frac{k_x}{k_\rho} B_y^{(1)}(\bar{k}_\perp - \bar{k}_{\perp mi})\Bigg]$$
$$+ ik_z F(\bar{k}_\perp - \bar{k}_{\perp mi}) \sum_{n=-\infty}^{\infty} Q_{mn}^{+}(k_z)\Bigg\{-\Bigg[\frac{k_z}{k}\frac{k_x}{k_\rho}$$
$$- \frac{k_\rho}{k}\frac{(n-m)\frac{2\pi}{p}}{k_z} + \frac{k_\rho}{k}\frac{(k_x - k_{xmi})}{k_z}\Bigg]\alpha_{xn}^{(0)}$$
$$- \left[\frac{k_z}{k}\frac{k_y}{k_\rho} + \frac{k_\rho}{k}\frac{(k_y - k_{yi})}{k_z}\right]\alpha_{yn}^{(0)} - \frac{k_y}{k_\rho}\beta_{xn}^{(0)} + \frac{k_x}{k_\rho}\beta_{yn}^{(0)}\Bigg\}\Bigg\} \qquad (B3b)$$

Appendix C. Second-Order Equation and Solution

The second-order equations are given as follows.

$$\sum_{m=-\infty}^{\infty} \frac{1}{2} Q_{m0}^{-}(k_{zl})\Bigg\{\frac{k_{yl}}{k_{\rho l}} A_x^{(2)}(\bar{k}_{\perp l} - \bar{k}_{\perp mi}) - \frac{k_{xl}}{k_{\rho l}} A_y^{(2)}(\bar{k}_{\perp l} - \bar{k}_{\perp mi})$$
$$+ \left(\frac{k_{zl}}{k}\frac{k_{xl}}{k_{\rho l}} + \frac{k_{\rho l}}{k}\frac{m\frac{2\pi}{p}}{k_{zl}}\right) B_x^{(2)}(\bar{k}_{\perp l} - \bar{k}_{\perp mi})$$
$$+ \frac{k_{zl}}{k}\frac{k_{yl}}{k_{\rho l}} B_y^{(2)}(\bar{k}_{\perp l} - \bar{k}_{\perp mi})\Bigg\}$$
$$= -\sum_{m=-\infty}^{\infty} F * F(\bar{k}_{\perp l} - \bar{k}_{\perp mi}) \sum_{n=-\infty}^{\infty} Q_{mn}^{-}(k_{zl})(-\frac{1}{2})k_{zl}^2$$
$$\Bigg\{\frac{k_{yl}}{k_{\rho l}}\alpha_{xn}^{(0)} - \frac{k_{xl}}{k_{\rho l}}\alpha_{yn}^{(0)} + \Bigg[\frac{k_{zl}}{k}\frac{k_{xl}}{k_{\rho l}} - \frac{k_{\rho l}}{k}\frac{(n-m)\frac{2\pi}{p}}{k_{zl}}$$
$$+ \frac{k_{\rho l}}{k}\frac{(k_{xl} - k_{xmi})}{k_{zl}}\Bigg]\beta_{xn}^{(0)} + \left[\frac{k_{zl}}{k}\frac{k_{yl}}{k_{\rho l}} + \frac{k_{\rho l}}{k}\frac{(k_{yl} - k_{yi})}{k_{zl}}\right]\beta_{yn}^{(0)}\Bigg\}$$
$$- \sum_{m=-\infty}^{\infty} ik_{zl} Q_{m0}^{-}(k_{zl})\Bigg\{\frac{k_{yl}}{k_{\rho l}} F * A_x^{(1)}(\bar{k}_{\perp l} - \bar{k}_{\perp mi})$$

$$
\begin{aligned}
&- \frac{k_{xl}}{k_{\rho l}} F * A_y^{(1)}(\overline{k}_{\perp l} - \overline{k}_{\perp mi}) \\
&+ \left(\frac{k_{zl}}{k} \frac{k_{xl}}{k_{\rho l}} + \frac{k_{\rho l}}{k} \frac{m\frac{2\pi}{p}}{k_{zl}} \right) F * B_x^{(1)}(\overline{k}_{\perp l} - \overline{k}_{\perp mi}) \\
&+ \frac{k_{zl}}{k} \frac{k_{yl}}{k_{\rho l}} F * B_y^{(1)}(\overline{k}_{\perp l} - \overline{k}_{\perp mi}) \\
&+ \frac{k_{\rho l}}{kk_{zl}} [(k_{xl} - k_{xmi}) F(\overline{k}_{\perp l} - \overline{k}_{\perp mi})] * B_x^{(1)}(\overline{k}_{\perp l} - \overline{k}_{\perp mi}) \\
&+ \left. \frac{k_{\rho l}}{kk_{zl}} [(k_{yl} - k_{ymi}) F(\overline{k}_{\perp l} - \overline{k}_{\perp mi})] * B_y^{(1)}(\overline{k}_{\perp l} - \overline{k}_{\perp mi}) \right\} \qquad (C1a)
\end{aligned}
$$

$$
\begin{aligned}
\sum_{m=-\infty}^{\infty} \frac{1}{2} Q_{m0}^{-}(k_{zl}) &\left\{ \left(\frac{k_{zl}}{k} \frac{k_{xl}}{k_{\rho l}} + \frac{k_{\rho l}}{k} \frac{m\frac{2\pi}{p}}{k_{zl}} \right) A_x^{(2)}(\overline{k}_{\perp l} - \overline{k}_{\perp mi}) \right. \\
&+ \frac{k_{zl}}{k} \frac{k_{yl}}{k_{\rho l}} A_y^{(2)}(\overline{k}_{\perp l} - \overline{k}_{\perp mi}) - \frac{k_{yl}}{k_{\rho l}} B_x^{(2)}(\overline{k}_{\perp l} - \overline{k}_{\perp mi}) \\
&+ \left. \frac{k_{xl}}{k_{\rho l}} B_y^{(2)}(\overline{k}_{\perp l} - \overline{k}_{\perp mi}) \right\} \\
= - \sum_{m=-\infty}^{\infty} & F * F(\overline{k}_{\perp l} - \overline{k}_{\perp mi}) \sum_{n=-\infty}^{\infty} Q_{mn}^{-}(k_{zl}) \left(-\frac{1}{2} \right) k_{zl}^2 \\
&\left\{ \left[\frac{k_{zl}}{k} \frac{k_{xl}}{k_{\rho l}} - \frac{k_{\rho l}}{k} \frac{(n-m)\frac{2\pi}{p}}{k_{zl}} + \frac{k_{\rho l}}{k} \frac{(k_{xl} - k_{xmi})}{k_{zl}} \right] \alpha_{xn}^{(0)} \right. \\
&+ \left. \left[\frac{k_{zl}}{k} \frac{k_{yl}}{k_{\rho l}} + \frac{k_{\rho l}}{k} \frac{(k_{yl} - k_{yi})}{k_{zl}} \right] \alpha_{yn}^{(0)} - \frac{k_{yl}}{k_{\rho l}} \beta_{xn}^{(0)} + \frac{k_{xl}}{k_{\rho l}} \beta_{yn}^{(0)} \right\} \\
&- \sum_{m=-\infty}^{\infty} i k_{zl} Q_{m0}^{-}(k_{zl}) \left\{ \left(\frac{k_{zl}}{k} \frac{k_{xl}}{k_{\rho l}} + \frac{k_{\rho l}}{k} \frac{m\frac{2\pi}{p}}{k_{zl}} \right) F * A_x^{(1)}(\overline{k}_{\perp l} - \overline{k}_{\perp mi}) \right. \\
&+ \frac{k_{zl}}{k} \frac{k_{yl}}{k_{\rho l}} F * A_y^{(1)}(\overline{k}_{\perp l} - \overline{k}_{\perp mi}) \\
&- \frac{k_{yl}}{k_{\rho l}} F * B_x^{(1)}(\overline{k}_{\perp l} - \overline{k}_{\perp mi}) + \frac{k_{xl}}{k_{\rho l}} F * B_y^{(1)}(\overline{k}_{\perp l} - \overline{k}_{\perp mi}) \\
&+ \frac{k_{\rho l}}{kk_{zl}} [(k_{xl} - k_{xmi}) F(\overline{k}_{\perp l} - \overline{k}_{\perp mi})] * A_x^{(1)}(\overline{k}_{\perp l} - \overline{k}_{\perp mi})
\end{aligned}
$$

$$+ \frac{k_{\rho l}}{kk_{zl}}[(k_{yl} - k_{ymi})F(\overline{k}_{\perp l} - \overline{k}_{\perp mi})] * A_y^{(1)}(\overline{k}_{\perp l} - \overline{k}_{\perp mi})\Bigg\} \qquad (C1b)$$

$$\sum_{m=-\infty}^{\infty} \frac{1}{2} Q_{m0}^{+}(k_{1zl}) \Bigg\{ \frac{k}{k_1}\frac{k_{yl}}{k_{\rho l}} A_x^{(2)}(\overline{k}_{\perp l} - \overline{k}_{\perp mi})$$

$$- \frac{k}{k_1}\frac{k_{xl}}{k_{\rho l}} A_y^{(2)}(\overline{k}_{\perp l} - \overline{k}_{\perp mi})$$

$$- \left(\frac{k_{1zl}}{k_1}\frac{k_{xl}}{k_{\rho l}} + \frac{k_{\rho l}}{k_1}\frac{m\frac{2\pi}{p}}{k_{1zl}} \right) B_x^{(2)}(\overline{k}_{\perp l} - \overline{k}_{\perp mi})$$

$$- \frac{k_{1zl}}{k_1}\frac{k_{yl}}{k_{\rho l}} B_y^{(2)}(\overline{k}_{\perp l} - \overline{k}_{\perp mi})\Bigg\}$$

$$= - \sum_{m=-\infty}^{\infty} F * F(\overline{k}_{\perp l} - \overline{k}_{\perp mi}) \sum_{n=-\infty}^{\infty} Q_{mn}^{+}(k_{1zl})(-\frac{1}{2})k_{1zl}^2$$

$$\Bigg\{ \frac{k}{k_1}\frac{k_{yl}}{k_{\rho l}} \alpha_{xn}^{(0)} - \frac{k}{k_1}\frac{k_{xl}}{k_{\rho l}} \alpha_{yn}^{(0)} - \Bigg[\frac{k_{1zl}}{k_1}\frac{k_{xl}}{k_{\rho l}} - \frac{k_{\rho l}}{k_1}\frac{(n-m)\frac{2\pi}{p}}{k_{1zl}}$$

$$+ \frac{k_{\rho l}}{k_1}\frac{(k_{xl} - k_{xmi})}{k_{1zl}} \Bigg] \beta_{xn}^{(0)} - \left[\frac{k_{1zl}}{k_1}\frac{k_{yl}}{k_{\rho l}} + \frac{k_{\rho l}}{k_1}\frac{(k_{yl} - k_{yi})}{k_{1zl}} \right] \beta_{yn}^{(0)} \Bigg\}$$

$$- \sum_{m=-\infty}^{\infty} -ik_{1zl} Q_{m0}^{+}(k_{1zl}) \Bigg\{ \frac{k}{k_1}\frac{k_{yl}}{k_{\rho l}} F * A_x^{(1)}(\overline{k}_{\perp l} - \overline{k}_{\perp mi})$$

$$- \frac{k}{k_1}\frac{k_{xl}}{k_{\rho l}} F * A_y^{(1)}(\overline{k}_{\perp l} - \overline{k}_{\perp mi})$$

$$- \left(\frac{k_{1zl}}{k_1}\frac{k_{xl}}{k_{\rho l}} + \frac{k_{\rho l}}{k_1}\frac{m\frac{2\pi}{p}}{k_{1zl}} \right) F * B_x^{(1)}(\overline{k}_{\perp l} - \overline{k}_{\perp mi})$$

$$- \frac{k_{1zl}}{k_1}\frac{k_{yl}}{k_{\rho l}} F * B_y^{(1)}(\overline{k}_{\perp l} - \overline{k}_{\perp mi})$$

$$- \frac{k_{\rho l}}{k_1 k_{1zl}}[(k_{xl} - k_{xmi})F(\overline{k}_{\perp l} - \overline{k}_{\perp mi})] * B_x^{(1)}(\overline{k}_{\perp l} - \overline{k}_{\perp mi})$$

$$- \frac{k_{\rho l}}{k_1 k_{1zl}}[(k_{yl} - k_{ymi})F(\overline{k}_{\perp l} - \overline{k}_{\perp mi})] * B_y^{(1)}(\overline{k}_{\perp l} - \overline{k}_{\perp mi})\Bigg\}$$

$$(C1c)$$

$$\sum_{m=-\infty}^{\infty} \frac{1}{2} Q^{+}_{m0}(k_{1zl}) \Bigg\{ -\frac{k}{k_1} \left(\frac{k_{1zl}}{k_1}\frac{k_{xl}}{k_{\rho l}} + \frac{k_{\rho l}}{k_1}\frac{m\frac{2\pi}{p}}{k_{1zl}} \right) A^{(2)}_x(\overline{k}_{\perp l} - \overline{k}_{\perp mi})$$
$$- \frac{k}{k_1}\frac{k_{1zl}}{k_1}\frac{k_{yl}}{k_{\rho l}} A^{(2)}_y(\overline{k}_{\perp l} - \overline{k}_{\perp mi}) - \frac{k_{yl}}{k_{\rho l}} B^{(2)}_x(\overline{k}_{\perp l} - \overline{k}_{\perp mi})$$
$$+ \frac{k_{xl}}{k_{\rho l}} B^{(2)}_y(\overline{k}_{\perp l} - \overline{k}_{\perp mi}) \Bigg\}$$
$$= -\sum_{m=-\infty}^{\infty} F * F(\overline{k}_{\perp l} - \overline{k}_{\perp mi}) \sum_{n=-\infty}^{\infty} Q^{+}_{mn}(k_{1zl}) \left(-\frac{1}{2}\right) k^2_{1zl}$$
$$\Bigg\{ -\frac{k}{k_1} \left[\frac{k_{1zl}}{k_1}\frac{k_{xl}}{k_{\rho l}} - \frac{k_{\rho l}}{k_1}\frac{(n-m)\frac{2\pi}{p}}{k_{1zl}} + \frac{k_{\rho l}}{k_1}\frac{(k_{xl} - k_{xmi})}{k_{1zl}} \right] \alpha^{(0)}_{xn}$$
$$- \frac{k}{k_1} \left[\frac{k_{1zl}}{k_1}\frac{k_{yl}}{k_{\rho l}} + \frac{k_{\rho l}}{k_1}\frac{(k_{yl} - k_{yi})}{k_{1zl}} \right] \alpha^{(0)}_{yn} - \frac{k_{yl}}{k_{\rho l}} \beta^{(0)}_{xn} + \frac{k_{xl}}{k_{\rho l}} \beta^{(0)}_{yn} \Bigg\}$$
$$- \sum_{m=-\infty}^{\infty} -ik_{1zl} Q^{+}_{m0}(k_{1zl}) \Bigg\{ -\frac{k}{k_1} \Bigg(\frac{k_{1zl}}{k_1}\frac{k_{xl}}{k_{\rho l}}$$
$$+ \frac{k_{\rho l}}{k_1}\frac{m\frac{2\pi}{p}}{k_{1zl}} \Bigg) F * A^{(1)}_x(\overline{k}_{\perp l} - \overline{k}_{\perp mi})$$
$$- \frac{k}{k_1}\frac{k_{1zl}}{k_1}\frac{k_{yl}}{k_{\rho l}} F * A^{(1)}_y(\overline{k}_{\perp l} - \overline{k}_{\perp mi}) - \frac{k_{yl}}{k_{\rho l}} F * B^{(1)}_x(\overline{k}_{\perp l} - \overline{k}_{\perp mi})$$
$$+ \frac{k_{xl}}{k_{\rho l}} F * B^{(1)}_y(\overline{k}_{\perp l} - \overline{k}_{\perp mi})$$
$$- \frac{k}{k_1}\frac{k_{\rho l}}{k_1 k_{1zl}} [(k_{xl} - k_{xmi}) F(\overline{k}_{\perp l} - \overline{k}_{\perp mi})] * A^{(1)}_x(\overline{k}_{\perp l} - \overline{k}_{\perp mi})$$
$$- \frac{k}{k_1}\frac{k_{\rho l}}{k_1 k_{1zl}} [(k_{yl} - k_{ymi}) F(\overline{k}_{\perp l} - \overline{k}_{\perp mi})] * A^{(1)}_y(\overline{k}_{\perp l} - \overline{k}_{\perp mi}) \Bigg\} \tag{C1d}$$

where $Q^{\pm}_{mn}$ are given by (A3).

The components of the second-order scattered field are given explicitly below.

$$E^{(2)}_{hsm} = \left(-\frac{k}{2k_z}\right) \Bigg\{ \sum_{m=-\infty}^{\infty} \frac{1}{2} Q^{+}_{m0}(k_z) \Bigg[\frac{k_y}{k_\rho} A^{(2)}_x(\overline{k}_\perp - \overline{k}_{\perp mi})$$

$$- \frac{k_x}{k_\rho} A_y^{(2)}(\overline{k}_\perp - \overline{k}_{\perp mi}) - \frac{k_z}{k}\frac{k_y}{k_\rho} B_y^{(2)}(\overline{k}_\perp - \overline{k}_{\perp mi})$$

$$- (\frac{k_z}{k}\frac{k_x}{k_\rho} + \frac{k_\rho}{k}\frac{m\frac{2\pi}{p}}{k_z}) B_x^{(2)}(\overline{k}_\perp - \overline{k}_{\perp mi})\Big]$$

$$+ \sum_{m=-\infty}^{\infty} F * F(\overline{k}_\perp - \overline{k}_{\perp mi}) \sum_{n=-\infty}^{\infty} Q_{mn}^{+}(k_z) \left(-\frac{1}{2}\right) k_z^2 \left[\frac{k_y}{k_\rho}\alpha_{xn}^{(0)}\right.$$

$$- \frac{k_x}{k_\rho}\alpha_{yn}^{(0)} - \left[\frac{k_z}{k}\frac{k_x}{k_\rho} - \frac{k_\rho}{k}\frac{(n-m)\frac{2\pi}{p}}{k_z} + \frac{k_\rho}{k}\frac{(k_x - k_{xmi})}{k_z}\right]\beta_{xn}^{(0)}$$

$$\left. - \left[\frac{k_z}{k}\frac{k_y}{k_\rho} + \frac{k_\rho}{k}\frac{(k_y - k_{yi})}{k_z}\right]\beta_{yn}^{(0)}\right]$$

$$- \sum_{m=-\infty}^{\infty} ik_z Q_{m0}^{+}(k_z) \left[\frac{k_y}{k_\rho} F * A_x^{(1)}(\overline{k}_\perp - \overline{k}_{\perp mi})\right.$$

$$- \frac{k_x}{k_\rho} F * A_y^{(1)}(\overline{k}_\perp - \overline{k}_{\perp mi})$$

$$- \left(\frac{k_z}{k}\frac{k_x}{k_\rho} + \frac{k_\rho}{k}\frac{m\frac{2\pi}{p}}{k_z}\right) F * B_x^{(1)}(\overline{k}_\perp - \overline{k}_{\perp mi})$$

$$- \frac{k_z}{k}\frac{k_y}{k_\rho} F * B_y^{(1)}(\overline{k}_\perp - \overline{k}_{\perp mi})$$

$$- \frac{k_\rho}{kk_z}[(k_x - k_{xmi})F(\overline{k}_\perp - \overline{k}_{\perp mi})] * B_x^{(1)}(\overline{k}_\perp - \overline{k}_{\perp mi})$$

$$\left. - \frac{k_\rho}{kk_z}[(k_y - k_{ymi})F(\overline{k}_\perp - \overline{k}_{\perp mi})] * B_y^{(1)}(\overline{k}_\perp - \overline{k}_{\perp mi})\right]\Bigg\} \qquad (C2a)$$

$$E_{vsm}^{(2)} = \left(-\frac{k}{2k_z}\right)\Bigg\{\sum_{m=-\infty}^{\infty} \frac{1}{2} Q_{m0}^{+}(k_z)\Big[-\Big(\frac{k_z}{k}\frac{k_x}{k_\rho}$$

$$+ \frac{k_\rho}{k}\frac{m\frac{2\pi}{p}}{k_z}\Big) A_x^{(2)}(\overline{k}_\perp - \overline{k}_{\perp mi}) - \frac{k_z}{k}\frac{k_y}{k_\rho} A_y^{(2)}(\overline{k}_\perp - \overline{k}_{\perp mi})$$

$$- \frac{k_y}{k_\rho} B_x^{(2)}(\overline{k}_\perp - \overline{k}_{\perp mi}) + \frac{k_x}{k_\rho} B_y^{(2)}(\overline{k}_\perp - \overline{k}_{\perp mi})\Big]$$

$$+ \sum_{m=-\infty}^{\infty} F * F(\overline{k}_\perp - \overline{k}_{\perp mi}) \sum_{n=-\infty}^{\infty} Q_{mn}^{+}(k_z) \left(-\frac{1}{2}\right) k_z^2$$

$$\left[-\left[\frac{k_z}{k}\frac{k_x}{k_\rho} - \frac{k_\rho}{k}\frac{(n-m)\frac{2\pi}{p}}{k_z} + \frac{k_\rho}{k}\frac{(k_x - k_{xmi})}{k_z} \right] \alpha_{xn}^{(0)} \right.$$

$$- \left[\frac{k_z}{k}\frac{k_y}{k_\rho} + \frac{k_\rho}{k}\frac{(k_y - k_{yi})}{k_z} \right] \alpha_{yn}^{(0)} - \frac{k_y}{k_\rho}\beta_{xn}^{(0)} + \frac{k_x}{k_\rho}\beta_{yn}^{(0)} \Bigg]$$

$$- \sum_{m=-\infty}^{\infty} ik_z Q_{m0}^{+}(k_z) \left[- \left(\frac{k_z}{k}\frac{k_x}{k_\rho} + \frac{k_\rho}{k}\frac{m\frac{2\pi}{p}}{k_z} \right) F * A_x^{(1)}(\bar{k}_\perp - \bar{k}_{\perp mi}) \right.$$

$$- \frac{k_z}{k}\frac{k_y}{k_\rho} F * A_y^{(1)}(\bar{k}_\perp - \bar{k}_{\perp mi})$$

$$- \frac{k_y}{k_\rho} F * B_x^{(1)}(\bar{k}_\perp - \bar{k}_{\perp mi}) + \frac{k_x}{k_\rho} F * B_y^{(1)}(\bar{k}_\perp - \bar{k}_{\perp mi})$$

$$- \frac{k_\rho}{kk_z}[(k_x - k_{xmi})F(\bar{k}_\perp - \bar{k}_{\perp mi})] * A_x^{(1)}(\bar{k}_\perp - \bar{k}_{\perp mi})$$

$$\left. - \frac{k_\rho}{kk_z}[(k_y - k_{ymi})F(\bar{k}_\perp - \bar{k}_{\perp mi})] * A_y^{(1)}(\bar{k}_\perp - \bar{k}_{\perp mi}) \right] \Bigg\} \qquad (C2b)$$

Appendix D. Narrow-Band Gaussian Random Process

The narrow-band Gaussian random process can be expressed as [Davenport and Root, 1958]

$$A(x)\cos\left[\frac{2\pi}{P}x + \psi(x)\right] = \xi_c(x)\cos\frac{2\pi}{P}x - \xi_s \sin\frac{2\pi}{P}x \qquad (D1)$$

where $\xi_c(x)$ and $\xi_s(x)$ are independent Gaussian

$$\langle \xi_c(x) \rangle = \langle \xi_s(x) \rangle = 0 \qquad (D2)$$

and

$$\langle \xi_c(x_1)\xi_c(x_2) \rangle = \langle \xi_s(x_1)\xi_s(x_2) \rangle = \sigma_x^2 C_x(|x_1 - x_2|) \qquad (D3)$$

$$\langle \xi_c(x_1)\xi_s(x_2) \rangle = 0 \qquad (D4)$$

where σ_x is the standard deviation of $\xi_c(x)$ and $\xi_s(x)$ and $C_x(|x_1 - x_2|)$ is the normalized correlation function. The covariance of narrow-band Gaussian random processes at x_1 and x_2 is given by

$$\left\langle A(x_1)\cos\left[\frac{2\pi}{P}x_1 + \psi(x_1)\right] A(x_2)\cos\left[\frac{2\pi}{P}x_2 + \psi(x_2)\right] \right\rangle$$

$$= \sigma_x^2 C_x(|x_1 - x_2|)\cos\frac{2\pi}{P}(x_1 - x_2) \qquad (D5)$$

The probability density function in terms of A and ψ is given by

$$P(A,\psi) = \begin{cases} \frac{A}{2\pi\sigma_x^2} \exp\left(-\frac{A^2}{2\sigma_x^2}\right) & \text{for } A > 0, \quad 0 \le \psi \le 2\pi \\ 0 & \text{otherwise} \end{cases} \tag{D6}$$

Appendix E. Calculations of $|\langle I\rangle|^2$ and D_I

The integral I is given by

$$I = \int_{A_o} \exp(i\bar{k}_d \cdot \bar{r}')d\bar{r}'_\perp \tag{E1}$$

The ensemble average of I is given by

$$\langle I\rangle = \int_{A_o} dx'dy' \exp(ik_{dx}x' + ik_{dy}y')\langle \exp[ik_{dz}f(x',y')]\rangle \tag{E2}$$

with

$$\begin{aligned}\langle \exp[ik_{dz}f(x',y')]\rangle = \exp&\left[-\frac{1}{2}k_{dz}^2(\sigma^2+\sigma_x^2)\right] \\ \times \exp&\left[ik_{dz}B\cos\left(\frac{2\pi}{P}x'+\phi\right)\right]\end{aligned} \tag{E3}$$

$$\sigma^2 = \langle \xi^2\rangle \tag{E4}$$

$$\sigma_x^2 = \langle \xi_c^2\rangle = \langle \xi_s^2\rangle \tag{E5}$$

Therefore,

$$\begin{aligned}\langle I\rangle = &4L_xL_y \exp\left[-\frac{1}{2}k_{dz}^2(\sigma^2+\sigma_x^2)\right] \sum_{n=-\infty}^{\infty} a_n \\ &\times \operatorname{sinc}\left[\left(k_{dz}+n\frac{2\pi}{P}\right)L_x\right] \operatorname{sinc}[k_{dy}L_y]\end{aligned} \tag{E6}$$

where

$$a_n = (-1)^n J_n(-k_{dz}B)\exp\left[in\left(\frac{\pi}{2}+\phi\right)\right] \tag{E7}$$

J_n is the nth-order Bessel function, and $2L_x$ and $2L_y$ are the lengths of the rough surface in the x and y directions, respectively, so that

$$A_o = 4L_xL_y \tag{E8}$$

Assuming that the area illuminated contains many periods $(L_x, L_y \gg P)$, we have

$$|\langle I\rangle|^2 \simeq 16 L_x^2 L_y^2 \exp\left[-k_{dz}^2(\sigma^2+\sigma_x^2)\right] \sum_{n=-\infty}^{\infty} |a_n|^2 \times \mathrm{sinc}^2\left[\left(k_{dz}+n\frac{2\pi}{P}\right)L_x\right] \mathrm{sinc}^2\left[k_{dy}L_y\right] \tag{E9}$$

By allowing L_x and L_y to approach infinity in the above equation, we obtain

$$|\langle I\rangle|^2 = 4\pi^2 A_o \exp\left[-k_{dz}^2(\sigma^2+\sigma_x^2)\right] \sum_{n=-\infty}^{\infty} |a_n|^2 \,\delta\left(k_{dx}+n\frac{2\pi}{P}\right)\delta\left(k_{dy}\right) \tag{E10}$$

where δ is the Dirac delta function.

$$\langle II^*\rangle = \int_{A_o} d\bar{r}_\perp \int_{A_O} d\bar{r}'_\perp \exp\left[i\bar{k}_{d\perp}\cdot(\bar{r}_\perp-\bar{r}'_\perp\right] \langle \exp\left(i\bar{k}_{dz}[f(x,y)-f(x',y')]\right)\rangle \tag{E11}$$

With the change of variables we have

$$\begin{aligned}\langle II^*\rangle = \frac{1}{4}\int_{-2L_x+|x|}^{2L_x-|x|} dx' \int_{-2L_y+|y|}^{2L_y-|y|} dy' \int_{-2L_x}^{2L_x} dx \int_{-2L_y}^{2L_y} dy \\ \times \exp\left[i\bar{k}_{d\perp}\cdot\bar{r}_\perp\right] \sum_{n=-\infty}^{\infty} a_n(x)\exp\left[i\frac{n}{2}\frac{2\pi}{P}x'\right] \\ \times \exp\left[-k_{dz}^2\sigma^2+k_{dz}^2\sigma^2 C(\bar{r}_\perp)\right] \\ \times \exp\left[-k_{dz}^2\sigma_x^2+k_{dz}^2\sigma_x^2\cos\frac{2\pi}{P}x\, C_x(x)\right]\end{aligned} \tag{E12}$$

where

$$a_n(x) = J_n\left[2k_{dz}B\sin\left(\frac{\pi}{P}x\right)\right]$$

Expanding $a_n(x)$ and carrying out the dx' and dy' integrations, we obtain

$$\langle II^*\rangle = \frac{1}{2}\int_{-2L_x}^{2L_x} dx \int_{-2L_y}^{2L_y} dy \sum_{n=-\infty}^{\infty}\sum_{\mu=-\infty}^{\infty} \beta_{n\mu}\exp\left[i\left(\bar{k}_{d\perp}+\hat{x}\mu\frac{2\pi}{P}\right)\cdot\bar{r}_\perp\right]$$

$$[2L_y - |y|] \frac{1}{in\frac{2\pi}{P}} \left\{ \exp\left[in\frac{2\pi}{P}(2L_x - |x|) \right] \right.$$

$$\left. - \exp\left[-in\frac{2\pi}{P}(2L_x - |x|) \right] \right\} \exp\left[-k_{dz}^2\sigma^2 + k_{dz}^2\sigma^2 C(\overline{r}_\perp) \right]$$

$$\times \exp\left[-k_{dz}^2\sigma_x^2 + k_{dz}^2\sigma_x^2 \cos\frac{2\pi}{P}x\, C_x(x) \right] \tag{E13}$$

where

$$\beta_{n\mu} = (-1)^\mu J_{n-\mu}(k_{dz}B)\, J_{n+\mu}(k_{dz}B)$$

It is clear from the above equation that the $n = 0$ term is proportional to L_x, while the $n \neq 0$ term is proportional to P. The argument of the Bessel function $k_{dz}B$ will dictate the number of terms that needs to be summed up. However, if $L_x \gg P$ (we eventually take the limit $L_x \to \infty$ later on), then the $n = 0$ term will make dominant contributions, and other terms will be negligible. Therefore, keeping only the $n = 0$ term, we obtain

$$\langle II^* \rangle = \int_{-2L_x}^{2L_x} dx \int_{-2L_y}^{2L_y} dy \sum_{\mu=-\infty}^{\infty} b_\mu \exp\left[i\left(\overline{k}_{d\perp} + \hat{x}\mu\frac{2\pi}{P} \right) \cdot \overline{r}_\perp \right]$$

$$[2L_x - |x|]\,[2L_y - |y|] \exp\left[-k_{dz}^2\sigma^2 + k_{dz}^2\sigma^2 C(\overline{r}_\perp) \right]$$

$$\times \exp\left[-k_{dz}^2\sigma_x^2 + k_{dz}^2\sigma_x^2 \cos\frac{2\pi}{P}x\, C_x(x) \right] \tag{E14}$$

where

$$b_\mu = J_\mu^2(k_{dz}B) \tag{E15}$$

We assume the correlation functions $C(\overline{r}_\perp)$ and $C_x(x)$ to have a Gaussian form

$$C(\overline{r}_\perp) = \exp\left[-(x^2 + y^2)/l^2 \right] \tag{E16}$$

$$C_x(x) = \exp\left[-x^2/l_x^2 \right] \tag{E17}$$

where l is the correlation length for the random variable $\xi(\overline{r}\perp)$ in the transverse plane, and l_x is the correlation length for the random variables $\xi_c(x)$ and $\xi_s(x)$ in the x direction.

The expressions for the standard deviation of the integral I can now be evaluated in a closed form. Assuming $L_x, L_y \gg l, l_x, P$, we obtain from $(E10)$, $(E14)$, $(E16)$, and $(E17)$,

$$D_I = \langle II^* \rangle - |\langle I \rangle| = \frac{4\pi A_o}{k^2}(D_{I_1} + D_{I_2}) \tag{E18}$$

with

$$D_{I_1} = \frac{k^2}{4} \sum_{\mu=-\infty}^{\infty} b_\mu \exp\left[-k_{dz}^2(\sigma^2 + \sigma_x^2)\right] \sum_{m=1}^{\infty} \frac{(k_{dz}^2\sigma^2)^m}{m!}$$
$$\times \frac{l^2}{m}\left(\exp\left\{-\left[\left(k_{dx} + \mu\frac{2\pi}{P}\right)^2 + k_{dy}^2\right]\frac{l^2}{4m}\right\}\right.$$
$$+ \sum_{m'=1}^{\infty}\sum_{n=0}^{m'} \frac{1}{2^{m'}}\binom{m'}{n}\frac{(k_{dz}^2\sigma_x^2)^{m'}}{m'!}$$
$$\left.\times \frac{1}{q}\exp\left\{-k_{dy}^2\frac{l^2}{4m} - \left[k_{dx} + (\mu + m' - 2n)\frac{2\pi}{P}\right]^2\frac{l^2}{4mq}\right\}\right) \qquad (E19)$$

where

$$q = \sqrt{1 + \frac{l^2 m'}{l_x^2 m}}$$

and

$$D_{I_2} = \frac{k^2}{2}\delta(k_{dy}) \sum_{\mu=-\infty}^{\infty} b_\mu \sum_{m=1}^{\infty}\sum_{n=0}^{m} \frac{1}{2^m}\binom{m}{n} \exp\left[-k_{dz}^2(\sigma^2 + \sigma_x^2)\right]$$
$$\times \sqrt{\frac{\pi}{m}}\, l_x \frac{(k_{dz}^2\sigma_x^2)^m}{m!} \exp\left\{-\left[k_{dx} + (\mu + m - 2n)\frac{2\pi}{P}\right]^2 \frac{l_x^2}{4m}\right\} \qquad (E20)$$

where we made use of

$$\lim_{L_y\to\infty} \frac{L_y}{\pi}\operatorname{sinc}^2(k_{dy}L_y) = \delta(k_{dy}) \qquad (E21)$$

We note that when $l \gg l_x, P$, the expressions for D_{I_1}, (E19), can be simplified to

$$D_{I_1} \simeq \frac{k^2}{4} \sum_{\mu=-\infty}^{\infty} b_\mu \exp\left[-k_{dz}^2(\sigma^2 + \sigma_x^2)\right] \sum_{m=1}^{\infty} \frac{(k_{dz}^2\sigma^2)^m}{m!}\frac{l^2}{m}$$
$$\exp\left\{-\left[\left(k_{dx} + \mu\frac{2\pi}{P}\right)^2 + k_{dy}^2\right]\frac{l^2}{4m}\right\} \qquad (E22)$$

After some manipulations, the above expression further simplifies to

$$D_{I_1} \simeq \frac{1}{P}\int_0^P dx\,\frac{k^2}{4}\,\exp\left[-k_{dz}^2(\sigma^2+\sigma_x^2)\right]\sum_{m=1}^{\infty}\frac{(k_{dz}^2\sigma^2)^m}{m!}\frac{l^2}{m}$$
$$\times\exp\left\{-\left[\left(k_{dx}+k_{dz}2\pi\frac{B}{P}\sin\frac{2\pi}{P}x\right)^2+k_{dy}^2\right]\frac{l^2}{4m}\right\} \quad (E23)$$

The above result is consistent with the result obtained using the incoherent model [Ulaby et al., 1982] where the physical optics solution is averaged over the local slopes. This is due to the fact that when the period P is much larger than the correlation length l, then within the correlation length the periodic component will appear to be planar with the local slope.

Acknowledgments

This work was supported by the NASA Grants NAG5-270 and NAWG-1272, the ARMY Corp of Engineers Contract DACA39-87-K-0022, the NSF Grant 8504381-ECS, and the ONR Contract N00014-83-K-0258.

References

[1] Barrick, D. E., "Relationship between slope probability density function and the physical optics integral in rough surface scattering," *Proc. IEEE,* **56**, 1728–1729, 1968.

[2] Batlivala, P. P. and F. T. Ulaby, "Radar look direction and row crops," *Photogrammetric Eng. Remote Sens.,* **42**, 233–238, 1976.

[3] Beckmann, P. and A. Spizzichino, *The Scattering of Electromagnetic Waves from Rough surfaces,* MacMillan, New York, 1963.

[4] Brown, G. S., "Backscattering from a Gaussian-distributed perfectly conducting rough surface," *IEEE Trans. Ant. Prop.,* **AP-26**, 472–482, 1978.

[5] Chuang, S. L. and J. A. Kong, "Scattering of waves from periodic surfaces," *IEEE Proc.,* **69**, 1132–1144, 1981.

[6] Chuang, S. L. and J. A. Kong, "Wave scattering from a periodic dielectric surface for a general angle of incidence," *Radio Science,* **17**, 545–557, 1982.

[7] Davenport, W. B., Jr. and W. L. Root, *An Introduction to the Theory of Random Signals and Noise*, McGraw-Hill, New York, 1958.

[8] Fenner, R. G., G. F. Pels, and S. C. Reid, "A parametric study of tillage effects on radar backscatter," *Proceedings of the Terrain and Sea Scatter Workshop (TASS),* George Washington Univ., Wash., DC., 1980.

[9] Guissard, A., C. Baufays, and P. Sobieski, "Sea surface description requirements for electromagnetic scattering calculations," *J. Geophys. Res.,* **91**, c2, 2477–2492, 1986.

[10] Hessel, A. and A. A. Oliner, "A New theory of Wood's anomalies on optical gratings," *Applied Optics,* **4**, No. 10, 1275–1297, 1965.

[11] Holzer, J. A. and C. C. Sung, "Scattering of electromagnetic waves from a rough surface II," *J. Appl. Phys.,* **49**, 1002–1011, 1978.

[12] Jordan, A. K. and R. H. Lang, "Electromagnetic scattering patterns from sinusoidal surfaces," *Radio Science,* **14**,1077–1088, 1975.

[13] Kong, J. A., *Electromagnetic Wave Theory,* John Wiley & Sons, New York, 1986.

[14] Leader, J. C., "Bidirectional scattering of electromagnetic waves from rough surfaces," *J. Appl. Phys.,* **42**, 4808–4816, 1971.

[15] Peake, W. H., "Interaction of electromagnetic waves with some natural surfaces," *IEEE Trans. Ant. Prop.,* **AP-7**, Special Supplement, S324–S329, 1959.

[16] Rice, S. O., "Reflection of EM waves by slightly rough surfaces," The Theory of Electromagnetic Waves, M. Kline, Ed., Interscience, NY, 1951.

[17] Sancer, M. I., "Shadow-corrected electromagnetic scattering from a randomly rough surface," *IEEE Trans. Ant. Prop.,* **AP-17**, 577–585, 1969.

[18] Shin, R. T. and J. A. Kong, "Scattering of electromagnetic waves from a randomly perturbed quasiperiodic surface," *J. Appl. Phys.*, **56**, 10–21, 1984.

[19] Stogryn, A., "Electromagnetic scattering from rough finitely conducting surfaces," *Radio Science,* **2**, 415–428, 1967.

[20] Tsang, L. and J. A. Kong, "Energy conservation for reflectivity and transmissivity at a very rough surface," *J. Appl. Phys.,* **51**, 673–680, 1980.

[21] Tsang, L., J. A. Kong, and R. T. Shin, *Theory of Microwave Remote Sensing*, Wiley & Sons, New York, 1985.

[22] Tsang, L. and R. W. Newton, "Microwave emissions from soils with rough surfaces," *J. Geophys. Res.,* **87**, 9017–9024, 1982.

[23] Ulaby, F. T., C. T. Allen, and A. K. Fung, "Method for retrieving the true backscattering coefficient from measurements with a real antenna," IEEE International Geoscience and Remote Sensing Symposium, Munich, **TA-6**, 6.1–6.7, Federal Republic of Germany, 1982.

[24] Ulaby, F. T. and J. E. Bare, "Look direction modulation function of the radar backscattering coefficient of agricultural fields," *Photogrammetric Eng. Remote Sens.,* **45**, 1495–1506, 1979.

[25] Ulaby, F. T., F. Kouyate, A. F. Fung, and A. J. Sieber, "A backscatter model for a randomly perturbed periodic surface," *IEEE Trans. Geosci. Remote Sens.,* **GE-20**, 518–528, 1982.

[26] Valenzuela, G. R., "Depolarization of EM waves by slightly rough surfaces," *IEEE Trans. Ant. Prop.,* **AP-15**, 552–557, 1967.

[27] Valenzeula, G. R., "Scattering of electromagnetic waves from a tilted slightly rough surface," *Radio Science,* **3**, 1057–1064, 1968.

[28] Waterman, P. C., "Scattering by periodic surfaces," *J. Acoust. Soc. Am.,* **57**, 791–802, 1975.

[29] Wright, J. W., "A new model for sea clutter," *IEEE Trans. Ant. Prop.,* **AP-16**, 217–223, 1969.

[30] Wu, S. T. and A. K. Fung, "A noncoherent model for microwave emissions an backscattering from the sea surface," *J. Geophys. Res.,* **77**, 5917–5929, 1972.

5

RADIATIVE TRANSFER THEORY FOR ACTIVE REMOTE SENSING OF TWO-LAYER RANDOM MEDIUM

R. T. Shin and J. A. Kong

5.1 Introduction

The remote sensing of the earth and the elements of its environment at microwave frequencies have been found to contain many practical applications. The primary advantage inherent in remote sensing at microwave frequencies over optical and infrared frequencies is in its all-weather, day-and-night operational capabilities. Active and passive microwave remote sensing with both radar and radiometer have been investigated in areas of snow and ice covered land or water [1–15], vegetation canopy [16–19], cloud and rainfall [20,21], and soil moisture studies [22–30]. While extensive effort has been concentrated in the measurement and collection of voluminous experimental data, theoretical models that are useful in interpreting these data have not been satisfactorily developed, especially where combinations of absorption, scattering, layering and rough surface are important factors. Although past theoretical emphasis has been largely concentrated on rough surface scattering, recent theoretical models have been proposed to account for volume scattering effects [31].

In the active and passive microwave remote sensing of earth terrain, the scattering effects due to medium inhomogeneities and rough interfaces play a dominant role in the determination of brightness temperatures and radar backscattering coefficients. The effects of volume scattering have been treated with two theoretical models for the terrain media: (1) the random medium model where scattering effects can be accounted for by introducing a randomly fluctuating part in the permittivities, and (2) the discrete scatterer model where discrete scatterers are imbedded in a homogeneous background medium.

In the theoretical developments for passive remote sensing the effect of volume scattering due to medium inhomogeneity was first accounted for by Gurvich et al., [32]. They derived expressions for the brightness temperature of a halfspace random medium with a laminar structure, assuming uniform temperature distribution. Tsang and Kong [33] solved the problem of thermal microwave emission from a halfspace random medium with a laminar structure and nonuniform temperature distribution using the radiative transfer theory. England [34] first examined thermal microwave emission from a uniform low-loss dielectric medium containing randomly distributed isotropic scatterers, with a radiative transfer approach. He [35] then considered the more general case of a scattering layer over a homogeneous halfspace, using the radiative transfer theory and a Rayleigh scattering model. Tsang and Kong [36] derived a more general result than that of England for both the halfspace and two-layer case, using a Mie scattering model. With the Born approximation, Tsang and Kong [37] obtained the emissivity of a halfspace random medium with a three-dimensional variation.

In active remote sensing, Stogryn [38] first calculated the bistatic scattering coefficients for a random medium with a spherical correlation function using a perturbation approach. Leader [39] studied scattering from Rayleigh scatterers imbedded in a dielectric slab using the matrix doubling method. Using the Born approximation, Tsang and Kong [36] studied scattering of electromagnetic waves by a halfspace random medium. They [40] also developed the radiative transfer theory to calculate the bistatic scattering coefficients from a halfspace random medium. An iterative approach is used to obtain results to the second order, in order to exhibit depolarizaton of backscattered power. Using the first-order renormalization method, Fung and Fung [41] obtained the bistatic scattering coefficients from a vegetation-like halfspace random medium. Fung [42] then extended the result to the case of a vegetation-like layer over a homogeneous halfspace. Zuniga and Kong [43] studied the scattering from a slab of random medium using the Born approximation. Then, Zuniga et al., [44] extended the result to the second order in albedo to show the depolarization effect in the backscattering direction.

The radiative transfer theory has been useful in the interpretation of remote sensing data [45]. Even though it deals only with the intensities of the field quantities and neglects their coherent nature, it accounts for the multiple scattering and obeys energy conservation. The modified radiative transfer (MRT) theory [46–48] which takes into account the partial coherent effects due to the boundaries has been derived for the cases when the interference effects become important [49]. The MRT equations have been developed for a two-layer random medium with laminar structure by applying the nonlinear approximation to Dyson's equation and the ladder approximation to the Bethe-Salpeter equation [46]. Then, the MRT equations for electromagnetic wave propogation in a two-layer medium with three-dimensional permittivity fluctuations are derived [47,48]. The MRT equations are then solved with the first order renormalization approximation to obtain the backscattering cross sections.

Most of the previous work on volume scattering all assumed planar boundaries, and the effect of rough surface scattering was neglected. However, in order to understand in a more meaningful way the problems of radar backscattering and thermal microwave emission from natural terrains, a composite model that can account for both the volume and surface scattering effects must be studied. Recently, the Rayleigh scattering model has been used with the radiative transfer equations to study the combined volume and rough surface scattering effects [50–52].

In this chapter, we solve the problem of scattering from a two-layer random medium using the radiative transfer theory. The volume scattering effect is first studied in section 5.2 by considering the two-layer random medium with planar interfaces. Using all four Stokes parameters, the bistatic scattering coefficients of two-layer random medium are first calculated using a numerical approach which provides a valid solution for both small and large albedos. A Fourier-series expansion in the azimuthal direction is used to eliminate the azimuthal ϕ-dependence from the radiative transfer equations. Then, the set of equations without the ϕ-dependence is solved using the method of Gaussian quadrature. The integrals in the radiative transfer equations are replaced by a Gaussian quadrature and the resulting system of first-order differential equations is solved by obtaining eigenvalues and eigenvectors and matching the boundary conditions. The order of the system of eigen-equations is reduced for more efficient computation by making use of the symmetry properties of the scattering function matrix. The numerical results are illustrated by plotting backscattering cross sections and the bistatic scattering coefficients as functions of frequency, incident angle, and the scattering angles.

In section 5.3, the combined volume and rough surface scattering effects are studied with a two-layer random medium model with rough interfaces. The rough surface effects are incorporated into the radiative transfer equations by modifying the boundary conditions. The reflected and transmitted bistatic scattering coefficients derived with the randomly rough surface models are used to derived the boundary conditions satisfied by the intensities at the top and bottom interfaces. The radiative transfer equations are again solved numerically using the Fourier-series expansion and the Gaussian quadrature method. The thoeretical results are illustrated by plotting the backscattering cross sections as functions of frequency and incident angle.

5.2 Two-Layer Random Medium with Planar Interfaces

a. Formulation

Consider a layer of random medium characterized by the permittivity $\epsilon_1 + \epsilon_f$, where ϵ_f stands for the randomly fluctuating part whose amplitude is very small and whose ensemble average is zero, on top of a

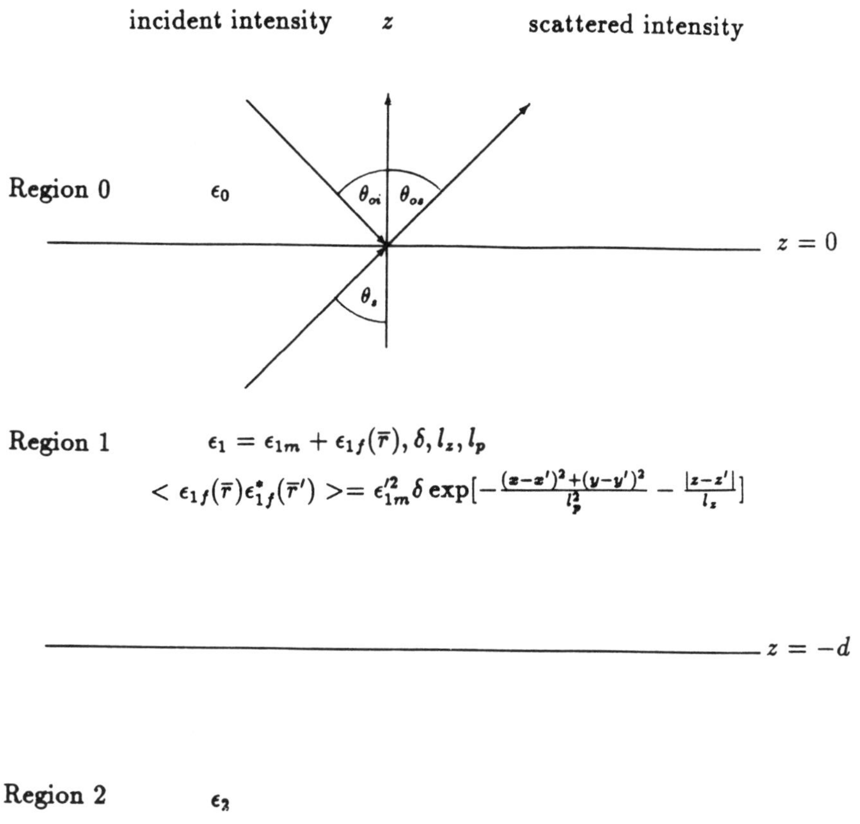

Figure 5.2.1 Geometrical configuration of the problem.

homogeneous medium with permittivity ϵ_2 [Fig. 5.2.1]. The radiative transfer equations which govern the propagation of intensities inside the scattering medium are, for $0 < \theta < \pi$,

$$\cos\theta \frac{d}{dz}\overline{I}(\theta,\phi,z) = -K_a\overline{I}(\theta,\phi,z) - \overline{\overline{K}}_s(\theta)\cdot\overline{I}(\theta,\phi,z) + \int_0^{\pi} d\theta' \sin\theta' \int_0^{2\pi} d\phi' \, \overline{\overline{P}}(\theta,\phi;\theta'\phi')\cdot\overline{I}(\theta',\phi',z) \quad (1)$$

where

$$\overline{I}(\theta,\phi,z)=\begin{bmatrix} I_v(\theta,\phi,z)\\ I_h(\theta,\phi,z)\\ U(\theta,\phi,z)\\ V(\theta,\phi,z)\end{bmatrix} \tag{2}$$

I_v is the vertically polarized specific intensity, I_h is the horizontally polarized specific intensity, and U and V represent the correlation between two polarizations [40,53], $\overline{\overline{P}}(\theta,\phi;\theta',\phi')$ is a 4×4 scattering function matrix, which relates scattered intensities into the direction (θ,ϕ) from the incident intensities in the direction (θ',ϕ'), K_a is the loss per unit length due to absorption, and $\overline{\overline{K}}_s(\theta)$ is the loss per unit length due to scattering. The random permittivity fluctuation is characterized by the variance of the fluctuation δ and the correlation function with lateral correlation length l_ρ and vertical correlation length l_z. The scattering function matrix and the scattering coefficient have been derived by applying Born approximation with the far-field solution and the explicit expressions for the correlation function with gaussian dependence laterally and exponential dependence vertically are given in Appendix A [40].

Consider an incident wave with specific intensity $\overline{I}_{oi}(\pi-\theta_o,\phi_o)$ impinging from region 0, which is assumed to be free space, upon the scattering layer. The incident beam in region 0 assumes the form

$$\overline{I}_{oi}(\pi-\theta_o,\phi_o)=\overline{I}_{oi}\,\delta(\cos\theta_o-\cos\theta_{oi})\,\delta(\phi_o-\phi_{oi}) \tag{3}$$

where the use of Dirac delta function is made.

The boundary conditions for the four Stokes parameters at a planar dielectric interface have been derived [40] from the continuity of tangential electric and magnetic fields. The results are, for $0<\theta<\pi/2$, at $z=0$,

$$\overline{I}(\pi-\theta,\phi,z=0)=\overline{\overline{T}}_{01}(\theta_o)\cdot\overline{I}_{oi}(\pi-\theta_o,\phi_o)+\overline{\overline{R}}_{10}(\theta)\cdot\overline{I}(\theta,\phi,z=0) \tag{4}$$

and, at $z=-d$,

$$\overline{I}(\theta,\phi,z=-d)=\overline{\overline{R}}_{12}(\theta)\cdot\overline{I}(\pi-\theta,\phi,z=-d) \tag{5}$$

where we have broken up intensities in the scattering layer into upward going intensities $\overline{I}(\theta,\phi,z)$ and downward going intensities $\overline{I}(\pi-$

$\theta, \phi, z)$. In the above equations, $\overline{\overline{T}}_{01}(\theta_o)$ represents the coupling from region 0 to region 1, $\overline{\overline{R}}_{10}(\theta)$ represents the coupling from upward going intensities into downward going intensities at the boundary of region 1 and region 0, and $\overline{\overline{R}}_{12}(\theta)$ represents similar coupling at the boundary of region 1 and region 2 [Appendix B].

Once the radiative transfer equations are solved subject to the boundary conditions (4) and (5), the intensity in the direction (θ_{os}, ϕ_{os}) in region 0 is

$$\overline{I}_o(\theta_{os}, \phi_{os}) = \overline{\overline{T}}_{10}(\theta_s) \cdot \overline{I}(\theta_s, \phi_s, z = 0) + \overline{\overline{R}}_{01}(\theta_o) \cdot \overline{I}_{oi}(\pi - \theta_o, \phi_o) \quad (6)$$

where $\overline{\overline{T}}_{10}(\theta)$ represents the coupling from region 1 to region 0. The bistatic scattering coefficients $\gamma_{\beta\alpha}(\theta_{os}, \phi_{os}; \theta_{oi}, \phi_{oi})$ are defined as the ratio of the scattered power of polarization β per unit solid angle in the direction (θ_{os}, ϕ_{os}) and the intercepted incident power of polarization α in the direction (θ_{oi}, ϕ_{oi}) averaged over 4π radians [54].

$$\gamma_{\beta\alpha}(\theta_{os}, \phi_{os}; \theta_{oi}, \phi_{oi}) = 4\pi \frac{\cos\theta_{os}\, I_{o\beta s}(\theta_{os}, \phi_{os})}{\cos\theta_{oi}\, I_{o\alpha i}} \quad (7)$$

where $\alpha, \beta = v$ or h with v denoting vertical polarization and h denoting horizontal polarization. In the backscattering direction $\theta_{os} = \theta_{oi}$ and $\phi_{os} = \pi - \phi_{oi}$. The backscattering cross sections per unit area are defined to be

$$\sigma_{\beta\alpha}(\theta_{oi}) = \cos\theta_{oi}\, \gamma_{\beta\alpha}(\theta_{oi}, \pi + \phi_{oi}; \theta_{oi}, \phi_{oi}) \quad (8)$$

b. Fourier Series Expansion

The radiative transfer equations can be solved using an iterative approach which gives closed form solutions [40,53] when the effect of scattering is small (small albedo). The radiative transfer equations and the boundary conditions are cast into the integral equation form, then an iterative process is applied to solve the integral equation to both the first and second order in albedo. The depolarization of the backscattered intensities has been shown to be the second-order effect. However, for the general cases when the effect of scattering is not small, we must resort to the numerical approach to solve the radiative transfer equations. We first use a Fourier-series expansion in the azimuthal direction to eliminate the ϕ-dependence from the radiative transfer equations. We let

$$\overline{I}(\theta,\phi,z) = \sum_{m=0}^{\infty} \left[\overline{I}^{mc}(\theta,z)\cos m(\phi-\phi_i) + \overline{I}^{ms}(\theta,z)\sin m(\phi-\phi_i) \right] \tag{9}$$

$$\overline{\overline{P}}(\theta,\phi;\theta'\phi') = \sum_{m=0}^{\infty} \frac{1}{(1+\delta_m)\pi} \times \left[\overline{\overline{P}}^{mc}(\theta,\theta')\cos m(\phi-\phi') + \overline{\overline{P}}^{ms}(\theta,\theta')\sin m(\phi-\phi') \right] \tag{10}$$

where superscript m indicates the order of harmonics in the azimuthal direction, superscripts c and s indicate the cosine and sine dependence, and Neumann number $\delta_m = 0$ for $m \neq 0$ and $\delta_o = 1$. Also note that the zeroth-order sine dependence terms are zero.

$$\overline{I}^{os}(\theta,z) = 0 \tag{11a}$$

$$\overline{\overline{P}}^{os}(\theta,\theta') = 0 \tag{11b}$$

Substituting (9) and (10) into the radiative transfer equations, the ϕ'-integration can be carried out. Then, by collecting terms with the same sine or cosine dependence, we obtain a set of equations without the ϕ dependence. For $m = 0,1,2,\cdots$

$$\cos\theta \frac{d}{dz}\overline{I}^{mc}(\theta,z) = -K_a\overline{I}^{mc}(\theta,z) - \overline{\overline{K}}_s(\theta)\cdot\overline{I}^{mc}(\theta,z) + \int_0^{\pi} d\theta' \sin\theta' \times \left[\overline{\overline{P}}^{mc}(\theta,\theta')\cdot\overline{I}^{mc}(\theta',z) - \overline{\overline{P}}^{ms}(\theta,\theta')\cdot\overline{I}^{ms}(\theta',z) \right] \tag{12a}$$

$$\cos\theta \frac{d}{dz}\overline{I}^{ms}(\theta,z) = -K_a\overline{I}^{ms}(\theta,z) - \overline{\overline{K}}_s(\theta)\cdot\overline{I}^{ms}(\theta,z) + \int_0^{\pi} d\theta' \sin\theta' \times \left[\overline{\overline{P}}^{ms}(\theta,\theta')\cdot\overline{I}^{mc}(\theta',z) + \overline{\overline{P}}^{mc}(\theta,\theta')\cdot\overline{I}^{ms}(\theta',z) \right] \tag{12b}$$

The closed form expressions for the Fourier-series expanded scattering function matrices $\overline{\overline{P}}^{mc}(\theta,\theta')$ and $\overline{\overline{P}}^{ms}(\theta,\theta')$ are obtained [Appendix C]. We note that for azimuthally isotropic media the scattering function matrix can be expanded as

$$\overline{\overline{P}}^{mc}(\theta,\theta') = \begin{bmatrix} P_{11}^{mc} & P_{12}^{mc} & 0 & 0 \\ P_{21}^{mc} & P_{22}^{mc} & 0 & 0 \\ 0 & 0 & P_{33}^{mc} & P_{34}^{mc} \\ 0 & 0 & P_{43}^{mc} & P_{44}^{mc} \end{bmatrix} \tag{13a}$$

$$\overline{\overline{P}}^{ms}(\theta,\theta') = \begin{bmatrix} 0 & 0 & P_{13}^{ms} & P_{14}^{ms} \\ 0 & 0 & P_{23}^{ms} & P_{24}^{ms} \\ P_{31}^{ms} & P_{32}^{ms} & 0 & 0 \\ P_{41}^{ms} & P_{42}^{ms} & 0 & 0 \end{bmatrix} \tag{13b}$$

Thus, the coupled equations (12a) and (12b) can be changed into two decoupled equations by defining

$$\overline{I}^{me}(\theta,z) = \begin{bmatrix} I_v^{mc}(\theta,z) \\ I_h^{mc}(\theta,z) \\ U^{ms}(\theta,z) \\ V^{ms}(\theta,z) \end{bmatrix} \tag{14a}$$

$$\overline{I}^{mo}(\theta,z) = \begin{bmatrix} I_v^{ms}(\theta,z) \\ I_h^{ms}(\theta,z) \\ U^{mc}(\theta,z) \\ V^{mc}(\theta,z) \end{bmatrix} \tag{14b}$$

where superscripts e and o stands for even or odd dependence in the first two Stokes parameters. Decoupled equations are given by, for $m = 0, 1, 2, \cdots$

$$\begin{aligned} \cos\theta \frac{d}{dz}\overline{I}^{me}(\theta,z) = & - K_a \overline{I}^{me}(\theta,z) - \overline{\overline{K}}_s(\theta)\cdot\overline{I}^{me}(\theta,z) \\ & + \int_0^\pi d\theta' \sin\theta' \overline{\overline{P}}^{me}(\theta,\theta')\cdot\overline{I}^{me}(\theta',z) \end{aligned} \tag{15a}$$

$$\begin{aligned} \cos\theta \frac{d}{dz}\overline{I}^{mo}(\theta,z) = & - K_a \overline{I}^{mo}(\theta,z) - \overline{\overline{K}}_s(\theta)\cdot\overline{I}^{mo}(\theta,z) \\ & + \int_0^\pi d\theta' \sin\theta' \overline{\overline{P}}^{mo}(\theta,\theta')\cdot\overline{I}^{mo}(\theta',z) \end{aligned} \tag{15b}$$

where

$$\overline{\overline{P}}^{me}(\theta,\theta') = \begin{bmatrix} P_{11}^{mc} & P_{12}^{mc} & -P_{13}^{ms} & -P_{14}^{ms} \\ P_{21}^{mc} & P_{22}^{mc} & -P_{23}^{ms} & -P_{24}^{ms} \\ P_{31}^{ms} & P_{32}^{ms} & P_{33}^{mc} & P_{34}^{mc} \\ P_{41}^{ms} & P_{42}^{ms} & P_{43}^{mc} & P_{44}^{mc} \end{bmatrix} \tag{16a}$$

$$\overline{\overline{P}}^{mo}(\theta,\theta') = \begin{bmatrix} P_{11}^{mc} & P_{12}^{mc} & P_{13}^{ms} & P_{14}^{ms} \\ P_{21}^{mc} & P_{22}^{mc} & P_{23}^{ms} & P_{24}^{ms} \\ P_{31}^{ms} & P_{32}^{ms} & P_{33}^{mc} & P_{34}^{mc} \\ P_{41}^{ms} & P_{42}^{ms} & P_{43}^{mc} & P_{44}^{mc} \end{bmatrix} \tag{16b}$$

In order to derive the boundary conditions for the Fourier-series expanded intensities, we first expand the incident intensity $\overline{I}_{oi}(\pi - \theta_o, \phi_{oi})$ into the Fourier series:

$$\begin{aligned} \overline{I}_{oi}(\pi - \theta_o, \phi_o) &= \overline{I}_{oi}\, \delta(\cos\theta_o - \cos\theta_{oi})\, \delta(\phi_o - \phi_{oi}) \\ &= \overline{I}_{oi}\, \delta(\cos\theta_o - \cos\theta_{oi}) \sum_{m=0}^{\infty} \frac{1}{(\delta_m + 1)\pi} \cos m(\phi_o - \phi_{oi}) \end{aligned} \tag{17}$$

Substituting the above equation along with the Fourier-series expanded intensities into the boundary conditions in (4) and (5) and collecting terms with the same azimuthal dependence, the boundary conditions for each harmonic can be obtained. The results are, for $\alpha = e$ or o, and $0 < \theta < \pi/2$, at $z = 0$

$$\overline{I}^{m\alpha}(\pi - \theta, z = 0) = \overline{\overline{T}}_{01}(\theta_o) \cdot \overline{I}_{oi}^{m\alpha}(\pi - \theta_o) + \overline{\overline{R}}_{10}(\theta) \cdot \overline{I}^{m\alpha}(\theta, z = 0) \tag{18}$$

and at $z = -d$

$$\overline{I}^{m\alpha}(\theta, z = -d) = \overline{\overline{R}}_{12}(\theta) \cdot \overline{I}^{m\alpha}(\pi - \theta, z = -d) \tag{19}$$

where

$$\overline{I}_{oi}^{m\alpha}(\pi - \theta_o) = \overline{I}_{oi}^{m\alpha} \frac{1}{(\delta_m + 1)\pi} \delta(\cos\theta_o - \cos\theta_{oi}) \tag{20}$$

with

$$\overline{I}_{oi}^{me} = \begin{bmatrix} I_{vi} \\ I_{hi} \\ 0 \\ 0 \end{bmatrix} \tag{21a}$$

$$\overline{I}_{oi}^{mo} = \begin{bmatrix} 0 \\ 0 \\ U_i \\ V_i \end{bmatrix} \tag{21b}$$

We define m_{max} to be the number of harmonics that has to be kept in the expansion of the scattering function matrix such that

$$\overline{\overline{P}}^{mc}(\theta,\theta') \simeq \overline{\overline{P}}^{ms}(\theta,\theta') \simeq 0 \qquad \text{for} \quad m > m_{max} \tag{22}$$

Then, for $m > m_{max}$ the radiative transfer equations simplify to

$$\cos\theta \frac{d}{dz}\overline{I}^{me}(\theta,z) = -K_a\overline{I}^{me} - \overline{\overline{K}}_s(\theta)\cdot\overline{I}^{me}(\theta,z) \tag{23a}$$

$$\cos\theta \frac{d}{dz}\overline{I}^{mo}(\theta,z) = -K_a\overline{I}^{mo} - \overline{\overline{K}}_s(\theta)\cdot\overline{I}^{mo}(\theta,z) \tag{23b}$$

and the solutions to these equations can be obtained analytically, without resorting to the numerical approach.

c. Gaussian Quadrature Method

The set of decoupled radiative transfer equations without the azimuthal dependence for each harmonic can be solved numerically using the Gaussian quadrature method. The integrals in the radiative transfer equations are replaced by a Gaussian quadrature, an appropriately weighted sum over $2n$ intervals between $2n$ zeroes of the even-order Legendre polynomial $P_{2n}(\theta)$. The resulting system of $8n$ first-order differential equations are solved by obtaining eigenvalues and eigenvectors and matching the boundary conditions. In obtaining the eigenvalues and eigenvectors, the order of system of equations can be reduced by a factor of two to $4n$ equations by making use of the symmetry properties of the scattering function matrix and noting that the eigenvalues occur in pairs such that if ξ is an eigenvalue, so is $-\xi$. We first break up the radiative transfer equations into two set of equations by defining, for $\alpha = e$, $\beta = o$ or $\alpha = o$, $\beta = e$, and $m = 0, 1, 2, \cdots$

$$\overline{I}_1(\theta,z) \equiv \begin{bmatrix} I_{1_1}(\theta,z) \\ I_{1_2}(\theta,z) \end{bmatrix} = \begin{bmatrix} I_v^{m\alpha}(\theta,z) \\ I_h^{m\alpha}(\theta,z) \end{bmatrix} \tag{24a}$$

$$\overline{I}_2(\theta, z) \equiv \begin{bmatrix} I_{2_1}(\theta, z) \\ I_{2_2}(\theta, z) \end{bmatrix} = \begin{bmatrix} U^{m\beta}(\theta, z) \\ V^{m\beta}(\theta, z) \end{bmatrix} \tag{24b}$$

$$\overline{\overline{P}}_{11}(\theta, \theta') \equiv \begin{bmatrix} P_{11_{11}} & P_{11_{12}} \\ P_{11_{21}} & P_{11_{22}} \end{bmatrix} = \begin{bmatrix} P_{11}^{m\alpha} & P_{12}^{m\alpha} \\ P_{21}^{m\alpha} & P_{22}^{m\alpha} \end{bmatrix} \tag{25a}$$

$$\overline{\overline{P}}_{12}(\theta, \theta') \equiv \begin{bmatrix} P_{12_{11}} & P_{12_{12}} \\ P_{12_{21}} & P_{12_{22}} \end{bmatrix} = \begin{bmatrix} P_{13}^{m\alpha} & P_{14}^{m\alpha} \\ P_{23}^{m\alpha} & P_{24}^{m\alpha} \end{bmatrix} \tag{25b}$$

$$\overline{\overline{P}}_{21}(\theta, \theta') \equiv \begin{bmatrix} P_{21_{11}} & P_{21_{12}} \\ P_{21_{21}} & P_{21_{22}} \end{bmatrix} = \begin{bmatrix} P_{31}^{m\alpha} & P_{32}^{m\alpha} \\ P_{41}^{m\alpha} & P_{42}^{m\alpha} \end{bmatrix} \tag{25c}$$

$$\overline{\overline{P}}_{22}(\theta, \theta') \equiv \begin{bmatrix} P_{22_{11}} & P_{22_{12}} \\ P_{22_{21}} & P_{22_{22}} \end{bmatrix} = \begin{bmatrix} P_{33}^{m\alpha} & P_{34}^{m\alpha} \\ P_{43}^{m\alpha} & P_{44}^{m\alpha} \end{bmatrix} \tag{25d}$$

We then have

$$\cos\theta \frac{d}{dz}\overline{I}_1(\theta, z) = -\overline{\overline{K}}_{e1}(\theta) + \int_0^\pi d\theta' \sin\theta' \times \left[\overline{\overline{P}}_{11}(\theta, \theta') \cdot \overline{I}_1(\theta', z) + \overline{\overline{P}}_{12}(\theta, \theta') \cdot \overline{I}_2(\theta', z)\right] \tag{26a}$$

$$\cos\theta \frac{d}{dz}\overline{I}_2(\theta, z) = -\overline{\overline{K}}_{e2}(\theta) + \int_0^\pi d\theta' \sin\theta' \times \left[\overline{\overline{P}}_{21}(\theta, \theta') \cdot \overline{I}_1(\theta', z) + \overline{\overline{P}}_{22}(\theta, \theta') \cdot \overline{I}_2(\theta', z)\right] \tag{26b}$$

where

$$\overline{\overline{K}}_{e1}(\theta) = K_a + \overline{\overline{K}}_{s1} \tag{27a}$$

$$\overline{\overline{K}}_{e2}(\theta) = K_a + \overline{\overline{K}}_{s2} \tag{27b}$$

with

$$\overline{\overline{K}}_{s1} = \begin{bmatrix} K_v & 0 \\ 0 & K_h \end{bmatrix} \tag{28a}$$

$$\overline{\overline{K}}_{s2} = \begin{bmatrix} \frac{K_v + K_h}{2} & 0 \\ 0 & \frac{K_v + K_h}{2} \end{bmatrix} \tag{28b}$$

The scattering function matrix can be shown to satisfy the following properties. For $\alpha, \beta = 1$ or 2,

$$\overline{\overline{P}}_{\alpha\alpha}(\theta, \theta') = \overline{\overline{P}}_{\alpha\alpha}(\pi - \theta, \pi - \theta') \tag{29a}$$

$$\overline{\overline{P}}_{\alpha\alpha}(\pi - \theta, \theta') = \overline{\overline{P}}_{\alpha\alpha}(\theta, \pi - \theta') \tag{29b}$$

and for $\alpha \neq \beta$

$$\overline{\overline{P}}_{\alpha\beta}(\theta, \theta') = -\overline{\overline{P}}_{\alpha\beta}(\pi - \theta, \pi - \theta') \tag{30a}$$

$$\overline{\overline{P}}_{\alpha\beta}(\pi - \theta, \theta') = -\overline{\overline{P}}_{\alpha\beta}(\theta, \pi - \theta') \tag{30b}$$

Further breaking up the intensities into upward and downward propagating intensities, denoted by superscripts $+$ and $-$, respectively, and applying the Gaussian quadrature method, we obtain the following set of equations by making use of the symmetry properties of the scattering function matrix:

$$\overline{\overline{\mu}} \cdot \frac{d}{dz}\overline{I}_1^+ = -\overline{\overline{K}}_{e1} \cdot \overline{I}_1^+ + \overline{\overline{F}}_{11} \cdot \overline{I}_1^+ + \overline{\overline{B}}_{11} \cdot \overline{I}_1^- + \overline{\overline{F}}_{12} \cdot \overline{I}_2^+ + \overline{\overline{B}}_{12} \cdot \overline{I}_2^- \tag{31a}$$

$$-\overline{\overline{\mu}} \cdot \frac{d}{dz}\overline{I}_1^- = -\overline{\overline{K}}_{e1} \cdot \overline{I}_1^- + \overline{\overline{B}}_{11} \cdot \overline{I}_1^+ + \overline{\overline{F}}_{11} \cdot \overline{I}_1^- - \overline{\overline{B}}_{12} \cdot \overline{I}_2^+ - \overline{\overline{F}}_{12} \cdot \overline{I}_2^- \tag{31b}$$

$$\overline{\overline{\mu}} \cdot \frac{d}{dz}\overline{I}_2^+ = -\overline{\overline{K}}_{e2} \cdot \overline{I}_2^+ + \overline{\overline{F}}_{21} \cdot \overline{I}_1^+ + \overline{\overline{B}}_{21} \cdot \overline{I}_1^- + \overline{\overline{F}}_{22} \cdot \overline{I}_2^+ + \overline{\overline{B}}_{22} \cdot \overline{I}_2^- \tag{32a}$$

$$-\overline{\overline{\mu}} \cdot \frac{d}{dz}\overline{I}_2^- = -\overline{\overline{K}}_{e2} \cdot \overline{I}_2^- - \overline{\overline{B}}_{21} \cdot \overline{I}_1^+ - \overline{\overline{F}}_{21} \cdot \overline{I}_1^- + \overline{\overline{B}}_{22} \cdot \overline{I}_2^+ + \overline{\overline{F}}_{22} \cdot \overline{I}_2^- \tag{32b}$$

where, for $\alpha, \beta = 1, 2$, I_α^+ and I_α^- are $2n \times 1$ matrices

$$\overline{I}_\alpha^+ = \begin{bmatrix} I_{\alpha_1}(\mu_1, z) \\ \vdots \\ I_{\alpha_1}(\mu_n, z) \\ I_{\alpha_2}(\mu_1, z) \\ \vdots \\ I_{\alpha_2}(\mu_n, z) \end{bmatrix} \qquad \overline{I}_\alpha^- = \begin{bmatrix} I_{\alpha_1}(-\mu_1, z) \\ \vdots \\ I_{\alpha_1}(-\mu_n, z) \\ I_{\alpha_2}(-\mu_1, z) \\ \vdots \\ I_{\alpha_2}(-\mu_n, z) \end{bmatrix} \tag{33}$$

and $\overline{\overline{F}}_{\alpha\beta}$ and $\overline{\overline{B}}_{\alpha\beta}$ are $2n \times 2n$ matrices

$$\overline{\overline{F}}_{\alpha\beta} = \begin{bmatrix} P_{\alpha\beta_{11}}(\mu_1,\mu_1) & \cdots & P_{\alpha\beta_{11}}(\mu_1,\mu_n) & P_{\alpha\beta_{12}}(\mu_1,\mu_1) & \cdots & P_{\alpha\beta_{12}}(\mu_1,\mu_n) \\ \vdots & \vdots & \vdots & \vdots & \vdots & \vdots \\ P_{\alpha\beta_{11}}(\mu_n,\mu_1) & \cdots & P_{\alpha\beta_{11}}(\mu_n,\mu_n) & P_{\alpha\beta_{12}}(\mu_n,\mu_1) & \cdots & P_{\alpha\beta_{12}}(\mu_n,\mu_n) \\ P_{\alpha\beta_{21}}(\mu_1,\mu_1) & \cdots & P_{\alpha\beta_{21}}(\mu_1,\mu_n) & P_{\alpha\beta_{22}}(\mu_1,\mu_1) & \cdots & P_{\alpha\beta_{22}}(\mu_1,\mu_n) \\ \vdots & \vdots & \vdots & \vdots & \vdots & \vdots \\ P_{\alpha\beta_{21}}(\mu_n,\mu_1) & \cdots & P_{\alpha\beta_{21}}(\mu_n,\mu_n) & P_{\alpha\beta_{22}}(\mu_n,\mu_1) & \cdots & P_{\alpha\beta_{22}}(\mu_n,\mu_n) \end{bmatrix} \tag{34}$$

$$\overline{\overline{B}}_{\alpha\beta} = \begin{bmatrix} P_{\alpha\beta_{11}}(\mu_1,-\mu_1) & \cdots & P_{\alpha\beta_{11}}(\mu_1,-\mu_n) & P_{\alpha\beta_{12}}(\mu_1,-\mu_1) & \cdots & P_{\alpha\beta_{12}}(\mu_1,-\mu_n) \\ \vdots & \vdots & \vdots & \vdots & \vdots & \vdots \\ P_{\alpha\beta_{11}}(\mu_n,-\mu_1) & \cdots & P_{\alpha\beta_{11}}(\mu_n,-\mu_n) & P_{\alpha\beta_{12}}(\mu_n,-\mu_1) & \cdots & P_{\alpha\beta_{12}}(\mu_n,-\mu_n) \\ P_{\alpha\beta_{21}}(\mu_1,-\mu_1) & \cdots & P_{\alpha\beta_{21}}(\mu_1,-\mu_n) & P_{\alpha\beta_{22}}(\mu_1,-\mu_1) & \cdots & P_{\alpha\beta_{22}}(\mu_1,-\mu_n) \\ \vdots & \vdots & \vdots & \vdots & \vdots & \vdots \\ P_{\alpha\beta_{21}}(\mu_n,-\mu_1) & \cdots & P_{\alpha\beta_{21}}(\mu_n,-\mu_n) & P_{\alpha\beta_{22}}(\mu_n,-\mu_1) & \cdots & P_{\alpha\beta_{22}}(\mu_n,-\mu_n) \end{bmatrix} \tag{35}$$

and $\overline{\overline{\mu}}$ and $\overline{\overline{a}}$ are the $2n \times 2n$ diagonal matrices

$$\overline{\overline{\mu}} = \mathrm{diag}[\mu_1, \cdots, \mu_n, \mu_1, \cdots, \mu_n] \tag{36}$$

$$\overline{\overline{a}} = \mathrm{diag}[a_1, \cdots, a_n, a_1, \cdots, a_n] \tag{37}$$

In the above equation $\pm\mu_i$ are the zeroes of the Legendre polynomial $P_{2n}(\mu)$ and a_i are the corresponding Christoffel weighting functions and we made use of the relations $a_i = a_{-i}$ and $\mu_i = -\mu_{-i}$.

The system of $8n$ first-order differential equations, (31) and (32), can be cast into more compact form by defining $4n \times 1$ matrices $\overline{I}_a$ and $\overline{I}_s$

$$\overline{I}_a = \begin{bmatrix} \overline{I}_{a1} \\ \overline{I}_{s2} \end{bmatrix} \qquad \overline{I}_s = \begin{bmatrix} \overline{I}_{s1} \\ \overline{I}_{a2} \end{bmatrix} \tag{38}$$

such that upward propagating intensity $\overline{I}^+$ is given by

$$\overline{I}^+ \equiv \begin{bmatrix} \overline{I}_1^+ \\ \overline{I}_2^+ \end{bmatrix} = \frac{1}{2}[\overline{I}_a + \overline{I}_s] \tag{39}$$

and where, for $l = 1, 2$, $\overline{I}_{al} = \overline{I}_l^+ + \overline{I}_l^-$ and $\overline{I}_{sl} = \overline{I}_l^+ - \overline{I}_l^-$. Then, from (31) and (32), we obtain

$$\overline{\overline{\mu}} \cdot \frac{d}{dz}\overline{I}_a = \overline{\overline{W}} \cdot \overline{I}_s \tag{40a}$$

$$\overline{\overline{\mu}} \cdot \frac{d}{dz}\overline{I}_s = \overline{\overline{A}} \cdot \overline{I}_a \tag{40b}$$

where $\overline{\overline{W}}$ and $\overline{\overline{A}}$ are the $4n \times 4n$ matrices

$$\overline{\overline{W}} = -\begin{bmatrix} \overline{\overline{K}}_{e1} & 0 \\ 0 & \overline{\overline{K}}_{e2} \end{bmatrix} + \begin{bmatrix} (\overline{\overline{F}}_{11} - \overline{\overline{B}}_{11}) & (\overline{\overline{F}}_{12} + \overline{\overline{B}}_{12}) \\ (\overline{\overline{F}}_{21} - \overline{\overline{B}}_{21}) & (\overline{\overline{F}}_{22} + \overline{\overline{B}}_{22}) \end{bmatrix} \cdot \overline{\overline{a}} \tag{41a}$$

$$\overline{\overline{A}} = -\begin{bmatrix} \overline{\overline{K}}_{e1} & 0 \\ 0 & \overline{\overline{K}}_{e2} \end{bmatrix} + \begin{bmatrix} (\overline{\overline{F}}_{11} + \overline{\overline{B}}_{11}) & (\overline{\overline{F}}_{12} - \overline{\overline{B}}_{12}) \\ (\overline{\overline{F}}_{21} + \overline{\overline{B}}_{21}) & (\overline{\overline{F}}_{22} - \overline{\overline{B}}_{22}) \end{bmatrix} \cdot \overline{\overline{a}} \tag{41b}$$

and the $\overline{\overline{\mu}}$ and $\overline{\overline{a}}$ are $4n \times 4n$ diagonal matrices

$$\overline{\overline{\mu}} = \text{diag}[\mu_1, \cdots, \mu_n, \mu_1, \cdots, \mu_n, \mu_1, \cdots, \mu_n, \mu_1, \cdots, \mu_n] \tag{42a}$$

$$\overline{\overline{a}} = \text{diag}[a_1, \cdots, a_n, a_1, \cdots, a_n, a_1, \cdots, a_n, a_1, \cdots, a_n] \tag{42b}$$

The homogeneous solution can be obtained in the form

$$\overline{I}_a = \overline{I}_{ao} e^{\alpha z} \tag{43a}$$

$$\overline{I}_s = \overline{I}_{so} e^{\alpha z} \tag{43b}$$

Substituting the above equations into (40a) and (40b), we now have $4n$ eigenvalue equations.

$$(\overline{\overline{\mu}}^{-1} \cdot \overline{\overline{W}} \cdot \overline{\overline{\mu}}^{-1} \cdot \overline{\overline{A}} - \alpha^2 \overline{\overline{I}}) \cdot \overline{I}_{ao} = 0 \tag{44}$$

$$\overline{I}_{so} = \alpha^{-1} \overline{\overline{\mu}}^{-1} \cdot \overline{\overline{A}} \cdot \overline{I}_{ao} \tag{45}$$

where $\overline{\overline{I}}$ is an identity matrix. Thus, if α is an eigenvalue, so is $-\alpha$. Once the eigenvalues α_i and the corresponding eigenvectors $\overline{I}_{ai}$ are obtained, we let $\overline{\overline{E}} = (\overline{I}_{a_1}, \overline{I}_{a_2}, \cdots, \overline{I}_{a_{4n}})$ be the $4n \times 4n$ eigenmatrix. Then the total solution for the upward propagating intensity is given by

$$\overline{I}^+ = (\overline{\overline{E}} + \overline{\overline{Q}}) \cdot \overline{\overline{D}}(z) \cdot \overline{x} + (\overline{\overline{E}} - \overline{\overline{Q}}) \cdot \overline{\overline{U}}(z + d) \cdot \overline{y} \tag{46}$$

where

$$\overline{\overline{Q}} = \overline{\overline{\mu}}^{-1} \cdot \overline{\overline{A}} \cdot \overline{\overline{E}} \cdot \overline{\overline{\alpha}}^{-1} \tag{47}$$

$$\overline{\overline{\alpha}} = \mathrm{diag}[\alpha_1, \alpha_2, \cdots, \alpha_{4n}] \tag{48}$$

$$\overline{\overline{D}}(z) = \mathrm{diag}\left[e^{\alpha_1 z}, e^{\alpha_2 z}, \cdots, e^{\alpha_{4n} z}\right] \tag{49}$$

$$\overline{\overline{U}}(z) = \mathrm{diag}\left[e^{-\alpha_1 z}, e^{-\alpha_2 z}, \cdots, e^{-\alpha_{4n} z}\right] \tag{50}$$

and $\overline{x}$ and $\overline{y}$ are the $4n \times 1$ matrices which represent $8n$ unknowns to be determined by the boundary conditions.

In a similar manner, the downward propagating intensity $\overline{I}^-$ can calculated. We obtain,

$$\overline{I}^- = (\overline{\overline{E}}' + \overline{\overline{Q}}') \cdot \overline{\overline{D}}(z) \cdot \overline{x} + (\overline{\overline{E}}' - \overline{\overline{Q}}') \cdot \overline{\overline{U}}(z+d) \cdot \overline{y} \tag{51}$$

where

$$\overline{\overline{E}}' = \overline{\overline{\mu}}^{-1} \cdot \overline{\overline{W}}' \cdot \overline{\overline{Q}} \cdot \overline{\overline{\alpha}}^{-1} \tag{52a}$$

$$\overline{\overline{Q}}' = \overline{\overline{\mu}}^{-1} \cdot \overline{\overline{A}}' \cdot \overline{\overline{E}} \cdot \overline{\overline{\alpha}}^{-1} \tag{52b}$$

and

$$\overline{\overline{W}}' = -\begin{bmatrix} \overline{\overline{K}}_{e1} & 0 \\ 0 & -\overline{\overline{K}}_{e2} \end{bmatrix} + \begin{bmatrix} (\overline{\overline{F}}_{11} - \overline{\overline{B}}_{11}) & (\overline{\overline{F}}_{12} + \overline{\overline{B}}_{12}) \\ -(\overline{\overline{F}}_{21} - \overline{\overline{B}}_{21}) & -(\overline{\overline{F}}_{22} + \overline{\overline{B}}_{22}) \end{bmatrix} \cdot \overline{\overline{a}} \tag{53a}$$

$$\overline{\overline{A}}' = -\begin{bmatrix} -\overline{\overline{K}}_{e1} & 0 \\ 0 & \overline{\overline{K}}_{e2} \end{bmatrix} + \begin{bmatrix} -(\overline{\overline{F}}_{11} + \overline{\overline{B}}_{11}) & -(\overline{\overline{F}}_{12} - \overline{\overline{B}}_{12}) \\ (\overline{\overline{F}}_{21} + \overline{\overline{B}}_{21}) & (\overline{\overline{F}}_{22} - \overline{\overline{B}}_{22}) \end{bmatrix} \cdot \overline{\overline{a}} \tag{53b}$$

In the random medium model, the eigenvalue equations can be simplified further since there is no coupling between the first three Stokes parameters, I_v, I_h, and U, and the last Stokes parameter V in the scattering function matrix. However, the Stokes parameters U and V are coupled together in the boundary conditions, and, in general, we have to keep all four Stokes parameters.

The boundary conditions, which are to be used to determine the constants $\overline{x}$ and $\overline{y}$ of the upward and downward propagating intensities given by (46) and (51), are, at $z = -d$,

$$\overline{I}^+(z=-d) = \overline{\overline{R}}_{12} \cdot \overline{I}^-(z=-d) \tag{54}$$

and at $z = 0$

$$\overline{I}^{-}(z=0) = \overline{\overline{R}}_{10} \cdot \overline{I}^{+}(z=0) + \overline{\overline{T}}_{01} \cdot \overline{I}_{oi}^{-} \tag{55}$$

where $\overline{\overline{R}}_{12}$ and $\overline{\overline{R}}_{10}$ are the $4n \times 4n$ matrices which are obtained by evaluating the 4×4 coupling matrices $\overline{\overline{R}}_{12}(\theta)$ and $\overline{\overline{R}}_{10}(\theta)$ at n discrete quadrature angles, and $\overline{\overline{T}}_{01}$ is the $4n \times 4n_f$ matrix which is obtained by evaluating the coupling matrix $\overline{\overline{T}}_{01}$ at the quadrature angles. The n_f is the number of quadrature angles in region 0, and the quadrature angles are related by the Snell's law. Since $\epsilon_1' \geq \epsilon_o$ we have $n_f \leq n$. Thus, for the quadrature angles in region 1 which are greater than the critical angle between regions 1 and 0, θ_i where $i > n_f$, there is no incident intensity. In the above equation (55), the incident intensity $\overline{I}_{oi}^{-}$ is obtained by discretizing the incident intensity given by (20), which is given in terms of the delta function. One way to bypass the problem of discretizing the delta function is to change the source term at the boundary into the source term in the volume by calculating the zeroth-order solution explicitly and using the radiative transfer equations for the higher order terms with the zeroth-order solution acting as the volume source [51]. In this section we discretize the delta function in a consistent manner and keep the source term at the boundary. This approach gives the same solution as the other approach, and also the formulation does not have to be changed when the boundary conditions are changed to incorporate the rough surface scattering [51].

Consider an integral given by

$$I = \int_0^{\pi} d\theta' \sin\theta' f(\theta, \theta') \, g(\theta') \tag{56}$$

Using the Gaussian quadrature method the integral I is approximated as

$$I \simeq \sum_{j=-n}^{n} a_j \, f(\theta, \theta_j) \, g_j \tag{57}$$

If we now let $g(\theta') = \delta(\cos\theta_o' - \cos\theta_{oi})$ where θ_{oi} is one of the quadrature angles in region 0, then the integral I can be evaluated exactly to give

$$I = f(\theta, \theta_i) \frac{\epsilon_o}{\epsilon_1'} \frac{\cos\theta_{oi}}{\cos\theta_i} \tag{58}$$

where θ_i is the corresponding incident angle in region 1 which is related to θ_{oi} by Snell's law, and we made use of [40]

$$d\theta' \sin\theta' d\phi' = d\theta'_o \sin\theta'_o \frac{\epsilon_o}{\epsilon'_1} \frac{\cos\theta_{oi}}{\cos\theta_i} d\phi'_o \tag{59}$$

Therefore, comparing (58) with (57), we obtain the discrete form for the delta function:

$$g_j = \delta_{ji} \frac{1}{a_j} \frac{\epsilon_o}{\epsilon'_1} \frac{\cos\theta_{oi}}{\cos\theta_i} \tag{60}$$

where

$$\delta_{ji} = \begin{cases} 1 & \text{if } j = i \\ 0 & \text{otherwise} \end{cases} \tag{61}$$

The incident intensities for even and odd terms in each harmonic, given by (20), can now be cast into the quadrature form by making use of the above relations.

Substituting in the expressions for the upward and downward propagating intensities into the boundary conditions (54) and (55), we obtain the following $8n$ equations for $8n$ unknowns $\overline{x}$ and $\overline{y}$:

$$\begin{aligned} &\left[(\overline{\overline{E}}' + \overline{\overline{Q}}') - \overline{\overline{R}}_{10} \cdot (\overline{\overline{E}} + \overline{\overline{Q}})\right] \cdot \overline{x} \\ &\qquad + \left[(\overline{\overline{E}}' - \overline{\overline{Q}}') - \overline{\overline{R}}_{10} \cdot (\overline{\overline{E}} - \overline{\overline{Q}})\right] \cdot \overline{\overline{D}}(-d) \cdot \overline{y} = \overline{\overline{T}}_{01} \cdot \overline{I}_{oi}^{-} \end{aligned} \tag{62a}$$

$$\begin{aligned} &\left[(\overline{\overline{E}} + \overline{\overline{Q}}) - \overline{\overline{R}}_{12} \cdot (\overline{\overline{E}}' + \overline{\overline{Q}}')\right] \cdot \overline{\overline{D}}(-d) \cdot \overline{x} \\ &\qquad + \left[(\overline{\overline{E}} - \overline{\overline{Q}}) - \overline{\overline{R}}_{12} \cdot (\overline{\overline{E}}' - \overline{\overline{Q}}')\right] \cdot \overline{y} = 0 \end{aligned} \tag{62b}$$

The above equations can be solved for the constants $\overline{x}$ and $\overline{y}$ for each case when the incident intensity is at one of the quadrature angles. Note that in the halfspace random medium case when $d \to \infty$, the equations for $\overline{x}$ and $\overline{y}$ become decoupled since $\overline{\overline{D}} \to 0$ and the matrix equation does not become singular [51]. This is due to the form of the solution assumed in (46) and (51).

Once the constants $\overline{x}$ and $\overline{y}$ are determined, the scattered intensities from region 1 to region 0, represented by the first term on the right-hand-side of (6), can be determined. We have

$$\begin{aligned} \overline{I}_{os} &= \overline{\overline{T}}_{10} \cdot \overline{I}^{+}(z=0) \\ &= \overline{\overline{T}}_{10} \cdot \left[(\overline{\overline{E}} + \overline{\overline{Q}}) \cdot \overline{x} + (\overline{\overline{E}} - \overline{\overline{Q}}) \cdot \overline{\overline{D}}(-d) \cdot \overline{y}\right] \end{aligned} \tag{63}$$

Thus, the complete solution can be obtained by solving the radiative transfer equations using the Gaussian quadrature method for each harmonic as outlined above and reintroducing the azimuthal dependence. The total scattered intensities in region 0 is given by

$$\overline{I}_{os}(\phi_o) = \left\{ \overline{\overline{R}}_{01} + \overline{\overline{T}}_{10} \cdot \left[\overline{\overline{I}} - \overline{\overline{R}}_{10} \cdot \overline{\overline{R}}_{12} \cdot \exp[-\overline{\overline{\mu}}^{-1} \cdot \overline{\overline{K}}_e d] \right]^{-1} \cdot \overline{\overline{T}}_{01} \right\}$$
$$\cdot \overline{I}_{oi}\, \delta(\phi_o - \phi_{oi}) + \sum_{m=0}^{m_{max}} \left\{ \overline{\overline{T}}_{10} \cdot \left[\overline{I}^{mc^+}(z=0) \right.\right.$$
$$\left. - \left[\overline{\overline{I}} - \overline{\overline{R}}_{10} \cdot \overline{\overline{R}}_{12} \cdot \exp[-\mu^{-1} \cdot \overline{\overline{K}}_e d] \right]^{-1} \cdot \overline{\overline{T}}_{01} \cdot \overline{I}_{oi}^{mc} \right]$$
$$\left. \cdot \cos m(\phi_o - \phi_{oi}) + \overline{\overline{T}}_{10} \cdot \overline{I}^{ms^+}(z=0) \sin m(\phi_o - \phi_{oi}) \right\} \tag{64}$$

where we have summed up the zeroth-order solution and $\overline{I}^{mc^+}(z = 0)$ and $\overline{I}^{ms^+}(z = 0)$ are the upward propagating m-th cosine and sine harmonics evaluated at $z = 0$. Once the scattered intensities in region 0 are obtained, the bistatic scattering coefficients and the backscattering cross sections can be obtained from (7) and (8). Note that if we are interested in calculating the scattering intensities for vertically or horizontally polarized intensities only, then the even series needs to be calculated. This is because the odd series, represented by (14b), is zero due to the fact the incident intensity for the odd series as given by (21b) is zero. In general, both the even and odd series have to be calculated, especially for fully polarimetric scattering calculations.

d. Results and Discussion

The backscattering cross sections and the bistatic scattering coefficients are calculated and illustrated for the various cases. In our calculations $n = 16$ is used. The backscattering cross sections are illustrated as functions of frequency and incident angle. The bistatic scattering coefficients are plotted as functions of scattering angles θ_s and ϕ_s.

In Fig. 5.2.2 the horizontally polarized and depolarized backscattering cross sections are plotted as a function of frequency for a 48cm thick random medium. Backscattering cross sections increase as frequency is increased. This is due to the fact that as frequency increases

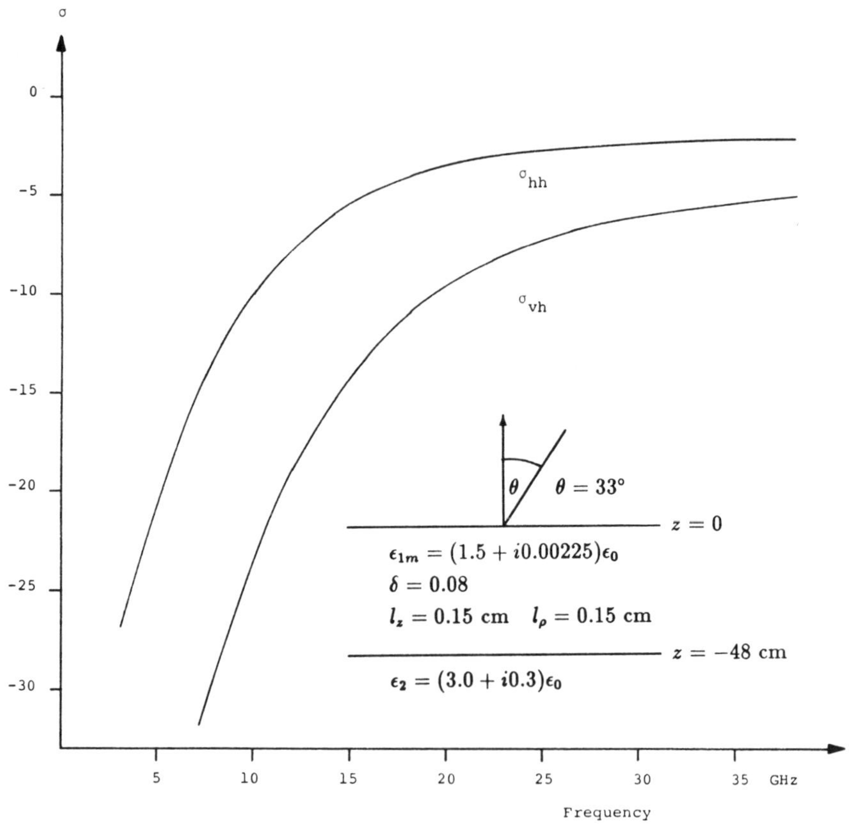

Figure 5.2.2 Backscattering cross sections as a function of frequency.

the albedo $[K_s/(K_a+K_s)]$ increases, and the scattering becomes dominant over the absorption. Also, the difference between the like-like polarized return and the depolarized return decreases. In Fig. 5.2.3 the backscattering cross sections are plotted as a function of incident angle at 10 GHz.

In Fig. 5.2.4, the bistatic scattering coefficients γ_{hh} and γ_{vh} are plotted as a function of scattering angle θ_s. The positive θ_s corresponds to the forward scattering case where $\phi_s = 0$ whereas negative θ_s corresponds to the backward scattering case with $\phi_s = 180°$. We note that there is symmetry about the $\theta_s = 0$ axis which is typical of Rayleigh scatterers. For correlation lengths small compared to the wavelength, the scattering pattern of the random medium is that of the Rayleigh scatterers [37]. The number of harmonics needed in this

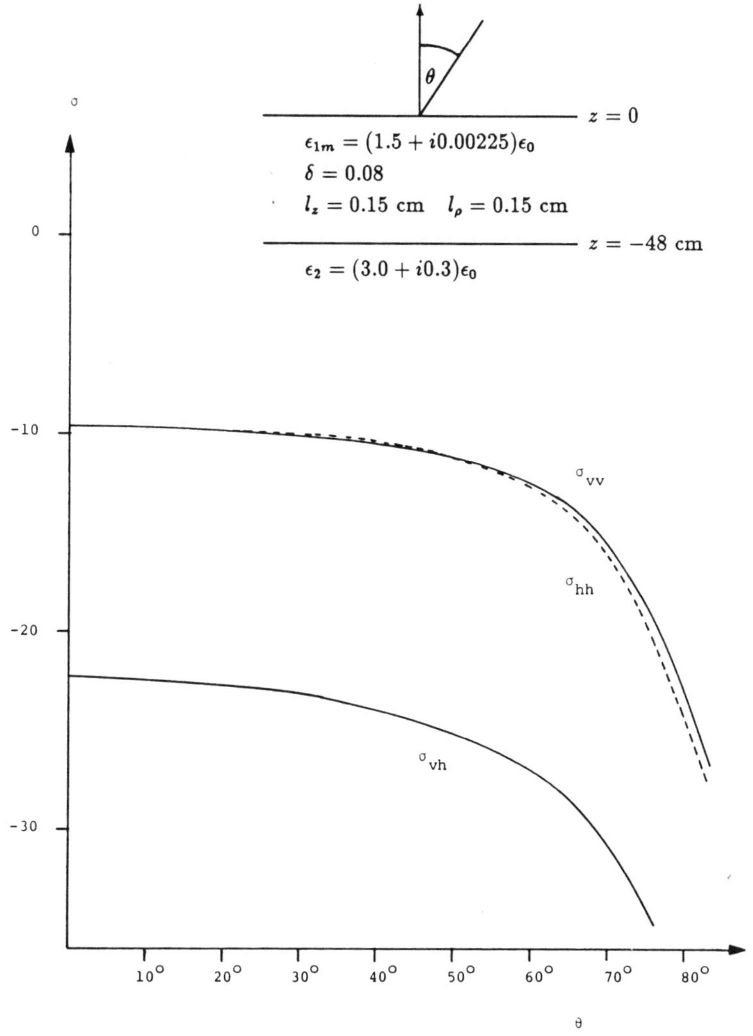

Figure 5.2.3 Backscattering cross sections as a function of incident angle at 10 GHz.

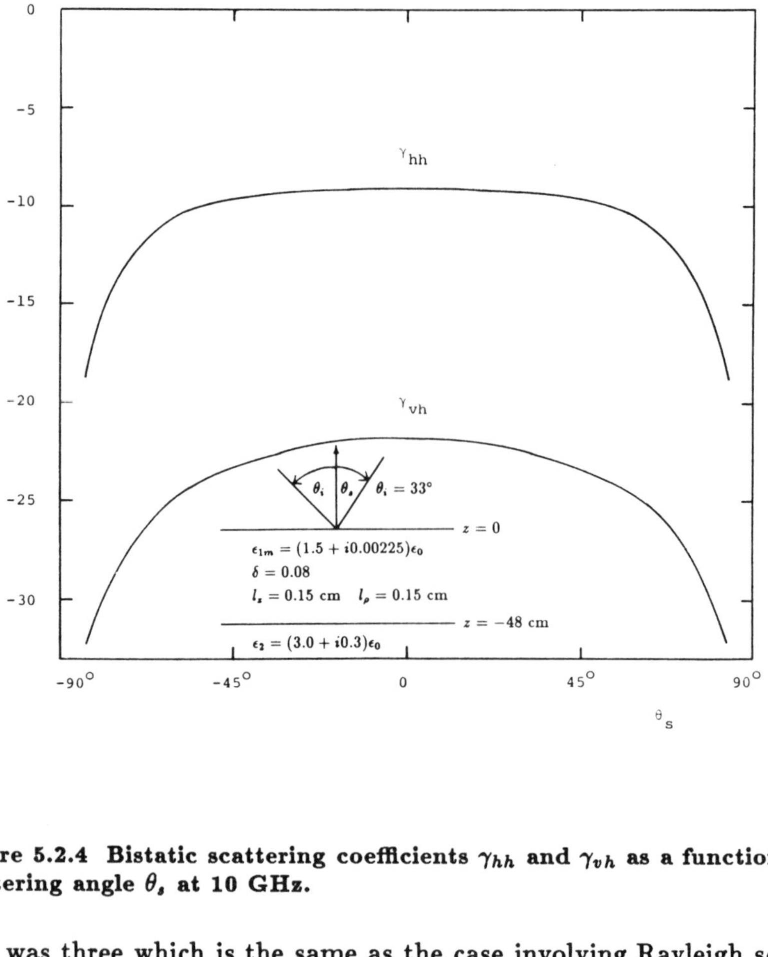

Figure 5.2.4 Bistatic scattering coefficients γ_{hh} and γ_{vh} as a function of scattering angle θ_s at 10 GHz.

case was three which is the same as the case involving Rayleigh scatterers [53]. In Fig. 5.2.5 we show the bistatic scattering coefficients for the larger correlation length l_ρ. Unlike the previous case there is no symmetry. The number of harmonics needed in the computation is also larger than the previous case.

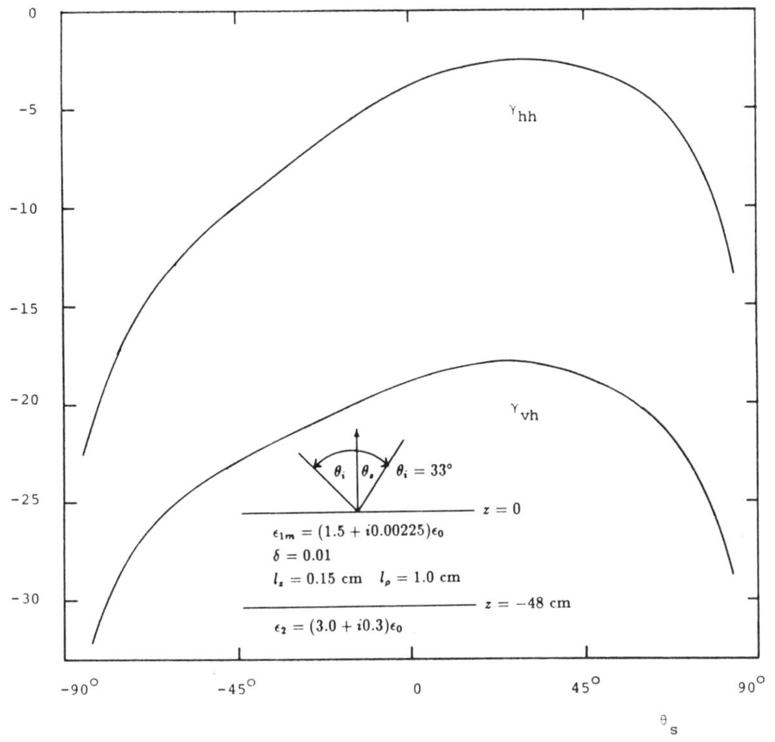

Figure 5.2.5 Bistatic scattering coefficients γ_{hh} and γ_{vh} as a function of scattering angle θ_s at 10 GHz.

In Figs. 5.2.6–5.2.9, the bistatic scattering coefficients are plotted as a function of azimuthal scattering angle ϕ_s for $\theta_i = \theta_s = 33°$. We only plot from $\phi_s = 0°$ to $\phi_s = 180°$ because the bistatc scattering coefficients are symmetrical. In Fig. 5.2.6, the bistatic scattering coefficients γ_{hh} and γ_{vh} are plotted for the case of small correlation length l_ρ. There is a symmetry in the bistatic scattering coefficients about $\phi_s = 90°$. In Fig. 5.2.7, γ_{hh} and γ_{vv} are compared. We note there is no symmetry for the bistatic scattering coefficient γ_{vv}. In Figs. 5.2.8 and 5.2.9, the bistatic scattering coefficients are plotted for the case of the large correlation length l_ρ. We note that the scattering coefficients are more peaked toward the forward scattering direction and that there is no symmetry about the $\phi_s = 90°$ axis.

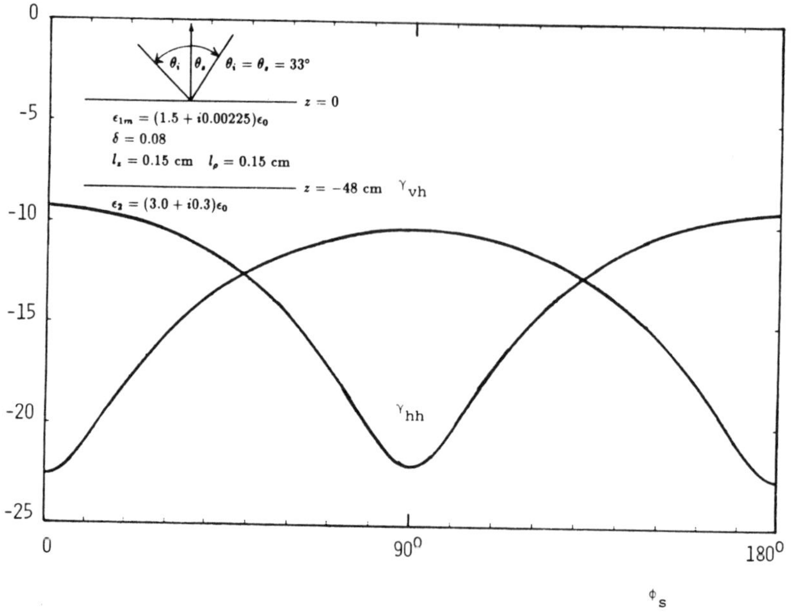

Figure 5.2.6 Bistatic scattering coefficients γ_{hh} and γ_{vh} as a function of azimuthal scattering angle ϕ_s at 10 GHz.

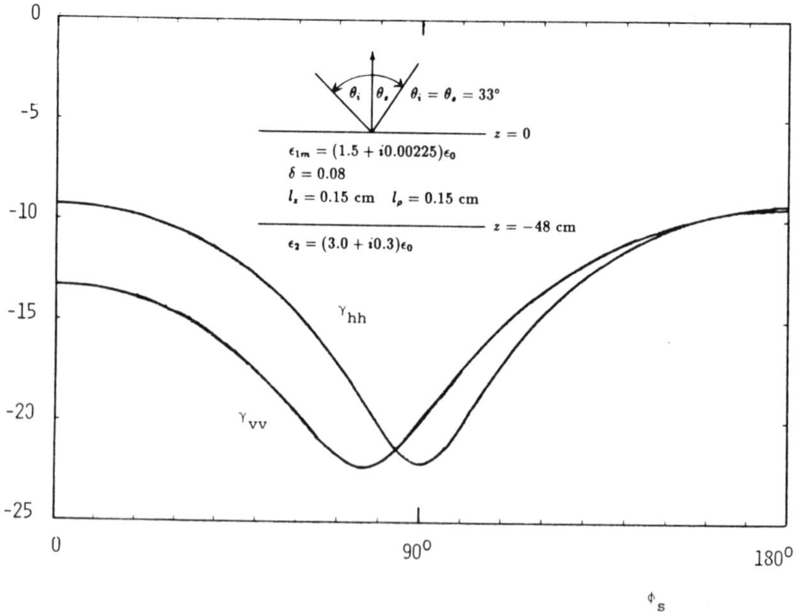

Figure 5.2.7 Bistatic scattering coefficients γ_{hh} and γ_{vh} as a function of azimuthal scattering angle ϕ_s at 10 GHz.

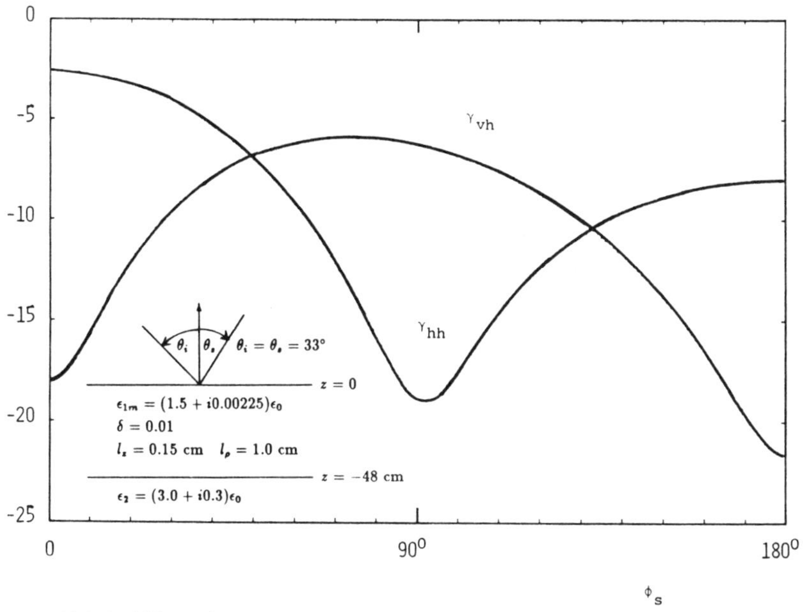

Figure 5.2.8 Bistatic scattering coefficients γ_{hh} and γ_{vh} as a function of azimuthal scattering angle ϕ_s at 10 GHz.

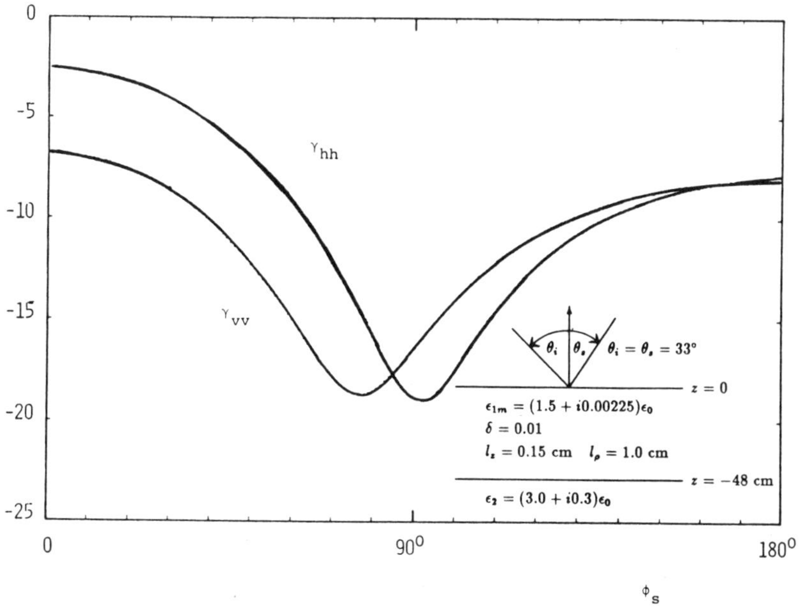

Figure 5.2.9 Bistatic scattering coefficients γ_{hh} and γ_{vh} as a function of azimuthal scattering angle ϕ_s at 10 GHz.

5.3 Two-Layer Random Medium with Rough Interfaces

a. Formulation

Consider a layer of random medium with rough interfaces characterized by the permittivity $\epsilon_1 + \epsilon_f$, where ϵ_f stands for the randomly fluctuating part whose amplitude is very small and whose ensemble average is zero, on top of a homogeneous medium with permittivity ϵ_2 [Fig. 5.3.1]. The radiative transfer equations which govern the propagation of intensities inside the scattering medium are given by (1) and (2).

For the case of rough interfaces, the boundary conditions are, for $0 < \theta < \pi/2$,

$$\begin{aligned}
&\overline{I}(\pi - \theta, \phi, z = 0) \\
&= \int_0^{2\pi} d\phi'_o \int_0^{\pi/2} d\theta'_o \sin\theta'_o \, \overline{\overline{T}}_{01}(\theta, \phi; \theta'_o, \phi'_o) \cdot \overline{I}_{oi}(\pi - \theta'_o, \phi'_o) \\
&\quad + \int_0^{2\pi} d\phi' \int_0^{\pi/2} d\theta' \sin\theta' \, \overline{\overline{R}}_{10}(\theta, \phi; \theta', \phi') \cdot \overline{I}(\theta', \phi', z = 0) \qquad (65)
\end{aligned}$$

$$\begin{aligned}
&\overline{I}(\theta, \phi, z = -d) \\
&= \int_0^{2\pi} d\phi' \int_0^{\pi/2} d\theta' \sin\theta' \, \overline{\overline{R}}_{12}(\theta, \phi; \theta', \phi') \cdot \overline{I}(\pi - \theta', \phi, z = -d) \qquad (66)
\end{aligned}$$

where we have broken up the intensities in the scattering layer into upward going intensities $\overline{I}(\theta, \phi, z)$ and downward going intensities $\overline{I}(\pi - \theta, \phi, z)$, and $\overline{I}_{oi}(\pi - \theta_o, \phi_o)$ is the incident intensity given by (3). In the above equations $\overline{\overline{T}}_{01}(\theta, \phi; \theta_o, \phi_o)$ represents the coupling from region 0 to region 1, $\overline{\overline{R}}_{10}(\theta, \phi; \theta', \phi')$ represents the coupling from upward going intensity in the direction (θ', ϕ') into downward going intensity in the direction $(\pi - \theta, \phi)$ at the boundary of region 1 and region 0, and $\overline{\overline{R}}_{12}(\theta, \phi; \theta', \phi')$ represents similar coupling at the boundary of region 1 and region 2.

Once the radiative transfer equations are solved subject to the boundary conditions, the scattered intensity in the direction (θ_{os}, ϕ_{os}) in region 0 is determined from

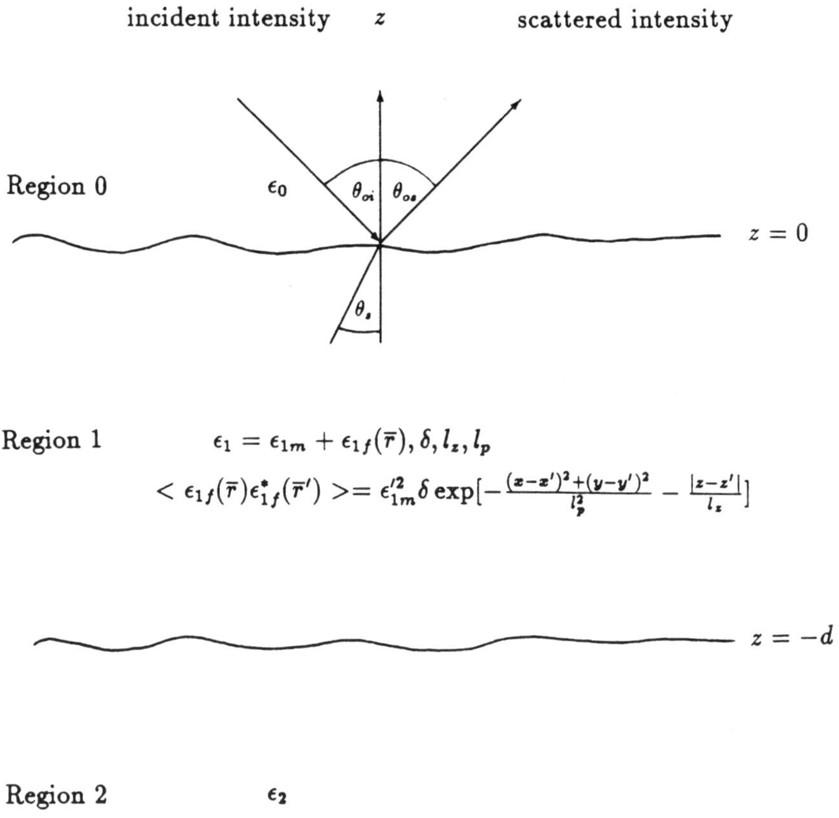

Figure 5.3.1 Geometrical configuration of the problem.

$$
\begin{aligned}
&\overline{I}_{os}(\theta_{os}, \phi_{os}) \\
&\quad = \int_0^{2\pi} d\phi'_o \int_0^{\pi/2} d\theta'_o \sin\theta'_o \, \overline{\overline{R}}_{01}(\theta_{os}, \phi_{os}; \theta'_o, \phi'_o) \cdot \overline{I}_{oi}(\pi - \theta'_o, \phi'_o) \\
&\quad + \int_0^{2\pi} d\phi' \int_0^{\pi/2} d\theta' \sin\theta' \, \overline{\overline{T}}_{10}(\theta_{os}, \phi_{os}; \theta', \phi') \cdot \overline{I}(\theta', \phi', z = 0) \quad (67)
\end{aligned}
$$

where $\overline{\overline{T}}_{10}(\theta_o, \phi_o; \theta', \phi')$ represents coupling from region 1 to region 0. The coupling coefficients used in the boundary conditions for the rough interfaces are derived in the next section.

b. Boundary Conditions

The boundary conditions satisfied by the specific intensities at rough dielectric interfaces are derived in this section. Unlike the planar interface case where the coupling at the boundary is only to the specular reflection and transmission directions, the incident intensity is coupled to all of the reflection and transmission directions. The reflection and transmission matrices are related to the bistatic scattering matrices, which is the generalization of the bistatic scattering coefficients to include the correlation between polarizations of the scattered fields.

Consider a plane wave incident from medium 1 onto medium 2 along the direction $\hat{k}_i$ upon a rough dielectric interface. The electric field of the incident wave is given by

$$\overline{E}_i = \hat{e}_i \, E_{ei} \, e^{i\overline{k}_i \cdot \overline{r}} \tag{68}$$

where $\overline{k}_i$ denotes the incident wave vector and $\hat{e}_i$ is the polarization of the electric field vector. The rough surface is characterized by a random height distribution $z = f(\overline{r}_\perp)$ where $f(\overline{r}_\perp)$ is a Gaussian random variable with zero mean, $\langle f(\overline{r}_\perp) \rangle = 0$. The incident field will generate the reflected and transmitted fields in medium 1 and 2, respectively. The solutions to the problem of scattering from a random rough surface have been extensively studied [55-67]. In general, the scattered and transmitted fields for vertical and horizontal polarizations for the incident field with vertical and horizontal polarizations are given by

$$\begin{bmatrix} E_{vs} \\ E_{hs} \end{bmatrix} = \frac{e^{ik_1 r}}{r} \begin{bmatrix} f_{vv}^r(\theta_s, \phi_s; \theta_i, \phi_i) & f_{vh}^r(\theta_s, \phi_s; \theta_i, \phi_i) \\ f_{hv}^r(\theta_s, \phi_s; \theta_i, \phi_i) & f_{hh}^r(\theta_s, \phi_s; \theta_i, \phi_i) \end{bmatrix} \cdot \begin{bmatrix} E_{vi} \\ E_{hi} \end{bmatrix} \tag{69}$$

$$\begin{bmatrix} E_{vt} \\ E_{ht} \end{bmatrix} = \frac{e^{ik_2 r}}{r} \begin{bmatrix} f_{vv}^t(\theta_t, \phi_t; \theta_i, \phi_i) & f_{vh}^t(\theta_t, \phi_t; \theta_i, \phi_i) \\ f_{hv}^t(\theta_t, \phi_t; \theta_i, \phi_i) & f_{hh}^t(\theta_t, \phi_t; \theta_i, \phi_i) \end{bmatrix} \cdot \begin{bmatrix} E_{vi} \\ E_{hi} \end{bmatrix} \tag{70}$$

Now the scattered specific intensities can be expressed in terms of the incident specific intensities.

$$\overline{I}_r = \overline{\overline{R}}_{12} \cdot \overline{I}_i \tag{71}$$

$$\overline{I}_t = \overline{\overline{T}}_{12} \cdot \overline{I}_i \tag{72}$$

where $\overline{I}_i$, $\overline{I}_s$, and $\overline{I}_t$ are the column matrices for the specific intensities containing the four Stokes parameters

$$\overline{I}_\alpha = \begin{bmatrix} I_{v\alpha} \\ I_{h\alpha} \\ U_\alpha \\ V_\alpha \end{bmatrix} \qquad \alpha = i,\, r,\, t \tag{73}$$

and

$$\overline{\overline{R}}_{12}(\theta_s, \phi_s; \theta_i, \phi_i) = \frac{1}{A_o \cos\theta_s} \overline{\overline{L}}^r \tag{74}$$

$$\overline{\overline{T}}_{12}(\theta_t, \phi_t; \theta_i, \phi_i) = \frac{1}{A_o \cos\theta_t} \frac{\eta_1}{\eta_2} \overline{\overline{L}}^t \tag{75}$$

with

$$\overline{\overline{L}}^\alpha = \begin{bmatrix} \langle |f^\alpha_{vv}|^2 \rangle & \langle |f^\alpha_{vh}|^2 \rangle & Re\langle (f^{\alpha^*}_{vh} f^\alpha_{vv}) \rangle & -Im\langle (f^{\alpha^*}_{vh} f^\alpha_{vv}) \rangle \\ \langle |f^\alpha_{hv}|^2 \rangle & \langle |f^\alpha_{hh}|^2 \rangle & Re\langle (f^{\alpha^*}_{hh} f^\alpha_{hv}) \rangle & -Im\langle (f^\alpha_{hv} f^{\alpha^*}_{hh}) \rangle \\ 2Re\langle (f^\alpha_{vv} f^{\alpha^*}_{hv}) \rangle & 2Re\langle (f^\alpha_{vh} f^{\alpha^*}_{hh}) \rangle & Re\langle (f^\alpha_{vv} f^{\alpha^*}_{hh} + f^\alpha_{vh} f^{\alpha^*}_{hv}) \rangle & -Im\langle (f^\alpha_{vv} f^{\alpha^*}_{hh} - f^\alpha_{vh} f^{\alpha^*}_{hv}) \rangle \\ 2Im\langle (f^\alpha_{vv} f^{\alpha^*}_{hv}) \rangle & 2Im\langle (f^\alpha_{vh} f^{\alpha^*}_{hh}) \rangle & Im\langle (f^\alpha_{vv} f^{\alpha^*}_{hh} + f^\alpha_{vh} f^{\alpha^*}_{hv}) \rangle & Re\langle (f^\alpha_{vv} f^{\alpha^*}_{hh} - f^\alpha_{vh} f^{\alpha^*}_{hv}) \rangle \end{bmatrix} \tag{76}$$

Thus, the boundary conditions for the specific intensities at a rough interface is given by

$$\overline{I}_1(\hat{k}_s) = \int_0^{2\pi} d\phi_i \int_0^{\pi/2} d\theta_i \sin\theta_i \, \overline{\overline{R}}_{12}(\theta_s, \phi_s; \theta_i, \phi_i) \cdot \overline{I}_1(\hat{k}_i) \tag{77}$$

$$\overline{I}_2(\hat{k}_t) = \int_0^{2\pi} d\phi_i \int_0^{\pi/2} d\theta_i \sin\theta_i \, \overline{\overline{T}}_{12}(\theta_t, \phi_t; \theta_i, \phi_i) \cdot \overline{I}_1(\hat{k}_i) \tag{78}$$

The reflected and transmitted intensities at the directions $\hat{k}_s$ and $\hat{k}_t$ are given by integration of all the scattered intensities which are coupled to that direction from the incident intensity.

The reflection and transmission coupling matrices can also be related to the rough surface bistatic scattering coefficients. First, the definition of bistatic scattering coefficients is generalized to include the correlation between polarizations. We define the bistatic scattering matrix $\overline{\overline{\gamma}}$ whose elements are given by

$$\gamma_{\alpha\beta}(\theta_s,\phi_s;\theta_i,\phi_i) = 4\pi \frac{\cos\theta_s\, I_{\alpha s}(\theta_s,\phi_s)}{\cos\theta_i\, I_{\beta i}} \qquad \alpha,\beta = 1,2,3,4 \tag{79}$$

where

$$\overline{I}_\alpha = \begin{bmatrix} I_{1\alpha} \\ I_{2\alpha} \\ I_{3\alpha} \\ I_{4\alpha} \end{bmatrix} = \begin{bmatrix} I_{v\alpha} \\ I_{h\alpha} \\ U_\alpha \\ V_\alpha \end{bmatrix} \qquad \alpha = i, s \tag{80}$$

The usual bistatic scattering coefficients are

$$\gamma_{vv} = \gamma_{11} \qquad \gamma_{vh} = \gamma_{12} \qquad \gamma_{hv} = \gamma_{21} \qquad \gamma_{hh} = \gamma_{22} \tag{81}$$

The reflection coupling matrix is related to the bistatic scattering matrix as follows:

$$\overline{\overline{R}}_{12}(\theta_s,\phi_s;\theta_i,\phi_i) = \frac{1}{4\pi}\frac{\cos\theta_i}{\cos\theta_s}\overline{\overline{\gamma}}(\theta_s,\phi_s;\theta_i,\phi_i) \tag{82}$$

In a similar manner, the transmission coupling matrix can be related to the bistatic scattering coefficients for the transmitted intensities:

$$\overline{\overline{T}}_{12}(\theta_t,\phi_t;\theta_i,\phi_i) = \frac{1}{4\pi}\frac{\cos\theta_i}{\cos\theta_t}\overline{\overline{\gamma}}^t(\theta_t,\phi_t;\theta_i,\phi_i) \tag{83}$$

The explicit expressions for the reflection and transmission matrices, obtained using the scattered and transmitted fields derived by a combination of Kirchhoff approximation and geometrical optics approach, are given in Appendix D [31,69]. The other solutions for the scattering from a rough dielectric interface, such as small perturbation method (SPM) [65,66] or modified SPM [67,68], can also be used to derive the coupling matrices to be used with the radiative transfer equations.

Note that, in general, the coupling matrices can be broken up into coherent and incoherent components. The coherent components only couple the incident intensity into the specular reflection and transmission directions while the incoherent components couple to all reflection and transmission directions.

$$\overline{\overline{R}}_{12}(\theta_s,\phi_s;\theta_i,\phi_i) = \overline{\overline{R}}^c_{12}(\theta_s,\phi_s;\theta_i,\phi_i) + \overline{\overline{R}}^i_{12}(\theta_s,\phi_s;\theta_i,\phi_i) \tag{84}$$

$$\overline{\overline{T}}_{12}(\theta_t, \phi_t; \theta_i, \phi_i) = \overline{\overline{T}}^c_{12}(\theta_t, \phi_t; \theta_i, \phi_i) + \overline{\overline{T}}^i_{12}(\theta_t, \phi_t; \theta_i, \phi_i) \tag{85}$$

c. Fourier Series Expansion

The radiative transfer equations and the associated boundary conditions can be solved using a numerical approach. We use a Fourier-series expansion in the azimuthal direction to eliminate the ϕ-dependence from the radiative transfer equations and obtain a set of equations without the ϕ-dependence [section 5.2.b] The ϕ-dependence from the boundary conditions can also be eliminated using the Fourier-series expansion. We let

$$\overline{I}_{oi}(\pi - \theta_o, \phi_o) = \overline{I}_{oi}\, \delta(\cos\theta_o - \cos\theta_{oi}) \sum_{m=0}^{\infty} \frac{1}{(\delta_m + 1)\pi} \cos m(\phi_o - \phi_{oi}) \tag{86}$$

and, for α, β = 0, 1, 2,

$$\overline{\overline{R}}^c_{\alpha\beta}(\theta, \phi; \theta', \phi') = \overline{\overline{R}}^c_{\alpha\beta}(\theta)\, \delta(\cos\theta - \cos\theta') \sum_{m=0}^{\infty} \frac{1}{(1 + \delta_m)\pi} \cos m(\phi - \phi') \tag{87a}$$

$$\overline{\overline{R}}^i_{\alpha\beta}(\theta, \phi; \theta', \phi') = \sum_{m=0}^{\infty} \frac{1}{(1 + \delta_m)\pi} \Big[\overline{\overline{R}}^{mc}_{\alpha\beta}(\theta, \theta') \cos m(\phi - \phi') + \overline{\overline{R}}^{ms}_{\alpha\beta}(\theta, \theta') \sin m(\phi - \phi')\Big] \tag{87b}$$

$$\overline{\overline{T}}^c_{01}(\theta, \phi; \theta'_o, \phi'_o) = \overline{\overline{T}}^c_{01}(\theta_o)\, \delta(\cos\theta_o - \cos\theta'_o) \sum_{m=0}^{\infty} \frac{1}{(1 + \delta_m)\pi} \cos m(\phi_o - \phi'_o) \tag{88a}$$

$$\overline{\overline{T}}^i_{01}(\theta, \phi; \theta'_o, \phi'_o) = \sum_{m=0}^{\infty} \frac{1}{(1 + \delta_m)\pi} \Big[\overline{\overline{T}}^{mc}_{01}(\theta, \theta'_o) \cos m(\phi_o - \phi'_o) + \overline{\overline{T}}^{ms}_{01}(\theta, \theta'_o) \sin m(\phi_o - \phi'_o)\Big] \tag{88b}$$

where θ_o and θ are related by the Snell's law.

Substituting (86), (87), and (88) into the boundary conditions (65) and (66) and carrying out the $d\phi'$ and $d\phi'_o$ integrations, we obtain the following set of equations. For $0 < \theta < \pi/2$ and $m = 0, 1, 2, \cdots$

$$\begin{aligned}
\overline{I}^{mc}(\pi-\theta, z=0) &= \int_0^{\pi/2} d\theta'_o \sin\theta'_o \left[\overline{\overline{T}}^{c}_{01}(\theta,\theta'_o)\cdot\overline{I}^{mc}_{oi}(\pi-\theta'_o)\right.\\
&\quad \left.+\overline{\overline{T}}^{mc}_{01}(\theta,\theta'_o)\cdot\overline{I}^{mc}_{oi}(\pi-\theta'_o)\right]\\
&\quad +\int_0^{\pi/2} d\theta' \sin\theta'\, \overline{\overline{R}}^{c}_{10}(\theta,\theta')\cdot\overline{I}^{mc}(\theta', z=0)\\
&\quad +\int_0^{\pi/2} d\theta' \sin\theta' \left[\overline{\overline{R}}^{mc}_{10}(\theta,\theta')\cdot\overline{I}^{mc}(\theta', z=0)\right.\\
&\quad \left.-\overline{\overline{R}}^{ms}_{10}(\theta,\theta')\cdot\overline{I}^{ms}(\theta', z=0)\right]
\end{aligned} \tag{89a}$$

$$\begin{aligned}
\overline{I}^{ms}(\pi-\theta, z=0) &= \int_0^{\pi/2} d\theta'_o \sin\theta'_o\, \overline{\overline{T}}^{ms}_{01}(\theta,\theta'_o)\cdot\overline{I}^{mc}_{oi}(\pi-\theta'_o)\\
&\quad +\int_0^{\pi/2} d\theta' \sin\theta'\, \overline{\overline{R}}^{c}_{10}(\theta,\theta')\cdot\overline{I}^{ms}(\theta', z=0)\\
&\quad +\int_0^{\pi/2} d\theta' \sin\theta' \left[\overline{\overline{R}}^{ms}_{10}(\theta,\theta')\cdot\overline{I}^{mc}(\theta', z=0)\right.\\
&\quad \left.+\overline{\overline{R}}^{mc}_{10}(\theta,\theta')\cdot\overline{I}^{ms}(\theta', z=0)\right]
\end{aligned} \tag{89b}$$

$$\begin{aligned}
\overline{I}^{mc}(\theta, z=-d) &= \int_0^{\pi/2} d\theta' \sin\theta'\, \overline{\overline{R}}^{c}_{12}(\theta,\theta')\cdot\overline{I}^{mc}(\pi-\theta', z=-d)\\
&\quad +\int_0^{\pi/2} d\theta' \sin\theta' \left[\overline{\overline{R}}^{mc}_{12}(\theta,\theta')\cdot\overline{I}^{mc}(\pi-\theta', z=-d)\right.\\
&\quad \left.-\overline{\overline{R}}^{ms}_{12}(\theta,\theta')\cdot\overline{I}^{ms}(\pi-\theta', z=-d)\right]
\end{aligned} \tag{90a}$$

$$\overline{I}^{ms}(\theta, z=-d) = \int_0^{\pi/2} d\theta' \sin\theta' \,\overline{\overline{R}}_{12}^{c}(\theta,\theta') \cdot \overline{I}^{ms}(\pi-\theta', z=-d)$$
$$+ \int_0^{\pi/2} d\theta' \sin\theta' \Big[\overline{\overline{R}}_{12}^{ms}(\theta,\theta') \cdot \overline{I}^{mc}(\pi-\theta', z=-d)$$
$$+\overline{\overline{R}}_{12}^{mc}(\theta,\theta') \cdot \overline{I}^{ms}(\pi-\theta', z=-d)\Big] \tag{90b}$$

where

$$\overline{I}_{oi}^{mc}(\pi-\theta_o) = \overline{I}_{oi} \frac{1}{(\delta_m+1)\pi} \delta(\cos\theta_o - \cos\theta_{oi}) \tag{91}$$

Note that the Fourier-series expanded coupling matrices for azimuthally isotropic rough boundaries do not couple the first two Stokes parameters to the last two Stokes parameters. The coupling matrices can be expanded as follows. For $A = R$ or T, and $\alpha, \beta = 0, 1, 2$,

$$\overline{\overline{A}}_{\alpha\beta}^{mc} = \begin{bmatrix} A^{mc}_{\alpha\beta_{11}} & A^{mc}_{\alpha\beta_{12}} & 0 & 0 \\ A^{mc}_{\alpha\beta_{21}} & A^{mc}_{\alpha\beta_{22}} & 0 & 0 \\ 0 & 0 & A^{mc}_{\alpha\beta_{33}} & A^{mc}_{\alpha\beta_{34}} \\ 0 & 0 & A^{mc}_{\alpha\beta_{43}} & A^{mc}_{\alpha\beta_{44}} \end{bmatrix} \tag{92a}$$

$$\overline{\overline{A}}_{\alpha\beta}^{ms} = \begin{bmatrix} 0 & 0 & A^{ms}_{\alpha\beta_{13}} & A^{ms}_{\alpha\beta_{14}} \\ 0 & 0 & A^{ms}_{\alpha\beta_{23}} & A^{ms}_{\alpha\beta_{24}} \\ A^{ms}_{\alpha\beta_{31}} & A^{ms}_{\alpha\beta_{32}} & 0 & 0 \\ A^{ms}_{\alpha\beta_{41}} & A^{ms}_{\alpha\beta_{42}} & 0 & 0 \end{bmatrix} \tag{92b}$$

Thus, the boundary conditions (89) and (90) can be decoupled by defining

$$\overline{I}^{me}(\theta, z) = \begin{bmatrix} I_v^{mc}(\theta,z) \\ I_h^{mc}(\theta,z) \\ U^{ms}(\theta,z) \\ V^{ms}(\theta,z) \end{bmatrix} \tag{93a}$$

$$\overline{I}^{mo}(\theta, z) = \begin{bmatrix} I_v^{ms}(\theta,z) \\ I_h^{ms}(\theta,z) \\ U^{mc}(\theta,z) \\ V^{mc}(\theta,z) \end{bmatrix} \tag{93b}$$

where superscripts e and o stand for even or odd dependence in the first two Stokes parameters. After carrying out the $d\theta_o$ integration, the decoupled boundary conditions are given by, for $\alpha = e$ or o,

$$\begin{aligned}&\overline{I}^{m\alpha}(\pi-\theta, z=0)\\&=\overline{\overline{T}}{}^{c}_{01}(\theta_o)\cdot\overline{I}^{m\alpha}_{oi}(\pi-\theta_o)+\overline{\overline{T}}{}^{m\alpha}_{01}(\theta,\theta_{oi})\cdot\overline{I}^{m\alpha}_{oi}\\&+\int_0^{\pi/2}d\theta'\sin\theta'\left[\overline{\overline{R}}{}^{c}_{10}(\theta,\theta')+\overline{\overline{R}}{}^{m\alpha}_{10}(\theta,\theta')\right]\cdot\overline{I}^{m\alpha}(\theta', z=0)\end{aligned} \tag{94a}$$

$$\begin{aligned}&\overline{I}^{m\alpha}(\theta, z=-d)\\&=\int_0^{\pi/2}d\theta'\sin\theta'\left[\overline{\overline{R}}{}^{c}_{12}(\theta,\theta')+\overline{\overline{R}}{}^{m\alpha}_{12}(\theta,\theta')\right]\cdot\overline{I}^{m\alpha}(\pi-\theta', z=-d)\end{aligned} \tag{94b}$$

where, for $A = R$ or T, and $\alpha, \beta = 0, 1, 2,$

$$\overline{\overline{A}}{}^{me}_{\alpha\beta}=\begin{bmatrix}A^{mc}_{\alpha\beta11} & A^{mc}_{\alpha\beta12} & -A^{ms}_{\alpha\beta13} & -A^{ms}_{\alpha\beta14}\\ A^{mc}_{\alpha\beta21} & A^{mc}_{\alpha\beta22} & -A^{ms}_{\alpha\beta23} & -A^{ms}_{\alpha\beta24}\\ A^{ms}_{\alpha\beta31} & A^{ms}_{\alpha\beta32} & A^{mc}_{\alpha\beta33} & A^{mc}_{\alpha\beta34}\\ A^{ms}_{\alpha\beta41} & A^{ms}_{\alpha\beta42} & A^{mc}_{\alpha\beta43} & A^{mc}_{\alpha\beta44}\end{bmatrix} \tag{95a}$$

$$\overline{\overline{A}}{}^{mo}_{\alpha\beta}=\begin{bmatrix}A^{mc}_{\alpha\beta11} & A^{mc}_{\alpha\beta12} & A^{ms}_{\alpha\beta13} & A^{ms}_{\alpha\beta14}\\ A^{mc}_{\alpha\beta21} & A^{mc}_{\alpha\beta22} & A^{ms}_{\alpha\beta23} & A^{ms}_{\alpha\beta24}\\ A^{ms}_{\alpha\beta31} & A^{ms}_{\alpha\beta32} & A^{mc}_{\alpha\beta33} & A^{mc}_{\alpha\beta34}\\ A^{ms}_{\alpha\beta41} & A^{ms}_{\alpha\beta42} & A^{mc}_{\alpha\beta43} & A^{mc}_{\alpha\beta44}\end{bmatrix} \tag{95b}$$

and, for $\alpha = e$ or o,

$$\overline{I}^{m\alpha}_{oi}(\pi-\theta_o)=\overline{I}^{m\alpha}_{oi}\frac{1}{(\delta_m+1)\pi}\delta(\cos\theta_o-\cos\theta_{oi}) \tag{96}$$

with

$$\overline{I}^{me}_{oi}=\begin{bmatrix}I_{vi}\\ I_{hi}\\ 0\\ 0\end{bmatrix} \tag{97a}$$

$$\overline{I}_{oi}^{mo} = \begin{bmatrix} 0 \\ 0 \\ U_i \\ V_i \end{bmatrix} \tag{97b}$$

The radiative transfer equations for the even and odd series can be obtained in a similar manner in section 5.2.b. We have, for $\alpha = e$ or o and $m = 0, 1, 2, \cdots$

$$\cos\theta \frac{d}{dz}\overline{I}^{m\alpha}(\theta, z) = -\ K_a \overline{I}^{m\alpha}(\theta, z) - \overline{\overline{K}}_s(\theta) \cdot \overline{I}^{m\alpha}(\theta, z) + \int_0^\pi d\theta' \sin\theta' \, \overline{\overline{P}}^{m\alpha}(\theta, \theta') \cdot \overline{I}^{m\alpha}(\theta', z) \tag{98}$$

where $\overline{\overline{P}}^{m\alpha}(\theta, \theta')$ is defined in a similar manner to (95) [section 5.2.b].

We define m_{max} and m'_{max} to be the number of harmonics that has to be kept in the expansions of the scattering function matrix and the coupling matrices, respectively, such that

$$\overline{\overline{P}}^{mc}(\theta, \theta') \simeq \overline{\overline{P}}^{ms}(\theta, \theta') \simeq 0 \qquad \text{for} \quad m > m_{max} \tag{99}$$

and

$$\overline{\overline{A}}_{\alpha\beta}^{mc} \simeq \overline{\overline{A}}_{\alpha\beta}^{ms} \simeq 0 \qquad \text{for} \quad m > m'_{max} \tag{100}$$

Then, for $m > m_{max}$ the radiative transfer equations simplify to

$$\cos\theta \frac{d}{dz}\overline{I}^{m\alpha}(\theta, z) = -K_a \overline{I}^{m\alpha} - \overline{\overline{K}}_s(\theta) \cdot \overline{I}^{m\alpha}(\theta, z) \tag{101}$$

where $\alpha = e$ or o. Similarly, for $m > m'_{max}$ the boundary conditions simplify to

$$\overline{I}^{m\alpha}(\pi - \theta, z = 0) = \overline{\overline{T}}_{01}^{c}(\theta_o) \cdot \overline{I}_{oi}^{m\alpha}(\pi - \theta_o) + \overline{\overline{R}}_{10}^{c}(\theta) \cdot \overline{I}^{m\alpha}(\theta, z = 0) \tag{102a}$$

and

$$\overline{I}^{m\alpha}(\theta, z = -d) = \overline{\overline{R}}_{12}^{c}(\theta) \cdot \overline{I}^{m\alpha}(\pi - \theta, z = -d) \tag{102b}$$

where $d\theta'$ integrations are carried out. Thus, for $m > \max[m_{max}, m'_{max}]$, we can use the simplified radiative transfer equations and the boundary conditions, given by (101) and (102), to obtain the solutions analytically without resorting to the numerical approach.

d. Numerical Solution

The set of decoupled radiative transfer equations without the azimuthal dependence for each harmonic can be solved numerically using the Gaussian quadrature method. The integrals in the radiative transfer equations are replaced by Gaussian quadratures, and the resulting system of first-order differential equations with constant coefficients are solved by obtaining eigenvalues and eigenvectors. The numerical solution for the specific intensity is given by, for each harmonic and for even or odd series,

$$\overline{I}^{+} = (\overline{\overline{E}} + \overline{\overline{Q}}) \cdot \overline{\overline{D}}(z) \cdot \overline{x} + (\overline{\overline{E}} - \overline{\overline{Q}}) \cdot \overline{\overline{U}}(z+d) \cdot \overline{y} \tag{103a}$$

$$\overline{I}^{-} = (\overline{\overline{E}}' + \overline{\overline{Q}}') \cdot \overline{\overline{D}}(z) \cdot \overline{x} + (\overline{\overline{E}}' - \overline{\overline{Q}}') \cdot \overline{\overline{U}}(z+d) \cdot \overline{y} \tag{103b}$$

where $\overline{I}^{+}$ and $\overline{I}^{-}$ represent the upward and downward propagating intensities and $\overline{x}$ and $\overline{y}$ are the unknown constants [section 5.2.c].

The boundary conditions, which are to be used to determine the unknown constants $\overline{x}$ and $\overline{y}$, can be obtained by discretizing the boundary conditions given by (94). Following the procedure outlined in the Appendix E, we obtain the following set of equations:

$$\overline{I}^{-}(z=0) = \overline{\overline{R}}_{10} \cdot \overline{I}^{+}(z=0) + \overline{\overline{T}}_{01} \cdot \overline{I}_{oi}^{-} \tag{104a}$$

$$\overline{I}^{+}(z=-d) = \overline{\overline{R}}_{12} \cdot \overline{I}^{-}(z=-d) \tag{104a}$$

Substituting in the expressions for the upward and downward propagating intensities into the above boundary conditions, we obtain the following set of equations for $\overline{x}$ and $\overline{y}$:

$$\begin{aligned} &\left[(\overline{\overline{E}}' + \overline{\overline{Q}}') - \overline{\overline{R}}_{10} \cdot (\overline{\overline{E}} + \overline{\overline{Q}})\right] \cdot \overline{x} \\ &\qquad + \left[(\overline{\overline{E}}' - \overline{\overline{Q}}') - \overline{\overline{R}}_{10} \cdot (\overline{\overline{E}} - \overline{\overline{Q}})\right] \cdot \overline{\overline{D}}(-d) \cdot \overline{y} = \overline{\overline{T}}_{01} \cdot \overline{I}_{oi}^{-} \end{aligned} \tag{105a}$$

$$\begin{aligned} &\left[(\overline{\overline{E}} + \overline{\overline{Q}}) - \overline{\overline{R}}_{12} \cdot (\overline{\overline{E}}' + \overline{\overline{Q}}')\right] \cdot \overline{\overline{D}}(-d) \cdot \overline{x} \\ &\qquad + \left[(\overline{\overline{E}} - \overline{\overline{Q}}) - \overline{\overline{R}}_{12} \cdot (\overline{\overline{E}}' - \overline{\overline{Q}}')\right] \cdot \overline{y} = 0 \end{aligned} \tag{105b}$$

The above equations can be solved for the constants $\overline{x}$ and $\overline{y}$ for each case when the incident intensity is at one of the quadrature angles.

Once the constants $\overline{x}$ and $\overline{y}$ are determined, the scattered intensities from region 1 to region 0, represented by the first term on the right-hand-side of (67), can be determined. We have

$$
\begin{aligned}
\overline{I}_{os} &= \overline{\overline{T}}_{10} \cdot \overline{I}^{+}(z=0) \\
&= \overline{\overline{T}}_{10} \cdot \left[(\overline{\overline{E}} + \overline{\overline{Q}}) \cdot \overline{x} + (\overline{\overline{E}} - \overline{\overline{Q}}) \cdot \overline{\overline{D}}(-d) \cdot \overline{y} \right]
\end{aligned} \tag{106}
$$

Thus, the complete solution can be obtained by solving the radiative transfer equations using the Gaussian quadrature method for each harmonic as outlined above and reintroducing the azimuthal dependence. The total scattered intensities in region 0 are given by

$$
\begin{aligned}
\overline{I}_{os}(\phi_o) = & \left\{ \overline{\overline{R}}{}^{c}_{01} + \overline{\overline{T}}{}^{c}_{10} \cdot \left[\overline{\overline{I}} - \overline{\overline{R}}{}^{c}_{10} \cdot \overline{\overline{R}}{}^{c}_{12} \cdot \exp[-\overline{\overline{\mu}}^{-1} \cdot \overline{\overline{K}}_e \, d] \right]^{\;1} \cdot \overline{\overline{T}}{}^{c}_{01} \right\} \\
& \cdot \overline{I}_{oi} \, \delta(\phi_o - \phi_{oi}) \\
& + \sum_{m=0}^{m''_{max}} \Bigg\{ \overline{\overline{R}}{}^{mc}_{01} \cdot \overline{I}^{mc}_{oi} \cos m(\phi_o - \phi_{oi}) + \overline{\overline{R}}{}^{ms}_{01} \cdot \overline{I}^{mc}_{oi} \sin m(\phi_o - \phi_{oi}) \\
& \quad + \overline{\overline{T}}{}^{c}_{10} \cdot \overline{I}^{mc^{+}}(z=0) \cos m(\phi_o - \phi_{oi}) \\
& \quad + \overline{\overline{T}}{}^{c}_{10} \cdot \overline{I}^{ms^{+}}(z=0) \sin m(\phi_o - \phi_{oi}) \\
& \quad + \left[\overline{\overline{T}}{}^{mc}_{10} \cdot \overline{I}^{mc^{+}}(z=0) - \overline{\overline{T}}{}^{ms}_{10} \cdot \overline{I}^{ms^{-}}(z=0) \right] \cos m(\phi_o - \phi_{oi}) \\
& \quad - \overline{\overline{T}}{}^{c}_{10} \cdot \left[\overline{\overline{I}} - \overline{\overline{R}}{}^{c}_{10} \cdot \overline{\overline{R}}{}^{c}_{12} \cdot \exp[-\overline{\overline{\mu}}^{-1} \cdot \overline{\overline{K}}_e \, d] \right]^{-1} \\
& \quad \cdot \overline{\overline{T}}{}^{c}_{01} \cdot \overline{I}^{mc}_{oi} \cos m(\phi_o - \phi_{oi}) \\
& \quad + \left[\overline{\overline{T}}{}^{ms}_{10} \cdot \overline{I}^{mc^{+}}(z=0) + \overline{\overline{T}}{}^{mc}_{10} \cdot \overline{I}^{ms^{-}}(z=0) \right] \sin m(\phi_o - \phi_{oi}) \Bigg\}
\end{aligned} \tag{107}
$$

where $m''_{max} = \max[m_{max}, m'_{max}]$ and for $m > m''_{max}$, we have evaluated the scattered intensities analytically and summed them up. Once the scattered intensities in region 0 are obtained, the bistatic scattering

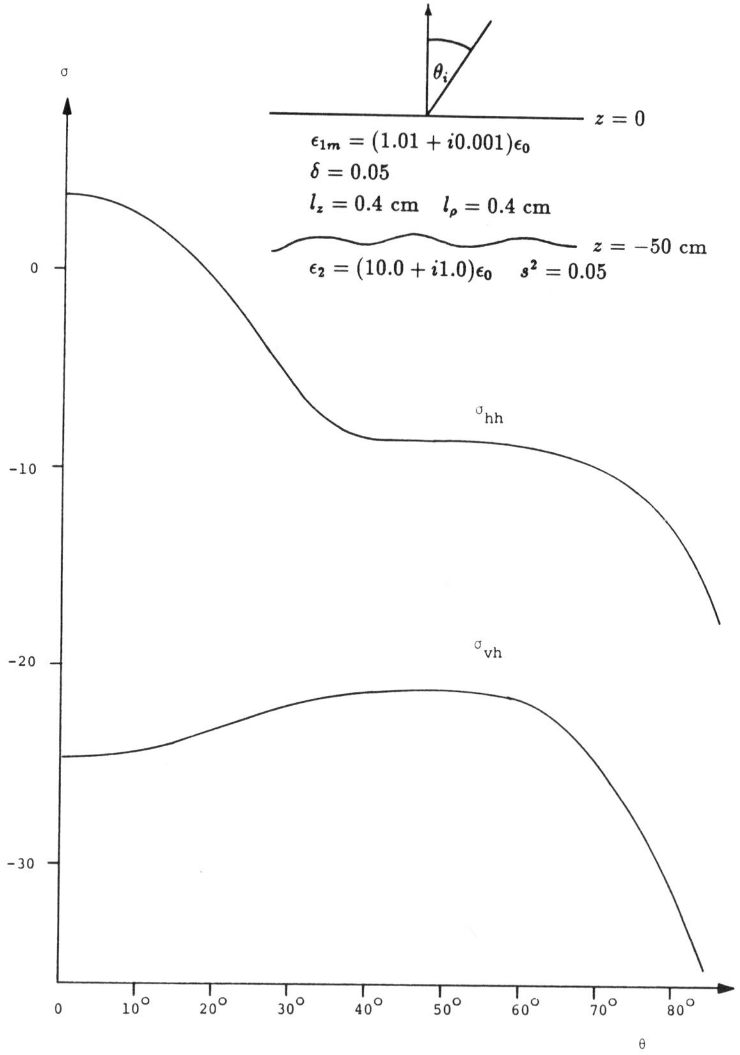

Figure 5.3.2 Backscattering cross sections as a function of incident angle for rough surface and volume scattering at 5 GHz.

coefficients and the backscattering cross sections can be obtained from (7) and (8). It should also be pointed out that the emissivity used in the passive remote sensing, which is related to the bistatic scattering coefficients, can be calculated in a similar manner [70].

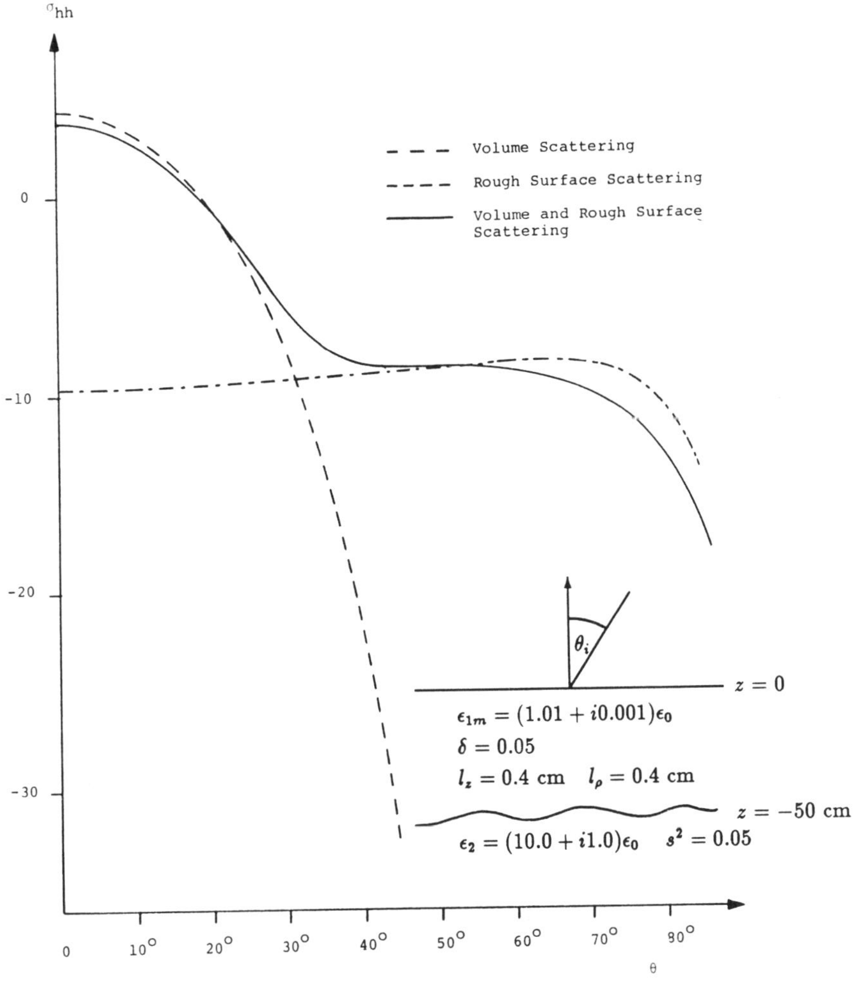

Figure 5.3.3 Backscattering cross sections as a function of incident angle for volume scattering, rough surface scattering, and combined volume and rough surface scattering at 5 GHz.

e. Results and Discussion

In this section we illustrate the theoretical results by plotting backscattering cross sections as functions of incident angle and frequency for various cases. In our calculations $n = 16$ is used. The combined volume and rough surface scattering model is illustrated us-

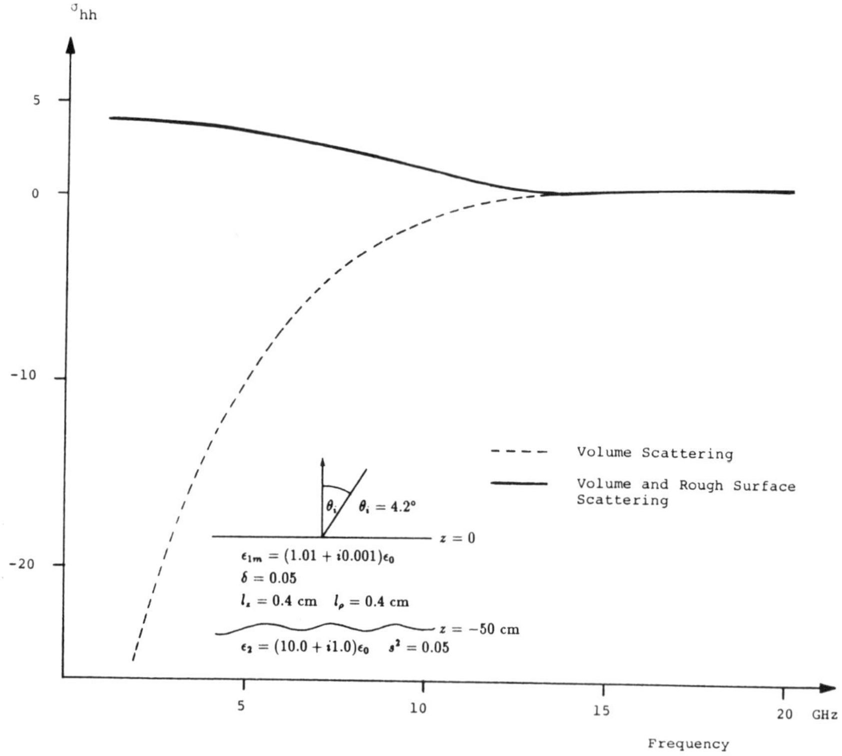

Figure 5.3.4 Backscattering cross sections as a function of frequency for $\theta_i = 4.2°$.

ing the geometrical optics solution for the rough surface modified with the shadowing function. In Fig. 5.3.2, the backscattering cross sections for like-like polarized return σ_{hh} and depolarized return σ_{vh} are plotted as a function of incident angle at 5 GHz. The bottom boundary is assumed to be rough with mean square surface slope $s^2 = 0.05$. We note that unlike the case of only volume scattering, which has a fairly smooth angular dependence, we have a peak near nadir. This is due to the contribution from the bottom rough surface. Also, depolarization return for the combined model is higher than the volume scattering model. In the volume scattering case, the depolarization of the backscattered power is due to the second-order and higher-order scattering effects. However, in the presence of a rough boundary, there

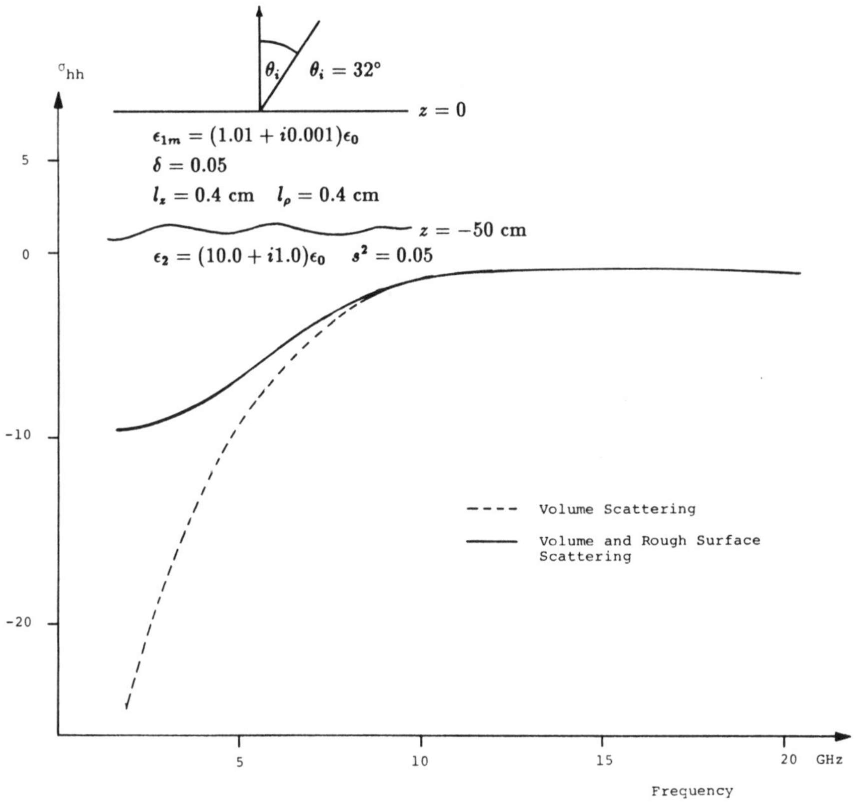

Figure 5.3.5 Backscattering cross sections as a function of frequency for $\theta_i = 32°$.

is the effect of interaction between the rough surface and volume scattering.

In Fig. 5.3.3, we compare the backscattering cross sections for volume scattering, rough surface scattering, and the combined volume and rough surface scattering. We can see that the backscattering cross section near nadir is dominated by the rough surface scattering whereas for larger angles of incidence volume scattering effects dominate.

In Figs. 5.3.4 and 5.3.5, we compare the volume scattering effects and the combined volume and rough surface scattering effects by plotting the backscattering cross sections as a function of frequency. In Fig. 5.3.4 the angle of incidence is $\theta_i = 4.2°$. At low frequencies the backscattered power due to volume scattering diminishes, and the rough surface effects dominate. As frequency is increased the effect of the bottom rough surface diminishes since the intensities do not penetrate the scattering layer as much as at lower frequencies, and the backscattered power is due to the volume scattering. In Fig. 5.3.5, we illustrate the same case for the angle of incidence $\theta_i = 32^\circ$. Again, the rough surface scattering dominates at low frequencies. However, the rough surface scattering effect diminishes faster as frequency is increased which is due to the fact that at higher incident angles the intensities have to travel a longer path before being affected by the bottom rough surface.

5.4 Summary

In the active microwave remote sensing of earth terrain, scattering effects due to medium inhomogeneities and surface roughness play a dominant role in the determination of radar backscattering coefficients. The volume scattering effects due to medium inhomogeneities have been studied by characterizing earth terrain with two-layer random medium model. The radiative transfer theory is used to calculate the backscattering and bistatic scattering coefficients from a two-layer random medium. Radiative transfer equations are solved numerically using the Fourier series expansion and the Gaussian quadrature method.

In order to more realistically model earth terrain, a composite model comprising an inhomogeneous layer over a homogeneous halfspace with rough boundaries has been developed. The random medium model is used to incorporate the volume scattering effects. To model rough top and bottom interfaces, the bistatic scattering coefficients for a randomly rough surface obtained using a combination of Kirchhoff theory and geometrical optics approach are used. The radiative transfer theory is used to solve for the scattering coefficients. Rough surface effects are incorporated into the radiative transfer theory by modifying the boundary conditions. Radiative transfer equations and the boundary conditions are solved numerically using the Fourier series expansion and the Gaussian quadrature method. It has been shown that the backscattering cross section for the angles of incidence near nadir is dominated by the rough surface effects whereas the large angle of incidence behavior is dominated by the volume scattering effects.

Appendix A: Scattering Function Matrix, Scattering Coefficient, and Absorption Coefficient for the Random Medium

The scattering function matrix has been derived for the random medium whose fluctuating permittivity is characterized by the correlation function

$$\langle \epsilon_f(\overline{r}')\epsilon_f^*(\overline{r}'')\rangle = \delta \epsilon_1'^2\, b(\overline{r}' - \overline{r}'') \tag{A1}$$

where δ is the variance of the fluctuation, and the function $b(\overline{r}' - \overline{r}'')$ is the normalized correlation function. The scattering function matrix is given by [40]

$$\overline{P}(\theta,\phi;\theta',\phi') = \frac{\pi k_1'^4 \delta}{2}\, \Phi(\theta,\phi;\theta',\phi') \begin{bmatrix} p_{11} & p_{12} & p_{13} & 0 \\ p_{21} & p_{22} & p_{23} & 0 \\ p_{31} & p_{32} & p_{33} & 0 \\ 0 & 0 & 0 & p_{44} \end{bmatrix} \tag{A2}$$

where $\Phi(\theta,\phi;\theta',\phi')$ is the spectral density of the fluctuation which is given by the Fourier transform of the correlation function and

$$\begin{aligned} p_{11} &= \sin^2\theta \sin^2\theta' + 2\sin\theta\sin\theta'\cos\theta\cos\theta'\cos(\phi-\phi') \\ &\quad + \cos^2\theta\cos^2\theta'\cos^2(\phi-\phi') \end{aligned} \tag{A3}$$

$$p_{12} = \cos^2\theta \sin^2(\phi-\phi') \tag{A4}$$

$$p_{13} = \cos\theta\sin\theta\sin\theta'\sin(\phi-\phi') + \cos^2\theta\cos\theta'\sin(\phi-\phi')\cos(\phi-\phi') \tag{A5}$$

$$p_{21} = \cos^2\theta'\sin^2(\phi-\phi') \tag{A6}$$

$$p_{22} = \cos^2(\phi-\phi') \tag{A7}$$

$$p_{23} = -\cos\theta'\sin(\phi-\phi')\cos(\phi-\phi') \tag{A8}$$

$$\begin{aligned} p_{31} &= -2\sin\theta\sin\theta'\cos\theta'\sin(\phi-\phi') \\ &\quad - 2\cos\theta\cos^2\theta'\cos(\phi-\phi')\sin(\phi-\phi') \end{aligned} \tag{A9}$$

$$p_{32} = 2\cos\theta\sin(\phi-\phi')\cos(\phi-\phi') \tag{A10}$$

$$p_{33} = \sin\theta\sin\theta'\cos(\phi-\phi') + \cos\theta\cos\theta'[\cos^2(\phi-\phi') - \sin^2(\phi-\phi')] \tag{A11}$$

$$p_{44} = \sin\theta\sin\theta'\cos(\phi-\phi') + \cos\theta\cos\theta' \tag{A12}$$

For the correlation function we assume gaussian dependence laterally and exponential dependence vertically

$$b(\bar{r}' - \bar{r}'') = \exp\left[-\frac{(x'-x'')^2 + (y'-y'')^2}{l_\rho^2} - \frac{|z'-z''|}{l_z}\right] \tag{A13}$$

Then the spectral density is given by

$$\Phi(\theta,\phi;\theta',\phi') = \frac{l_z l_\rho^2}{4\pi^2[1 + k_1'^2 l_z^2(\cos\theta' - \cos\theta)^2]} \cdot \exp\left[-\frac{k_1'^2 l_\rho^2}{4}[\sin^2\theta' + \sin^2\theta - 2\sin\theta'\sin\theta\cos(\phi'-\phi)]\right] \tag{A14}$$

The scattering coefficient $\overline{\overline{K}}_s(\theta)$ is given by

$$\overline{\overline{K}}_s(\theta) = \begin{bmatrix} K_v & 0 & 0 & 0 \\ 0 & K_h & 0 & 0 \\ 0 & 0 & \frac{K_v+K_h}{2} & 0 \\ 0 & 0 & 0 & \frac{K_v+K_h}{2} \end{bmatrix} \tag{A15}$$

where

$$K_v(\theta') = \int_{4\pi} d\Omega \, \frac{\pi k_1'^4 \delta}{2} \Phi(\theta,\phi;\theta',\phi')[p_{11}(\theta,\phi;\theta',\phi') + p_{21}(\theta,\phi;\theta',\phi')] \tag{A16}$$

$$K_h(\theta') = \int_{4\pi} d\Omega \frac{\pi k_1'^4 \delta}{2} \Phi(\theta,\phi;\theta',\phi')[p_{12}(\theta,\phi;\theta',\phi') + p_{22}(\theta,\phi;\theta',\phi')] \tag{A17}$$

and the $d\Omega$ integration is carried over a 4π solid angle. The absorption coefficient K_a is given by

$$K_a = 2k_1'' \tag{A18}$$

where k_1'' is the imaginary part of the wave number in region 1.

Appendix B: Coupling Matrices for Stokes Parameters at Planar Dielectric Interface

The coupling matrices in the boundary conditions take the form [40,53], for $\alpha,\ \beta = 0, 1, 2,$

$$\overline{\overline{T}}_{\alpha\beta}(\theta_\alpha) = \frac{\epsilon'_\beta}{\epsilon'_\alpha}\begin{bmatrix} t_{v\alpha\beta}(\theta_\alpha) & 0 & 0 & 0 \\ 0 & t_{h\alpha\beta}(\theta_\alpha) & 0 & 0 \\ 0 & 0 & g_{\alpha\beta}(\theta_\alpha) & -h_{\alpha\beta}(\theta_\alpha) \\ 0 & 0 & h_{\alpha\beta}(\theta_\alpha) & g_{\alpha\beta}(\theta_\alpha) \end{bmatrix} \qquad (B1)$$

and

$$\overline{\overline{R}}_{\alpha\beta}(\theta_\alpha) = \begin{bmatrix} r_{v\alpha\beta}(\theta_\alpha) & 0 & 0 & 0 \\ 0 & r_{h\alpha\beta}(\theta_\alpha) & 0 & 0 \\ 0 & 0 & W_{\alpha\beta}(\theta_\alpha) & -Z_{\alpha\beta}(\theta_\alpha) \\ 0 & 0 & Z_{\alpha\beta}(\theta_\alpha) & W_{\alpha\beta}(\theta_\alpha) \end{bmatrix} \qquad (B2)$$

where

$$t_{v\alpha\beta}(\theta_\alpha) = 1 - r_{v\alpha\beta}(\theta_\alpha) \qquad (B3)$$

$$t_{h\alpha\beta}(\theta_\alpha) = 1 - r_{h\alpha\beta}(\theta_\alpha) \qquad (B4)$$

$$g_{\alpha\beta}(\theta_\alpha) = (\cos\theta_\beta / \cos\theta_\alpha) Re(Y_{\alpha\beta} X^*_{\alpha\beta}) \qquad (B5)$$

$$h_{\alpha\beta}(\theta_\alpha) = (\cos\theta_\beta / \cos\theta_\alpha) Im(Y_{\alpha\beta} X^*_{\alpha\beta}) \qquad (B6)$$

for θ_α less than the critical angle, otherwise

$$g_{\alpha\beta}(\theta_\alpha) = h_{\alpha\beta}(\theta_\alpha) = 0 \qquad (B7)$$

and

$$X_{\alpha\beta}(\theta_\alpha) = 1 + R_{\alpha\beta}(\theta_\alpha) \qquad (B8)$$

$$Y_{\alpha\beta}(\theta_\alpha) = 1 + S_{\alpha\beta}(\theta_\alpha) \qquad (B9)$$

where $R_{\alpha\beta}(\theta_\alpha)$ and $S_{\alpha\beta}(\theta_\alpha)$ are the TE and TM Fresnel reflection coefficients, and

$$r_{v\alpha\beta}(\theta_\alpha) = |S_{\alpha\beta}(\theta_\alpha)|^2 \qquad (B10)$$

$$r_{h\alpha\beta}(\theta_\alpha) = |R_{\alpha\beta}(\theta_\alpha)|^2 \qquad (B11)$$

$$W_{\alpha\beta}(\theta_\alpha) = Re(S_{\alpha\beta} R^*_{\alpha\beta}) \qquad (B12)$$

$$Z_{\alpha\beta}(\theta_\alpha) = Im(S_{\alpha\beta} R^*_{\alpha\beta}) \qquad (B13)$$

Appendix C: Fourier Series Expansion of the Scattering Function Matrix

The scattering function matrix given in Appendix A can be expanded into the cosine and sine series as follows:

$$\overline{\overline{P}}(\theta,\phi:\theta'\phi') = \sum_{m=0}^{\infty} \frac{1}{(\delta_m+1)\pi}\left[\overline{\overline{P}}^{mc}(\theta,\theta')\cos m(\phi-\phi') + \overline{\overline{P}}^{ms}(\theta,\theta')\sin m(\phi-\phi')\right] \tag{C1}$$

where

$$\overline{\overline{P}}^{mc}(\theta,\theta') = q(\theta,\theta')\begin{bmatrix} p_{11}^{mc} & p_{12}^{mc} & 0 & 0 \\ p_{21}^{mc} & p_{22}^{mc} & 0 & 0 \\ 0 & 0 & p_{33}^{mc} & p_{34}^{mc} \\ 0 & 0 & p_{43}^{mc} & p_{44}^{mc} \end{bmatrix} \tag{C2}$$

$$\overline{\overline{P}}^{ms}(\theta,\theta') = q(\theta,\theta')\begin{bmatrix} 0 & 0 & p_{13}^{ms} & p_{14}^{ms} \\ 0 & 0 & p_{23}^{ms} & p_{24}^{ms} \\ p_{31}^{ms} & p_{32}^{ms} & 0 & 0 \\ p_{41}^{ms} & p_{42}^{ms} & 0 & 0 \end{bmatrix} \tag{C3}$$

with

$$q(\theta,\theta') = \frac{\delta k_1'^4}{4}\frac{l_\rho^2 l_z}{1+k_1'^2 l_z^2(\cos\theta-\cos\theta')^2} \tag{C4}$$

$$p_{11}^{mc} = e^{-y}\Big[\sin^2\theta\sin^2\theta' I_m(x) + \sin\theta\sin\theta'\cos\theta\cos\theta'(I_{m-1}(x)+I_{m+1}(x)) + \cos^2\theta\cos^2\theta'\left(\frac{1}{2}I_m(x)+\frac{1}{4}I_{m-2}(x)+\frac{1}{4}I_{m+2}(x)\right)\Big] \tag{C5}$$

$$p_{12}^{mc} = e^{-y}\cos^2\theta\frac{1}{2}\left[I_m(x)-\frac{1}{2}(I_{m-2}(x)+I_{m+2}(x))\right] \tag{C6}$$

$$p_{21}^{mc} = e^{-y}\cos^2\theta'\frac{1}{2}\left[I_m(x)-\frac{1}{2}(I_{m-2}(x)+I_{m+2}(x))\right] \tag{C7}$$

$$p_{22}^{mc} = e^{-y}\frac{1}{2}\left[I_m(x)+\frac{1}{2}(I_{m-2}(x)+I_{m+2}(x))\right] \tag{C8}$$

$$p_{33}^{mc} = e^{-y}\frac{1}{2}\Big[\sin\theta\sin\theta'(I_{m-1}(x) + I_{m+1}(x))$$
$$+ \cos\theta\cos\theta'(I_{m-2}(x) + I_{m+2}(x))\Big] \qquad (C9)$$

$$p_{44}^{mc} = e^{-y}\left[\cos\theta\cos\theta' I_m(x) + \sin\theta\sin\theta'\frac{1}{2}(I_{m-1}(x) + I_{m+1}(x))\right] \qquad (C10)$$

$$p_{34}^{mc} = p_{43}^{mc} = 0 \qquad (C11)$$

$$p_{13}^{ms} = e^{-y}\cos\theta\frac{1}{2}\Big[\sin\theta\sin\theta'(I_{m-1}(x) - I_{m+1}(x))$$
$$+\frac{1}{2}\cos\theta\cos\theta'(I_{m-2}(x) - I_{m+2}(x))\Big] \qquad (C12)$$

$$p_{23}^{ms} = -e^{-y}\cos\theta'\frac{1}{4}(I_{m-2}(x) - I_{m+2}(x)) \qquad (C13)$$

$$p_{31}^{ms} = -e^{-y}\cos\theta'\Big[\sin\theta\sin\theta'(I_{m-1}(x) - I_{m+1}(x))$$
$$+\frac{1}{2}\cos\theta\cos\theta'(I_{m-2}(x) - I_{m+2}(x))\Big] \qquad (C14)$$

$$p_{32}^{ms} = e^{-y}\cos\theta\frac{1}{2}(I_{m-2}(x) - I_{m+2}(x)) \qquad (C15)$$

$$p_{14}^{ms} = p_{24}^{ms} = p_{41}^{ms} = p_{42}^{ms} = 0 \qquad (C16)$$

$$y = \frac{1}{4}k_1'^2 l_\rho^2(\sin^2\theta + \sin^2\theta') \qquad (C17)$$

$$x = \frac{1}{2}k_1'^2 l_\rho^2 \sin\theta\sin\theta' \qquad (C18)$$

and I_m is the m-th order modified Bessel function.

Appendix D: Expressions for Coupling Matrices Obtained using Geometrical Optics Solution

The reflection and transmission coupling matrices at the rough dielectric interface can be obtained using the scattered and the transmitted fields derived by a combination of Kirchhoff approximation and geometrical optics approach. The explicit expressions for the coupling matrices $\overline{\overline{R}}_{12}$ and $\overline{\overline{T}}_{12}$ are given by [31]

$$\overline{\overline{R}}_{12}(\theta_s,\phi_s;\theta_i,\phi_i)=\frac{1}{\cos\theta_s}\frac{|\overline{k}_{1d}|^4}{4|\hat{k}_i\times\hat{k}_s|^4 k_{1dz}^4}\frac{1}{2\pi s^2}\cdot\exp\left[-\frac{k_{1dx}^2+k_{1dy}^2}{2k_{1dz}^2 s^2}\right]\overline{\overline{C}}{}_{12}^{R}(\theta_s,\phi_s;\theta_i,\phi_i) \quad (D1)$$

$$\overline{\overline{T}}_{12}(\theta_t,\phi_t;\theta_i,\phi_i)=\frac{1}{\cos\theta_t}\frac{k_2^2|\overline{k}_{2d}|^2(\hat{n}\cdot\hat{k}_t)^2}{|\hat{k}_i\times\hat{k}_t|^4 k_{2dz}^4}\frac{\eta_1}{\eta_2}\frac{1}{2\pi s^2}\cdot\exp\left[-\frac{k_{2dx}^2+k_{2dy}^2}{2k_{2dz}^2 s^2}\right]\overline{\overline{C}}{}_{12}^{T}(\theta_t,\phi_t;\theta_i,\phi_i) \quad (D2)$$

where s^2 is the mean square surface slope

$$\overline{k}_{1d}=\overline{k}_i-\overline{k}_s \quad (D3)$$

$$\overline{k}_{2d}=\overline{k}_i-\overline{k}_t \quad (D4)$$

$$\overline{\overline{C}}{}_{12}^{\alpha}=\begin{bmatrix} |f_{vv}^{\alpha}|^2 & |f_{vh}^{\alpha}|^2 & Re(f_{vh}^{\alpha *}f_{vv}^{\alpha}) & -Im(f_{vh}^{\alpha *}f_{vv}^{\alpha}) \\ |f_{hv}^{\alpha}|^2 & |f_{hh}^{\alpha}|^2 & Re(f_{hh}^{\alpha *}f_{hv}^{\alpha}) & -Im(f_{hv}^{\alpha}f_{hh}^{\alpha *}) \\ 2Re(f_{vv}^{\alpha}f_{hv}^{\alpha *}) & 2Re(f_{vh}^{\alpha}f_{hh}^{\alpha *}) & Re(f_{vv}^{\alpha}f_{hh}^{\alpha *}+f_{vh}^{\alpha}f_{hv}^{\alpha *}) & -Im(f_{vv}^{\alpha}f_{hh}^{\alpha *}-f_{vh}^{\alpha}f_{hv}^{\alpha *}) \\ 2Im(f_{vv}^{\alpha}f_{hv}^{\alpha *}) & 2Im(f_{vh}^{\alpha}f_{hh}^{\alpha *}) & Im(f_{vv}^{\alpha}f_{hh}^{\alpha *}+f_{vh}^{\alpha}f_{hv}^{\alpha *}) & Re(f_{vv}^{\alpha}f_{hh}^{\alpha *}-f_{vh}^{\alpha}f_{hv}^{\alpha *}) \end{bmatrix}$$

(D5)

with

$$f_{vv}^{r}=(\hat{h}_s\cdot\hat{k}_i)(\hat{h}_i\cdot\hat{k}_s)R_h+(\hat{v}_s\cdot\hat{k}_i)(\hat{v}_i\cdot\hat{k}_s)R_v \quad (D6)$$

$$f_{hv}^{r}=-(\hat{v}_s\cdot\hat{k}_i)(\hat{h}_i\cdot\hat{k}_s)R_h+(\hat{h}_s\cdot\hat{k}_i)(\hat{v}_i\cdot\hat{k}_s)R_v \quad (D7)$$

$$f_{vh}^{r}=(\hat{h}_s\cdot\hat{k}_i)(\hat{v}_i\cdot\hat{k}_s)R_h-(\hat{v}_s\cdot\hat{k}_i)(\hat{h}_i\cdot\hat{k}_s)R_v \quad (D8)$$

$$f_{hh}^{r}=(-\hat{v}_s\cdot\hat{k}_i)(\hat{v}_i\cdot\hat{k}_s)R_h+(\hat{h}_s\cdot\hat{k}_i)(\hat{h}_i\cdot\hat{k}_s)R_v \quad (D9)$$

and

$$f_{vv}^{t}=(\hat{h}_t\cdot\hat{k}_i)(\hat{h}_i\cdot\hat{k}_t)(1+R_h')+(\hat{v}_t\cdot\hat{k}_i)(\hat{v}_i\cdot\hat{k}_t)\frac{\eta_2}{\eta_1}(1+R_v') \quad (D10)$$

$$f^t_{hv} = -(\hat{v}_t \cdot \hat{k}_i)(\hat{h}_i \cdot \hat{k}_t)(1 + R'_h) + (\hat{h}_t \cdot \hat{k}_i)(\hat{v}_i \cdot \hat{k}_t)\frac{\eta_2}{\eta_1}(1 + R'_v) \quad (D11)$$

$$f^t_{vh} = -(\hat{h}_t \cdot \hat{k}_i)(\hat{v}_i \cdot \hat{k}_t)(1 + R'_h) + (\hat{v}_t \cdot \hat{k}_i)(\hat{h}_i \cdot \hat{k}_t)\frac{\eta_2}{\eta_1}(1 + R'_v) \quad (D12)$$

$$f^t_{hh} = (\hat{v}_t \cdot \hat{k}_i)(\hat{v}_i \cdot \hat{k}_t)(1 + R'_h) + (\hat{h}_t \cdot \hat{k}_i)(\hat{h}_i \cdot \hat{k}_t)\frac{\eta_2}{\eta_1}(1 + R'_v) \quad (D13)$$

R_v and R_h and R'_v and R'_h are the local reflection coefficients for the vertical and horizontal polarizations evaluated at the stationary phase points (α_o, β_o) and (α'_o, β'_o), respectively.

The geometrical optics solution used to derive the boundary conditons for rough dielectric interface satisfies the principle of reciprocity but violates the principle of energy conservation. This is due to the neglect of the effects of multiple scattering and shadowing. The shadowing effects can be incorporated to modify the boundary conditions as follows

$$\overline{\overline{R}}^m_{12}(\hat{k}_s; \hat{k}_i) = S(\hat{k}_s, \hat{k}_i)\,\overline{\overline{R}}_{12}(\hat{k}_s; \hat{k}_i) \quad (D14)$$

$$\overline{\overline{T}}^m_{12}(\hat{k}_t; \hat{k}_i) = S(\hat{k}_t, \hat{k}_i)\,\overline{\overline{T}}_{12}(\hat{k}_t; \hat{k}_i) \quad (D15)$$

where $S(\hat{k}_\alpha, \hat{k}_\beta)$ is the probability that a point will be illuminated by rays having the directions $\hat{k}_\beta$ and $-\hat{k}_\alpha$, given the values of the slope at the point [31].

Appendix E: Application of Legendre Quadrature Formula to the Boundary Conditions

The boundary conditions given by (94) can be cast into the matrix form using the quadrature formula. In this appendix we will illustrate the application of the Legendre quadrature formula to the boundary conditions. The boundary conditions are approximated in a manner such that the formulation does not have to be changed when applied to the flat surface case.

Consider the following scalar version of the boundary condition at $z = -d$:

$$I(\theta, z = -d) = \int_0^{\pi/2} d\theta' \sin\theta'\, r_{12}(\theta, \theta')\, I(\pi - \theta', z = -d) \quad (E1)$$

One way to approximate the above equations is to apply the Gaussian quadrature method. We obtain, for $i, j = 1, 2, \cdots, n$,

$$I(\theta_i, z = -d) = \sum_{j=1}^{n} a_j\, r_{12}(\theta_i, \theta_j)\, I(\pi - \theta_j, z = -d) \quad (E2)$$

This approach is justified as long as the approximation of changing the integration to the summation is accurate. This means the number of quadrature points n has to be large enough so that the above approximation is valid. Note that as $r_{12}(\theta, \theta')$ becomes a more sharply peaked function at the specular direction, the number of quadrature points has to be increased. Thus, it would be difficult to use the above approach for the case of near specular surface. In the limit of specular surface the coupling function is given by

$$r_{12}(\theta, \theta') = r_{12}(\theta)\, \delta(\cos\theta' - \cos\theta) \tag{E3}$$

and the boundary condition simplifies to

$$I(\theta, z = -d) = r_{12}(\theta) I(\pi - \theta, z = -d) \tag{E4}$$

In this limit we note that the number of quadrature points does not have to be large as long as $r_{12}(\theta)$ is a fairly smooth function.

One way to overcome the above problem is to use the Legendre quadrature formula. We let $\mu = \cos\theta$. Then, for $i = 1, 2, \cdots, n$ and $j = -n, \cdots, -1, 1, 2, \cdots, n$,

$$I(\mu_i, z = -d) = \sum_{j=-n}^{n} w_{ij}\, I(\mu_j, z = -d) \tag{E5}$$

where

$$w_{ij} = \frac{1}{\Pi'(\mu_j)} \int_0^1 d\mu\, r_{12}(\mu_i, \mu) \frac{\Pi(\mu)}{\mu - \mu_j} \tag{E6}$$

$$\Pi(\mu) = (\mu - \mu_1)(\mu - \mu_2) \cdots (\mu - \mu_n)(\mu + \mu_1)(\mu + \mu_2) \cdots (\mu + \mu_n) \tag{E7}$$

$$\Pi'(\mu_j) = \left. \frac{d}{d\mu} \Pi(\mu) \right|_{\mu = \mu_j} \tag{E8}$$

$$\mu_j = \cos\theta_j \qquad \mu_{-j} = \cos(\pi - \theta_j) \tag{E9}$$

In the above formulation, we note that as $r_{12}(\mu, \mu')$ becomes a sharply peaked function around the specular direction, the number of quadrature angles n does not have to be increased as long as the coefficients w_{ij} are evaluated accurately. In the limit of specular surface, we have

$$w_{ij} = r_{12}(\mu_j)\, \delta_{ij} \tag{E10}$$

where

$$\delta_{ij} = \begin{cases} 1 & \text{if } i = j \\ 0 & \text{otherwise} \end{cases} \tag{E11}$$

Therefore, in the limit of specular surface, we have

$$I(\theta_i, z = -d) = r_{12}(\theta_i) I(\pi - \theta_i, z = -d) \tag{E12}$$

Thus, if we use the Legendre quadrature formula to discretize the boundary conditions, then as the surface becomes more specular the number of quadrature angles does not have to be increased, and also the formulation does not have to be changed.

The boundary condition $(E5)$ can be cast into the following matrix equation:

$$\overline{I}^{+}(z = -d) = \overline{\overline{W}}_{12} \cdot \overline{I}^{-}(z = -d) + \overline{\overline{U}}_{12} \cdot \overline{I}^{+}(z = -d) \tag{E13}$$

where

$$\overline{I}^{+}(z = -d) = \begin{bmatrix} I(\mu_1, z = -d) \\ \vdots \\ I(\mu_n, z = -d) \end{bmatrix} \qquad \overline{I}^{-}(z = -d) = \begin{bmatrix} I(\mu_{-1}, z = -d) \\ \vdots \\ I(\mu_{-n}, z = -d) \end{bmatrix} \tag{E14}$$

$$\overline{\overline{W}}_{12} = \begin{bmatrix} w_{11} & \cdots & w_{1n} \\ \vdots & \vdots & \vdots \\ w_{n1} & \cdots & w_{nn} \end{bmatrix} \tag{E15}$$

$$\overline{\overline{U}}_{12} = \begin{bmatrix} w_{1(-1)} & \cdots & w_{1(-n)} \\ \vdots & \vdots & \vdots \\ w_{n(-1)} & \cdots & w_{n(-n)} \end{bmatrix} \tag{E16}$$

Thus, the appropriate boundary condition is given by

$$\overline{I}^{+}(z = -d) = \overline{\overline{R}}_{12} \cdot \overline{I}^{-}(z = -d) \tag{E17}$$

where

$$\overline{\overline{R}}_{12} = [\overline{\overline{I}} - \overline{\overline{U}}_{12}]^{-1} \cdot \overline{\overline{W}}_{12} \tag{E18}$$

We note that

$$\overline{\overline{U}}_{12} \simeq 0 \tag{E19}$$

and the coupling matrix is given by

$$\overline{\overline{R}}_{12} = \overline{\overline{W}}_{12} \tag{E20}$$

The boundary condition at $z = 0$ is next considered. Unlike the boundary condition at $z = -d$, there is an additional source term, which is the incident intensity being transmitted from the upper region. Consider the following scalar version of the boundary condition at $z = 0$:

$$I(\pi - \theta, z = 0) = g(\theta, \theta_{oi}) + \int_0^{\pi/2} d\theta' \sin\theta' \, r_{10}(\theta, \theta') \, I(\theta', z = 0) \tag{E21}$$

where $g(\theta, \theta_{oi})$ represents the incident intensity at θ_{oi} transmitted from region 0 to region 1. The second term in the left-side of the above equation can be approximated following the same procedure outlined above for the boundary condition at $z = -d$. Thus, we will concentrate on approximating the source term $g(\theta, \theta_{oi})$.

We note that for $g(\theta, \theta_{oi})$ which is a smooth function, there is no problem as long as the number of quadrature angles is sufficiently large. Then, the boundary condition can be discretized in a straightforward manner. In the limit of a specular surface, the source term is given by the delta function, and another approach must be used. One way to bypass the problem of discretizing the delta function is to change the source term at the boundary into the source term in the volume by calculating the zeroth-order solution explicitly and using the radiative transfer equations for the higher order terms with the zeroth-order solution acting as the volume source [51]. However, this approach requires two different formulations for the rough and planar boundaries. Also, the case of near specular surface where the source term is very sharply peaked in the specular direction cannot be treated easily.

In section 5.2, we outlined the procedure for discretizing the delta function and keeping the source term at the boundary. This approach also gives the same solution as the other approach of using the volume source terms. We will now generalize that procedure and discretize the sharply peaked incident intensity. Consider an integral given by

$$I = \int_0^{\pi/2} f(\theta, \theta') \, g(\theta', \theta_{oi}) \tag{E22}$$

where $f(\theta, \theta')$ is a smooth function. Using the Gaussian quadrature method, the integral I is approximated as

$$I \simeq \sum_{j=-n}^{j=n} a_j\, f(\theta, \theta_j)\, g_j \tag{E23}$$

Our task is to come up with a set of coefficients g_j, such that the above approximation is accurate for an arbitrary function $g(\theta', \theta_{oi})$. Using the Legendre quadrature formula, the integral I is accurately approximated as

$$I \simeq \sum_{j=-n}^{j=n} f(\theta, \theta_j)\, w_j \tag{E24}$$

where

$$w_j = \frac{1}{\Pi'(\mu_j)} \int_{-1}^{1} d\mu\, g(\mu, \mu_{oi}) \frac{\Pi(\mu)}{\mu - \mu_j} \tag{E25}$$

Comparing $(E23)$ and $(E24)$, we obtain

$$g_j = \frac{1}{a_j} w_j \tag{E26}$$

If we now let $g(\theta', \theta_{oi}) = \delta(\cos\theta'_o - \cos\theta_{oi})$ where θ_{oi} is one of the quadrature angles in region 0, the coefficient w_j is given by

$$w_j = \delta_{ji} \frac{\epsilon_o}{\epsilon'_1} \frac{\cos\theta_{oi}}{\cos\theta_i} \tag{E27}$$

Thus, the discrete form for the delta function is given by

$$g_j = \delta_{ji} \frac{1}{a_j} \frac{\epsilon_o}{\epsilon'_1} \frac{\cos\theta_{oi}}{\cos\theta_i} \tag{E28}$$

which is the same as the result given in Section II.3, Eq. (60). If $g(\theta', \theta_{oi})$ is a smooth function, then the coefficient w_j can be approximated as

$$\begin{aligned} w_j &\simeq g(\mu_j, \mu_{oi}) \frac{1}{\Pi'(\mu_j)} \int_{-1}^{1} d\mu \frac{\Pi(\mu)}{\mu - \mu_j} \\ &= g(\mu_j, \mu_{oi}) a_j \end{aligned} \tag{E29}$$

Therefore,

$$g_j = g(\mu_j, \mu_{oi}) \tag{E30}$$

which is also a consistent result. Thus, the discretization of the source term by $(E24)$ and $(E25)$ gives the correct results in both limits of very sharply peaked and smooth incident intensities. We also note that

$$w_j \simeq 0 \qquad \text{for} \quad j = -1, -2, \cdots, -n \tag{E31}$$

Acknowledgments

This work was supported by the NASA Grants NAG5-270 and NAWG-1272, the ARMY Corp of Engineers Contract DACA39-87-K-0022, the NSF Grant 8504381-ECS, and the ONR Contract N00014-83-K-0258.

References

[1] Rouse, J. W., Jr., "Arctic ice type identification by radar," *Proc. IEEE,* **57**, 605–611, 1969.

[2] Johnson, J. D. and L. D. Farmer, "Use of side-looking airborne radar for sea ice identification," *J. Geophys. Res.,* **76**, 2138–2155, 1971.

[3] Edgerton, A. T., A. Stogryn, and G. Poe, "Microwave radiometric investigations of snowpacks," Final Report 1285R-4, Aerojet-General Corp., El Monte, CA, 1971.

[4] Meier, M. F., "Application of remote sensing techniques to the study of seasonal snow cover," *J. Glaciology,* **15**, 251–265, 1975.

[5] Gloersen, P., W. Nordberg, T. J. Schmugge, T. T. Wilheit, and W. J. Campbell, "Microwave signatures of first-year and multiyear sea ice," *J. Geophys. Res.,* **78**, 3564–3572, 1973.

[6] Ketchum, R. D., Jr. and S. G. Tooma, Jr., "Analysis and interpretation of air-borne multifrequency side-looking radar sea ice imagery," *J. Geophys. Res.,* **3**, 520–538, 1973.

[7] Elachi, C., M. L. Bryaand, and W. F. Weeks, "*Imaging radar observations of frozen Arctic lakes*," *Rem. Sens. Environ.*, **5**, 169–175, 1976.

[8] Kunzi, K. F., A. D. Fisher, D. H. Staelin, and J. W. Waters, "Snow and ice surfaces measured by the Nimbus-5 microwave spectrometer," *J. Geophys. Res.*, **81**, 4965–4980, 1976.

[9] Parashar, S. K., R. M. Haralick, R. K. Moore, and A. W. Biggs, "Radar scatterometer discrimination of sea-ice types," *IEEE Trans. Geosci. Electron.*, **GE-15**, 83–87, 1977.

[10] Ulaby, F. T., W. H. Stiles, L. F. Dellwig, and B. C. Hanson, "Experiments on the radar backscatter of snow," *IEEE Trans. Geosci. Electron.*, **GE-15**, 185–189, 1977.

[11] Zwally, H. J., "Microwave emissivity and accumulation rate of polar firn," *J. Glaciology,* **18**, 195–215, 1977.

[12] Hofer, R. and E. Schanda, "Signatures of snow in the 5 to 94 GHz range, " Radio Science, **13**, 365–369, 1978.

[13] Shiue, J. C., A. T. C. Chang, H. Boyne, and D. Ellerbruch, "Remote sensing of snowpack with microwave radiometers for hydrologic applications," *Proceedings of the 12th International Symposium on Remote Sensing of Environment,* **2**, 877–886, Univ. of Michigan, Ann Arbor, MI, 1978.

[14] Stiles, W. H. and F. T. Ulaby, "The active and passive microwave response to snow parameters: 1. Wetness," *J. Geophys. Res.*, **85**, 1037–1044, 1980.

[15] Rango, A., A. T. C. Chang, and J. L. Foster, "The utilization of spaceborne microwave radiometers for monitoring snowpack properties," *Nordic Hydrol.*, **10**, 25–40, 1979.

[16] Ulaby, F. T., "Radar response to vegetation," *IEEE Trans. Ant. Prop.*, **AP-23**, 36–45, 1975.

[17] Ulaby, F. T., T. F. Bush and P. P. Batlivala, "Radar response to vegetation II: 8-18 GHz band," *IEEE Trans. Ant. Prop.*, **AP-23**, 608–618, 1975.

[18] Bush, T. F. and F. T. Ulaby, "Radar return from a continuous vegetation canopy," *IEEE Trans. Ant. Prop.*, **AP-24**, 269–276, 1976.

[19] Bush, T. F. and F. T. Ulaby, "An evaluation of radar as a crop classifier," *Rem. Sens. Environ.,* **7**, 15–36, 1978.

[20] Grody, N. C., "Remote sensing of atmospheric water content from satellites using microwave radiometry," *IEEE Trans. Ant. Prop.,* **AP-24**, 1976.

[21] Tsang, L., J. A. Kong, E. Njoku, D. H. Staelin, and J. Waters, "Theory for microwave thermal emission from a layer of cloud or rain," *IEEE Trans. Ant. Prop.,* **AP-25**, 650–657, 1977.

[22] Dickey, F. M., C. King, J.C. Holtzman, and R. K. Moore, "Moisture dependency of radar backscatter from irrigated and non-irrigated fields at 400 MHz and 13.3 GHz," *IEEE Trans. Geosci. Electron.,* **GE-12**, 19–22, 1974.

[23] Schmugge, T. J., P. Gloerson, T. Wilheit, and F. Geiger, "Remote sensing of soil moisture with microwave radiometers," *J. Geophys. Res.,* **9**, 317–323, 1974.

[24] Ulaby, F. T. and P. P. Batlivala, "Optimum radar parameters for mapping soil moisture," *IEEE Trans. Geosci. Electron.,* **GE-14**, 81–93, 1976.

[25] Njoku, E. G. and J. A. Kong, "Theory for passive sensing of near-surface soil moisture," *J. Geophys. Res.,* **82**, 3108–3118, 1977.

[26] Newton, R. W. and J. W. Rouse, Jr., "Microwave radiometer measurements of soil moisture constant," *IEEE Trans. Ant. Prop.,* **AP-28**, 680–686, 1980.

[27] Wang, J. R., R. W. Newton, and J. W. Rouse, "Passive microwave remote sensing of soil moisture: The effect of tilled row structure," *IEEE Trans. Geosci. Remote Sens.,* **GE-18**, 296–302, 1980.

[28] Jackson, T. J. and T. J. Schmugge, "Aircraft active microwave measurements for estimating soil moisture," *Photogrammetr. Eng. and Remote Sens.,* **47**, 801–805, 1981.

[29] Njoku, E. G. and P. E. O'Neill, "Multifrequency microwave radiometer measurements of soil moisture," *IEEE Trans. Geosci. Remote Sens.,* **GE-20**, 468–475, 1981.

[30] Wang, J. R., P. E. O'Neill, T. J. Jackson, and E. T. Engman, "Multifrequency measurements of the effects of soil moisture, soil texture and surface roughness," *IEEE Trans. Geosci. Remote Sens.,*

GE-21, 44–51, 1983.

[31] Tsang, L., J. A. Kong, and R. T. Shin, *Theory of Microwave Remote Sensing,* Wiley-Interscience, NY, 1985.

[32] Gurvich, A. S., V. L. Kalinin, and D. T. Matveyer, "Influence of internal structure of glaciers on their thermal radio emission," *Atm. Oceanic Phys.* USSR, **9**, 713–717, 1973.

[33] Tsang, L. and J. A. Kong, "The brightness temperature of a half-space random medium with nonuniform temperature profile," *Radio Science,* **10**, 1025–1033, 1975.

[34] England, A. W., "Thermal microwave emission from a halfpace containing scatterers," *Radio Science,* **9**, 447–454, 1974.

[35] England, A. W., "Thermal microwave emission from a scattering layer," *J. Geophys. Res.,* **80**, 4484–4496, 1975.

[36] Tsang, L. and J. A. Kong, "Emissivity of half-space random media," *Radio Science,* **11**, 593–598, 1976a.

[37] Tsang, L. and J. A. Kong, "Thermal microwave emission from a halfpace random media," *Radio Science,* **11**, 599–610, 1976b.

[38] Stogryn, A., "Electromagnetic scattering by random dielectric constant fluctuations in a bounded medium," *Radio Science,* **9**, 509–518, 1970.

[39] Leader, J. C., "Polarization dependence in EM scattering from Rayleigh scatterers embedded in a dielectric slab. I. Theory," *J. Appl. Phys.,* **46**, 4371–4385, 1975.

[40] Tsang, L. and J. A. Kong, "Radiative transfer theory for active remote sensing of half space random media," *Radio Science,* **13**, 763–774, 1978.

[41] Fung, A. K. and H. S. Fung, "Application of first order renormalization method to scattering from a vegetated-like half space," *IEEE Trans. Geosci. Electron.,* **GE-15**, 189–195, 1977.

[42] Fung, A. K., "Scattering from a vegetation layer," *IEEE Trans. Geosci. Electron.,* **GE-17**, 1–5, 1979.

[43] Zuniga, M. and J. A. Kong, "Active remote sensing of random media," *J. Appl. Phys.,* **51**, 74–79, 1980.

[44] Zuniga, M., J. A. Kong, and L. Tsang, "Depolarization effects in

the remote sensing of random media," *J. Appl. Phys.,* **51**, 2315–2325, 1980.

[45] Kong, J. A., R. Shin, J. C. Shiue, and L. Tsang, "Theory and experiment for passive remote sensing of snowpacks," *J. Geophys. Res.,* **84**, 5669–5673, 1979.

[46] Tsang, L. and J. A. Kong, "Microwave remote sensing of a two-layer random medium," *IEEE Trans. Ant. Prop.,* **AP-24**, 283–287, 1976c.

[47] Zuniga, M. A. and J. A. Kong, "Modified radiative transfer theory for a two-layer random medium," *J. Appl. Phys.,* **51**, 5228–5244, 1980.

[48] Zuniga, M. and J. A. Kong, "Mean dyadic Green's function for a two-layer random medium," *Radio Science,* **16**, 1255–1270, 1981.

[49] Blinn, J. C., III, J. E. Conel, and J. G. Quade, "Microwave emission from geological materials: Observations of interference effects," *J. Geophys. Res.,* **77**, 4366–4378, 1972.

[50] Fung, A. K. and M. F. Chen, "Emission from an inhomogeneous layer with irregular interfaces," *Radio Science,* **16**, 289–298, 1981a.

[51] Fung, A. K. and M. F. Chen, "Scattering from a Rayleigh layer with an irregular interface," *Radio Science,* **16**, 1337–1347, 1981b.

[52] Fung, A. K. and H. J. Eom, "A theory of wave scattering from an inhomogeneous layer with an irregular interface," *IEEE Trans. Ant. Prop.,* **AP-29**, 899–910, 1981.

[53] Shin, R. T. and J. A. Kong, "Radiative transfer theory for active remote sensing of homogeneous layer containing spherical scatterers," *J. Appl. Phys.,* **52**, 4221–4230, 1981.

[54] Peake, W. H., "Interaction of electromagnetic waves with some natural surfaces," *IEEE Trans. Ant. Prop.,* **AP-7**, Special Supplement, S324–S329, 1959.

[55] Bass, F. G. and I. M. Fuks, *Wave Scattering from Statistically Rough Surface,* translated by C. B. Vesecky and J. F. Vesecky, Rezamon Press, Oxford, 1979.

[56] Beckmann, P. and A. Spizzichino, *The Scattering of Electromagnetic Waves from Rough Surfaces,* MacMillan, New York, 1963.

[57] Kodis, R. D., "A note on the theory of scattering from an irregular

surface," *IEEE Trans. Ant. Prop.*, **AP-14**, 77–82, 1966.

[58] Lynch, P. J. and R. J. Wagner, "Energy conservation for rough surface scattering," *J. Acoust. Soc. Am.*, **47**, 816–821, 1968.

[59] Lynch, P. L. and R. J. Wagner, "Rough-surface scattering: Shadowing, multiple scatterer and energy conservation," *J. Math. Phys.*, **11**, 3032–3042, 1970.

[60] Sancer, M. I., "Shadow-corrected electromagnetic scattering from a randomly rough surface," *IEEE Trans. Ant. Prop.*, **17**, 577–585, 1969.

[61] Semenov, B., "Approximate computation of scattering electromagnetic waves by rough surface contours," *Radio Eng. Electron. Phys.*, **11**, 1179–1187, 1966.

[62] Stogryn, A., "Electromagnetic scattering from rough finitely conducting surfaces," Radio Science, **2**, 415–428, 1967.

[63] Barrick, D. E., "Relationship between slope probability density function and the physical optics integral in rough surface scattering," *Proc. IEEE*, **56**, 1728–1729, 1968.

[64] Burrows, M. L., "On the composite model for rough surface scattering," *IEEE Trans. Ant. Prop.*, **21**, 241–243, 1973.

[65] Rice, S. O., "Reflection of EM waves by slightly rough surfaces," *The Theory of Electromagnetic Waves*, M. Kline, Ed., Interscience, NY, 1963.

[66] Valenzuela, G. R., "Depolarization of EM waves by slightly rough surfaces," *IEEE Trans. Ant. Prop.*, **AP-15**, 552–557, 1967.

[67] Shen, J. and A. A. Maradudin, "Multiple scattering of waves from random rough surfaces," *Phys. Rev. B*, **22**, 4234–4240, 1980.

[68] Shin, R. T., "Theoretical models for microwave remote sensing of earth terrain," Ph.D. Thesis, MIT, 1984.

[69] Kong, J. A., *Electromagnetic Wave Theory*, Wiley-Interscience, New York, 1986.

[70] Shin, R. T. and J. A. Kong, "Theory for thermal microwave emission from a homogeneous layer with rough surfaces containing spherical scatterers" *J. Geophys. Res.*, **87**, 5566–5576, 1982.

INDEX